Smart Nanomaterials Technology

Series Editors

Azamal Husen, Department of Biotechnology, Smt. S. S. Patel Nootan Science and Commerce College, Sankalchand Patel University, Visnagar, Gujarat, India

Mohammad Jawaid, Department of Chemical and Petroleum Engineering, United Arab Emirates University, Al Ain, United Arab Emirates

Nanotechnology is a rapidly growing scientific field and has attracted a great interest over the last few years because of its abundant applications in different fields like biology, physics and chemistry. This science deals with the production of minute particles called nanomaterials having dimensions between 1 and 100 nm which may serve as building blocks for various physical and biological systems. On the other hand, there is the class of smart materials where the material that can stimuli by external factors and results a new kind of functional properties. The combination of these two classes forms a new class of smart nanomaterials, which produces unique functional material properties and a great opportunity to larger span of application. Smart nanomaterials have been employed by researchers to use it effectively in agricultural production, soil improvement, disease management, energy and environment, medical science, pharmaceuticals, engineering, food, animal husbandry and forestry sectors.

This book series in Smart Nanomaterials Technology aims to comprehensively cover topics in the fabrication, synthesis and application of these materials for applications in the following fields:

- Energy Systems—Renewable energy, energy storage (supercapacitors and electrochemical cells), hydrogen storage, photocatalytic water splitting for hydrogen production
- Biomedical—controlled release of drugs, treatment of various diseases, biosensors,
- Agricultural—agricultural production, soil improvement, disease management, animal feed, egg, milk and meat production/processing,
- Forestry—wood preservation, protection, disease management
- Environment—wastewater treatment, separation of hazardous contaminants from wastewater, indoor air filters.

Khalid Mahmood Zia

Chitosan

Green Derivatization and Applications

Springer

Khalid Mahmood Zia
Department of Chemistry
Government College University Faisalabad
Faisalabad, Pakistan

ISSN 3004-8273 ISSN 3004-8281 (electronic)
Smart Nanomaterials Technology
ISBN 978-981-96-2628-1 ISBN 978-981-96-2629-8 (eBook)
https://doi.org/10.1007/978-981-96-2629-8

This Springer imprint is published by the registered company Springer Nature Singapore Pte Ltd.
The registered company address is: 152 Beach Road, #21-01/04 Gateway East, Singapore 189721, Singapore

In the Honor of the Late Prof. Dr. Makshoof Athar
(25-09-1958 to 05-10-2021)

A wonderful mentor, passionate researcher and kind person
You are an inspiration for every chemist; your services for humanity will be alive

Foreword by Prof. John F. Kennedy

The definition of the fortune of the mankind has varied from an era to another. To embrace the civilization, to discover the new planets, to maintain the life stream, and to sustain the development is challenging in every respect. The world has faced the turbulence in the innovations and advancements. But it has never stopped the urge to discover new and innovative materials for the betterment of mankind. The hazardous effect of non-degradable materials is no longer a puzzle, and hence, there are continuous efforts to find the new materials to eradicate the existing issues. The renewable resources are the best option in this connection which are a cheap and abundant raw material with the countless hidden benefits. One such effort has been uncovered in the book in your hands about *Chitosan: Green Derivatization and Applications*. In writing/editing this book, the authors have selected a material which is emerging and has a lot of potential to be an important ingredient in diversified applications. This book is certainly a unique addition in the material of renewable resources and its diverse applications. It not only provides a comprehensive overview of chitosan but the utilization of this material in the most recent application is an indicator of the efforts of the researchers to probe the new materials. This book is a comprehensive review of chitosan's nanoparticle and its derivatives in all respects. It is certainly a novel addition in the biocompatible and biodegradable materials. It is an elaborated version for students and researchers. It provides the information about the chitosan, its derivatives, nanoparticles, special cases, and applications.

Prof. Khalid Mahmood Zia is an enthusiastic and outstanding researcher who is always keen to deliver the best of his research outcomes to the readers. Despite publishing his research envision on the utilization of the renewable resources in reputable journals, he has been able to compile the books dealing primarily with the biodegradability and biocompatibility of polymers. In the recent years, he has authored/edited three books in the areas of his expertise as *Algae Based Polymers Blends and Composites*; *Bionanocomposites*; and *Processing Technology for Bio-based Polymers*. All these books have been published by "Elsevier" listing Prof. Zia among highly professional authors. All these books have been internationally indexed and recognized by international academic and research institutions.

The book in hand and the previously authored three books by Dr. Zia are updated and highly structured material for undergraduate and postgraduate teachers and students in Chemistry and Applied Chemistry, Biotechnology, Material Sciences, Environmental Sciences, safety issues, and quality control domains. This book is valuable resource of recent scientific progress, along with the most known applications of the progress (including biological, physical and chemical engineering progress) on the green derivatization and modification of chitosan nanoparticles for the biomedical industry to be used by researchers, chemical and bio-engineers, and academic individuals. There are numerous novel opportunities and ideas for developing or improving technologies for the billion dollars biopolymer industry, biotechnologies, and industry.

This subject of the book has a potential relevance to community needs and is suitable for the intended audience. It is very relevant to other referred and textbooks on the subject as this book has been authored by the well-reputed signified, skilled competent expert of the field. In continuation of his efforts to amuse the readers with the recent updates in top-notch areas of research in bio-based polymers, this book is a significant addition.

This is a multifarious and explanatory volume which has meticulously comprised numerous important features of chitosan and its nanoparticles. Its fluent narration in every section amuses the reader. All the contents of the book are well connected starting from Chap. 1 which has discussed the chitosan as biomaterials, featuring the various synthetic modes for the nanoparticles. It has also discussed the limitations of the chitosan nanoparticles. Chapter 2 is an extension of preparation modes of chitosan nanoparticles, and it has provided the details of various chitosan nanoparticles, preparation, and functionalization of chitosan-derived nanoparticles. Chapter 3 has focused on the green derivatization of chitosan nanoparticles elaborating on how the new functionalities can be induced, and it has given the details of various chemical derivatization methods for the development of trimethyl chitosan, glycol chitosan, and chitosan catechol. Chapter 4 has discussed the modification process of chitosan nanoparticles, and it has discussed the physical and chemical modifications. Moreover, this chapter has also highlighted how these modifications can be induced by organic and polymeric compounds or inorganic compounds. Chapter 5 has provided the very interesting details about regulatory status and clinical trials of chitosan and derivatives. It has given the details of FDA approval of chitosan and how it has developed a prominent position in the twenty-first century. The chapter has also highlighted important details on the clinical trial of the chitosan nanoparticles for body weight reduction. Chapter 6 has provided the detailed overview of all the conventional and updated characterization details such as dynamic light scattering, dark field microscopy, aerodynamic particle sizer, and scanning mobility particle sizer. Chapter 7 has provided the details on biomedical application of modified chitosan nanoparticles, and the chapter has discussed a number of CSNPs derived from various sources such as hyaluronic acid coated, silver sulfadiazine-loaded, urocanic acid modified, or rutin-encapsulated chitosan nanoparticles. Chapter 8 has provided details about a very important topic of chitosan for the application of chemotherapy. It has given the details about the general chemotherapy and localized chemotherapy.

It has best described for cervical and brain cancer treatment. The widespread applications in food, water, and agriculture have been given in Chap. 9. Chapter 10 has provided the very interesting and important details of the chitosan to combat COVID-19. The chapter has been organized in a sequential manner while giving the details systematically about the emergence of virus SARS-CoV-2 responsible for the deadliest disease and how chitosan can be used to improve the immune system, this chapter has provided the details about antiviral role of chitosan and its potential use to surmount the SARS-CoV-2 virus. The last Chapter 11 has addressed the limitations of the chitosan, the recent challenges, and the practical aspects to utilize this efficient material. It has also addressed the future aspects and the probabilities of the utilization of the chitosan derivatives and its nanoparticles.

It is a complete package of updated knowledge and information on the topic. It is a great effort to restore the reader's trust to find a comprehensive and updated knowledge about the chitosan and its derivatives. It is a marvelous effort specifically for scientific personnel to motivate them for reading. Additionally, the book widens reader's perceptions by providing illustrations and tables summarizing the latest research on processing and characterization techniques of chitosan and its derivatives. By offering a detailed account of this field of research and describing real-world applications, it enables researchers, scientists, and professionals in industry to probe new materials to uncover the potential of biomaterials. The publication of this book is a landmark for the biomaterials in the form of chitosan. It will undoubtedly be of enormous value to students, researchers, teachers, and commoners. The publishers, editors, and authors are congratulated on a splendid effort. I hope the reader will enjoy every detail provided in this book.

Prof. John F. Kennedy
Editor-in-Chief of *Carbohydrate Polymers*
Editor-in-Chief of *Carbohydrate Polymer Technologies and Applications*
Editor-in-Chief of *International Journal of Biological Macromolecules*
Principal Director of Chembiotech Ltd.
Tenbury Wells, UK

Foreword by Prof. Atta-ur-Rahman

Chitosan is a biodegradable polymer derived from chitin. It is one of the most abundant natural polymers found in the exoskeletons of crustaceans and the cell walls of fungi. Chitosan has gained significant attention in recent years due to its versatile applications across various fields, including agriculture, biomedicine, food industry, wastewater treatment, and more. Furthermore, chitosan's eco-friendliness and sustainability have made it an appealing choice for researchers and industries aiming to develop sustainable products and processes.

Its unique structure imparts several notable properties to chitosan, including biocompatibility, biodegradability, non-toxicity, and antimicrobial activity. These characteristics make it a promising material for various applications. Chitosan can be obtained by using green methods from renewable sources like shrimp and crab shells, avoiding the use of hazardous chemicals and promoting resource efficiency. The process typically involves deproteinization, demineralization, and deacetylation of chitin, resulting in chitosan.

Chitosan has shown promise in agriculture as a natural and biodegradable alternative to synthetic pesticides. Its biodegradability ensures that it does not persist in the environment, making it an environmentally friendly soil conditioner. Chitosan-based nanoparticles and hydrogels have been extensively studied for drug delivery applications. Chitosan's biocompatibility, biodegradability, and mucoadhesive properties make it an excellent candidate for controlled drug release, targeting specific tissues, and improving the efficacy of therapeutic agents.

Chitosan wound dressings have been developed to promote wound healing. Its antibacterial properties, along with its ability to accelerate tissue regeneration, have made it valuable in managing chronic wounds and burn injuries, reducing the risk of infection and supporting tissue repair. In tissue engineering, chitosan scaffolds serve as a substrate for cell adhesion and proliferation. They can be tailored to mimic the extracellular matrix, providing a supportive environment for tissue regeneration. These biodegradable scaffolds can be used for bone, cartilage, and skin tissue engineering, offering a sustainable solution to tissue repair and replacement. Chitosan's antimicrobial properties make it a valuable natural preservative. Chitosan coatings can be applied to fruits, vegetables, and seafood to extend their shelf life by inhibiting the growth of spoilage microorganisms and pathogens. This reduces food waste and the need for synthetic preservatives.

Chitosan has also found applications in water and wastewater treatment due to its ability to remove heavy metals, organic pollutants, and dyes from contaminated water. The adsorption properties of chitosan, combined with its eco-friendly nature, make it a sustainable solution for water purification. Chitosan-based adsorbents have been developed for the removal of pollutants from industrial effluents and contaminated sites. These adsorbents are effective in reducing environmental contamination and can be regenerated or disposed of without causing harm to the environment.

Despite the numerous advantages of chitosan, challenges exist in scaling up its production from sustainable sources. Sourcing sufficient chitin-rich materials and developing cost-effective extraction methods are areas that require further research and innovation. Continued research into green synthesis methods for chitosan is crucial to further enhance its environmental credentials. Developing energy-efficient and eco-friendly processes for chitosan production will contribute to the sustainable use of this valuable material.

The book *Chitosan: Green Derivatization and Applications* presents the exciting developments in the field of chitosan chemistry. Several ways to obtain chitosan nanoparticles are presented in Chaps. 1 and 2. In Chap. 3, the topic of green derivatization of chitosan nanoparticles has been discussed in detail, along with how additional functionalities may be produced. In Chap. 4, the physical and chemical changes of chitosan nanoparticles have been covered. This chapter also presents how organic, polymeric, and inorganic substances might cause these alterations. The information in Chap. 5 on the regulatory situation and clinical trials with chitosan and its derivatives is particularly intriguing. Dynamic light scattering, dark field microscopy, aerodynamic particle sizers, and scanning mobility particle sizers are only a few of the traditional and modern characterization details covered in depth in Chap. 6. Information on the biomedical use of modified chitosan nanoparticles is offered in Chap. 7. Details on chitosan's use in chemotherapy are covered in Chap. 8, whereas Chap. 9 lists its numerous uses in food, water, and agriculture. The highly interesting and significant information on the use of chitosan to fight COVID-19 is found in Chap. 10. The last chapter has covered the limits, current difficulties, and practical considerations for using chitosan.

The book represents a comprehensive collection of current facts and expertise on the subject. It should help academics, scientists, and experts in industry to explore novel materials to unearth the promise of biomaterials by providing a thorough explanation of this area of study and outlining practical applications.

I hope that the readers will greatly enjoy the various aspects of chitosan chemistry and applications covered in the book.

Prof. Atta-ur-Rahman, FRS, N.I., H.I., S.I., T.I.
UNESCO Science Laureate
Academician Chinese Academy of Sciences, Professor Emeritus
International Centre for Chemical and Biological Sciences
University of Karachi
Karachi, Pakistan

Preface

This book is written and organized for students, researchers, scientists, industrialists, and everyone else who is specifically interested in the discovery of new biomaterials as a source to surpass the conventional materials, either due to their unavailability, scarcity, hazardous effects, or just possibly in the quest to find something novel and exciting. This book was developed with the intention of bringing the finer aspects of the creation, processing, range, and uses of chitosan, its derivatives, and nanoparticles to the surface. To accomplish the objectives of clarity, accuracy, and depth of the subject, we have placed enormous emphasis. The sequential arrangement, orderly presentation of ideas, modern terminology, and thorough analysis of literature with suitable citations will delight the reader. Each chapter is divided into an abstract and the specifics in the primary context, which are reinforced with current information, thorough tables, and illustrative figures. Our goal is to authentically inform the reader on the benefits and marvels of chitosan while also inspiring them to look for similar compounds in other parts of nature. Introduction to chitosan, methods of its synthesis, derivatization, functionalization, and modification, procedures for its characterization, and biomedical and nonbiomedical uses, notably in the areas of food, water, and agriculture, have all been covered by the writers. The role of chitosan and its derivatives to cope with COVID-19 has been addressed as well. Additionally, it has successfully addressed current issues and prospects for green derivatization of chitosan.

However, we selected this topic in the wake of green derivatization of chitosan as a useful tool to develop a number of products which can be effectively used in multidiscipline based on its structural and functional modifications. The vibrant features of this book will appeal to numerous researchers in diverse fields including; agriculture, chemistry, environmental science, biochemistry, polymer engineering, materials science, and biotechnology. The main composition of this book is as follows: Chap. 1, Chitosan Nanoparticles, is an introductory chapter which has given the details about the origin of chiton, its various features, limitations, and potential targets of this material. It has elaborated on the history of chitosan. The methods for the preparation of chitosan nanoparticles have also been listed. It has been described that chitosan nanoparticles possess a number of biological properties, which have attracted considerable attention in the present scenario. Chapter 2 demonstrates the procedures for preparing and functionalizing of chitosan derivatives. Chemical and enzymatic techniques for the production of chitosan derivatives have been used. It is narrated how chitosan is functionalized through cross-linking, grafting, carboxymethylation, phosphorylation, and etherification. Chapter 3 provides an overview about the derivatives of chitosan using the green technology. It has demonstrated the functionalization of the chitosan and its significance. This chapter describes that chitosan specification depends upon isolation process. It has also provided a comprehensive analysis of the chitosan derivatization.

Chapter 4 provides the prime focus of this chapter about the physical and chemical modifications of CSNPs with various organic, inorganic, and polymeric compounds. CSNPs' modification has been done in order to enhance its properties. Modified CSNPs have been utilized in various applications such as in biomedical field, wastewater treatment, agricultural field, and cosmetic industries and also in food packaging. In Chap. 5, chitosan has been described as a vehicle for delivering potential drugs and biologics like gene products. Chitosan has been studied as a potential drug for neuronal disorders, blood pressure control, and body weight loss. In addition, a number of animal studies and clinical trials of chitosan for cancer treatment have been examined. Chapter 6 has provided a brief summary of various characterization techniques involved in the structural confirmation and to demonstrate the various properties of the chitosan derivatives and its nanoparticles. It has provided an overview of the primitive techniques such as UV-visible spectroscopy and Fourier transform infrared spectroscopy (FTIR). The importance of zeta potential measurements for the chitosan nanoparticles has been evaluated. Darkfield microscopy has been used for the characterization of chitosan-capped gold nanoparticles. Chapter 7 aims to provide a comprehensive review of literature regarding the biomedical applications of modified chitosan nanoparticles. Several examples have been quoted in this connection including: chitosan tripolyphosphate nanoparticles in drug delivery system. Herbal therapy has lately been suggested as an alternate technique for developing drugs for a number of neurological illnesses, such as Alzheimer's disease and epilepsy for the remedy; the piperine-loaded CSNPs have been extensively investigated. For cancer treatment, hyaluronic acid-coated CSNPs have been utilized. Chapter 8 has provided an overview of modified chitosan nanoparticles for the chemotherapy. In this connection, the review about doxorubicin hydrochloride (Dox-HCl) has been reported as an antibiotics. In order to address the issues of the localized chemotherapy, collage/chitosan film-containing biofinylated glycol CSNPs have been reported. Chapter 9 has highlighted the applications of chitosan nanoparticles in various fields such as water, food, and agriculture. Polyacrylamide-grafted CSNPs have been reported for the removal of (Cu) from water. Chitosan-alginate nanoparticles (CANPs) have been reported for the removal of Hg from industrial wastewater. The biomass of Cu and Co-doped alginate-coated CSNPs has been used for the removal of heavy metals. The chitosan nanoparticles have also been used in the food industry for various applications, particularly in food packaging. In this connection, novel nanofibrillated cellulose/chitosan nanoparticles for food packaging have also been reported. Synthesized fish gelatin/CSNPs composite films for food packaging have also been reported. Chapter 10 has provided the details regarding the overview of a novel application of chitosan derivative and nanoparticle to combat COVID-19. The chapter has provided a detailed taxonomy and background of the disease COVID-19 and the virus SARS-CoV-2 responsible for the outbreak in 2019. Chapter 11 has discussed the advantages and limitations of CSNPs; it has also highlighted some future prospects of CSNPs.

I am humbly obliged to "Almighty Allah" who enabled me to learn and to spread knowledge to other humans. I express my deep gratitude to the entire Springer team for helping us to publish this book. In particular, am thankful to Loyola D'Silva, Ph.D. (Springer, Executive Editor) for providing with the exciting opportunity to work on this book. Special acknowledgments are made to Vinothini Elango who always encouraged and pushed up me to complete this task with her generous guidance to accomplish the publication of this book. Very special thanks to Tariq Mahmood Rehmani (PMS) Special Secretary (Operations), Specialized Healthcare & Medical Education Department, Government of the Punjab Pakistan, Prof. Dr. Rauf-I-Azam Vice Chancellor Government College University Faisalabad, Dr. Nadia Akram,Dr. Fatima Mumtaz, Dr. Shahzad Zafar Iqbal, and Rai Umar Farooq Abid for their continued support, and cooperation at every step to accomplish this task. I also would like to acknowledge the cooperation of all my lab scholars, especially Mohsin, Fizza, Rimsha, Sehrish, Sapna Anam, Aneeza, Adiba, Fatima, Amina, and others. I would also like to thank Nusrat Kahlid, Muhammad Abdullah Zia, and Harram Fatima Zia, who helped me a lot in finalizing this project within the limited time frame.

For comments and suggestions, please write to: drkmzia@gcuf.edu.pk

Dr. Khalid Mahmood Zia
Chairman/Professor of Chemistry
Department of Chemistry
Government College University
Faisalabad
Faisalabad, Pakistan

Contents

1 Chitosan Nanoparticles 1
- 1.1 Chitosan 1
  - 1.1.1 Introduction 1
  - 1.1.2 History 1
  - 1.1.3 Physicochemical Properties 2
  - 1.1.4 Molecular Weight 3
  - 1.1.5 Methods for Conversion of Chitin to Chitosan 3
  - 1.1.6 Structure of Cellulose, Chitin, and Chitosan 5
  - 1.1.7 Derivatives of Chitosan 6
- 1.2 Chitosan Nanoparticles (CSNPs) 7
- 1.3 Applications and Limitations of Chitosan Nanoparticles 9
  - 1.3.1 Applications of Chitosan Nanoparticles 9
  - 1.3.2 Limitations 14
- 1.4 Preparation of Chitosan Nanoparticles (CSNPs) 14
  - 1.4.1 Ionotropic Gelation Method 14
  - 1.4.2 Complex Co-acervation Method 15
  - 1.4.3 Co-precipitation Method 15
  - 1.4.4 Microemulsion Method 15
  - 1.4.5 Emulsion-Based Solvent Diffusion Method 17
  - 1.4.6 Reverse Micellar Method 17
  - 1.4.7 Polyelectrolyte Complexes 18
  - 1.4.8 Emulsion-Based Solvent Evaporation Method 18
- 1.5 Recent Advances in Preparations of CSNPs 20
  - 1.5.1 Emulsification and Cross-linking 20
  - 1.5.2 Emulsion Droplet Coalescence 21
  - 1.5.3 Ionic Gelation Through Radical Polymerization 22

1.5.4 Desolvation 23
1.6 Recent Development in Synthesis of Biocompatible Polymers Functionalized by Chitosan and Chitosan Nanoparticles 24
1.7 Conclusion 25
References 26

2 Preparation and Functionalization of the Chitosan Derivatives 31
2.1 Introduction 31
2.2 Chitosan and Its Derivatives 31
2.3 Chitosan Isolation 34
2.4 Preparation of Chitosan Derivatives by Chemical Method 34
2.4.1 Synthesis of *N*-Schiff Base Type *N*-Secondary Amino Type Chitosan Crown Ethers Derivatives 36
2.4.2 Synthesis of Schiff Base Type *N*-Heterocyclic Chitosan Derivatives 38
2.4.3 Synthesis of Chitosan Derivatives with Quaternary Ammonium Salt 39
2.4.4 Synthesis of Sugar-Bearing Chitosan Derivatives 42
2.4.5 Preparation of Chitosan Derivatives from Gladius of Squid *Sepioteuthis lessoniana* 43
2.4.6 Synthesis of Novel Quaternary Chitosan Derivatives via *N*-Chloroacyl-6-*O*-Triphenylmethyl Chitosans 43
2.4.7 Enzymatic Method 46
2.4.8 Enzymatic Synthesis of Chitosan Derivatives 47
2.5 Functionalization of Chitosan 52
2.5.1 Chemical Functionalization 52
2.5.2 Carboxymethylation of Chitosan 52
2.5.3 Phosphorylation of Chitosan 53
2.5.4 Graft Copolymerization of Chitosan 54
2.5.5 Esterification of Chitosan 54
2.5.6 Etherification of Chitosan 56
2.5.7 Cross-linking of Chitosan 56
2.6 Summary 58
References 60

3 Green Derivatization of Chitosan Nanoparticles 63
3.1 Chitosan Derivatives: New Functionalities 63
3.2 Chemical Derivation of Chitosan 64
3.3 Trimethyl Chitosan (TMC) 66
3.3.1 Synthesis of Trimethyl Chitosan 67
3.4 Carboxymethyl Chitosan 68
3.4.1 Synthesis of Carboxymethyl Chitosan 68
3.5 Thiolated Chitosan 69

3.5.1 Synthesis of Thiolated Chitosan 69
3.5.2 Properties of Thiolated Chitosan 70
3.6 Chitosan Enzyme Inhibitors 71
3.7 Chitosan-Ethylenediaminetetraacetic Acid (Chitosan-EDTA) 72
3.7.1 Preparation Method of Chitosan-EDTA 72
3.8 Half-Acetylated Chitosan 73
3.8.1 Synthesis of Half-Acetylated Chitosan 73
3.9 Acrylated Chitosan 73
3.9.1 Synthesis of Acrylated Chitosan 74
3.10 Glycol Chitosan (GCS) 75
3.11 Chitosan Catechol 75
3.11.1 Synthesis of Chi-C 76
3.12 Methyl Pyrrolidinone Chitosan (MPC) 77
3.13 Cyclodextrin Chitosan (CDCS) 79
3.13.1 Preparation of CD-g-CS 79
3.14 Oleoyl-Quaternized Chitosan 80
3.14.1 Synthesis of Oleoyl-Quaternized-Chitosan 81
3.14.2 Synthesis of Chitin/Chitosan-Based Shape Memory Polyurethanes Bionanocomposites 81
3.15 Conclusion 82
References 82

4 Modification of Chitosan Nanoparticles (CSNPs) 89
4.1 Modification of CSNPs 89
4.2 Physical Modification 89
4.3 Chemical Modification and Applications of CSNPs 90
4.3.1 Applications 91
4.4 Modification of CSNPs with Polymeric Compounds 99
4.5 Modification of CSNPs with Organic Compounds 100
4.5.1 Modification with Aldehydes 100
4.5.2 Modification with Acids 101
4.6 Modification of CSNPs with Inorganic Compounds 102
4.6.1 Sulfated CS Derivatives 102
4.6.2 Phosphorylated CS Derivatives 103
4.6.3 N-Substitutions of CS Without Protecting Agent 103
4.6.4 N-Substitutions Utilizing Protective Agents 104
4.6.5 O-Substitutions Utilizing Protective Agents 105
4.6.6 N,O-Substitutions 107
4.7 Chitosan Diacetate (CDA) and Chitosan Triacetate (CTA) Nanoparticles (NPs) 108
4.8 Conclusion 109
References 110

5 Regulatory Status and Clinical Trials of Chitosan and Derivatives ... 115
5.1 Introduction ... 115
5.2 Regulatory Status of Chitosan and Derivatives ... 115
5.2.1 Regulatory Aspects ... 116
5.2.2 Chitosan Food and Drug Administration (FDA) Approval ... 118
5.2.3 Regulatory Status: As a Pharmaceutical Excipient ... 119
5.2.4 Evidence for Possible Carcinogenic Activity ... 120
5.2.5 The Twenty-First Century Revival of Chitosan in Service to Bio-Organic Chemistry ... 120
5.3 Clinical Studies on Chitin, Chitosan, and Their Derivatives: Case Studies ... 122
5.3.1 A Chart of the Clinical Studies on Chitosan and Its Derivatives ... 122
5.3.2 An Overview of the Protective Effects of Chitosan and Acetylated Chitosan Oligosaccharides Against Neuronal Disorders ... 122
5.3.3 Clinical Trial Evaluating the Safety and Efficacy of the Use of Chitosan Derivatives ... 126
5.3.4 Quantitative Assessment of the Effects of Chitosan Intervention on Blood Pressure Control ... 127
5.3.5 Study of Chitosan for Pharmacologic Manipulation ... 127
5.3.6 A Meta-Analysis on Randomized Controlled Clinical Trials Evaluating the Effect of the Dietary Supplement Chitosan on Weight Loss, Lipid Parameters, and Blood Pressure ... 128
5.3.7 Single-Blind, Placebo-Controlled Randomized Clinical Study of Chitosan for Body Weight Reduction ... 129
5.3.8 Animal Experiments and Clinical Trials of ^{166}Ho-Chitosan for Various Cancers ... 130
5.4 Conclusion ... 131
References ... 132

6 Characterization Techniques for the Chitosan Nanoparticles and Derivatives ... 135
6.1 Chitosan Nanoparticles and Derivatives ... 135
6.2 Chitosan-Based Nanomaterials and Their Physicochemical Properties ... 138
6.3 Characterization of Chitosan Nanoparticles ... 139
6.4 Spectroscopic Analysis (UV–Visible Spectroscopy) ... 140
6.5 Fourier Transform Infrared Spectroscopy (FTIR) ... 140

6.5.1 Chitosan Infrared (CS-IR) Spectrum 142
6.5.2 Spectrum of Tripolyphosphate (TPP) 143
6.5.3 Spectrum of Triazoles (III–V) 143
6.5.4 The Spectrum of Chitosan-Tripolyphosphate (CS-TPP) Nanoparticles 144
6.6 X-Rays Diffraction (XRD) 144
6.7 Scanning Electron Microscopy (SEM) 145
6.8 Transmission Electron Microscopy (TEM) 146
6.8.1 Characterization of CS-TM-PEG Copolymer-Insulin Nanoparticles by TEM 147
6.9 Nuclear Magnetic Resonance (NMR) 150
6.10 Dynamic Light Scattering (DLS) 151
6.11 Inductively Coupled Plasma Mass Spectrometry (ICP-MS) 154
6.11.1 Characterization of Gd (TPyP) Conjugated Chitosan Nanoparticles 154
6.11.2 Determination of Loading Capacity, the Yield of Gd (TPyP)/CNs 156
6.12 Zeta Potential 157
6.13 Dark Field Microscopy 158
6.13.1 Darkfield and Chitosan-Capped Gold Nanoparticles 158
6.14 Aerodynamic Particle Sizer (APS) 159
6.15 Scanning Mobility Particle Sizer (SMPS) 159
6.15.1 Size Distribution and Scanning Mobility Particle Sizer (SMPS) 160
6.16 Matrix-Assisted Laser Desorption/Ionization Mass Spectrometry (MALDI-MS) 161
6.17 Size Exclusion Chromatography 163
6.18 Summary 166
References 167

7 Biomedical Applications of Modified Chitosan Nanoparticles 177
7.1 Pickering Emulsions-Chitosan Tripolyphophate Nanoparticles in Drug Delivery System 177
7.2 Delivery of Salazosulfapyridine 179
7.3 Polyelectrolyte Complex CSNPs and Poly(2-Acrylomido-2-Methylpropane Sulfonic Acid) 180
7.4 Chitosan-5-Flurouracil NPs 182
7.5 Propranolol CSNPs 184
7.6 Desvenlafaxine-Loaded PLGA CSNPs 186
7.7 Quetiapine Fumarate-Loaded CSNPs 187
7.8 Rutin-Encapsulated CSNPs 189
7.9 Piperine-Loaded CSNPs 190
7.10 Hyaluronic Acid-Coated CSNPs 191

7.11 Naringenin-Loaded Sulfobutylether-β-Cyclodextrin CSNPs 193
7.12 Primaquine Phosphate-Loaded CSNPs 194
7.13 Urocanic Acid Modified CSNPs 195
7.14 PVA Film Containing Tetracycline Hydrochloride-Loaded Quaternized CSNPs for Wound Treatment 196
7.15 Spanish Broom (*Spartium junceum* L.) Fiber with Vancomycin-Loaded CSNPs 199
7.16 Silver Sulfadiazine-Loaded CSNPs 200
7.17 Summary 202
References 202

8 Applications of Modified Chitosan Nanoparticles for Chemotherapy 213
8.1 Applications of Modified Chitosan Nanoparticles (CSNPs) for General Chemotherapy 213
8.1.1 Doxorubicin-Loaded *Auricularia auricula* Polysaccharide CSNPs 213
8.1.2 Collage/Chitosan Film Containing Biofinylated Glycol CSNPs 215
8.1.3 Doxorubicin-Loaded Polyethylene Glycoylated and Poloxemer-Modified CSNPs 215
8.1.4 Doxorubicin-Loaded Magnetic CSNPs 217
8.1.5 Paclitaxel (PTX)-Loaded CSNPs 218
8.1.6 Vincristine-Loaded CSNPs 221
8.1.7 Methotrexate Polyamido Amine-Loaded CSNPs 221
8.1.8 Docetaxel-Loaded CSNPs 221
8.1.9 PEGylated CS-Poly(Ethylene Imine) NPs 222
8.1.10 Small Interfering RNA Glycol CSNPs 223
8.1.11 Hydrotropic Oligomer-Conjugated Glycol CSNPs 223
8.2 Applications of Modified CSNPs for Localized Chemotherapy 224
8.2.1 Cervical Cancer Treatment 225
8.2.2 Curcumin-Loaded CSNPs 227
8.2.3 Fore Brain Cancer Treatment 227
8.2.4 Docetaxel-Loaded D-α-Tocopherol Polyethylene Glycol 1000 Succinate-Loaded CSNPs 229
8.2.5 Temozolomide (TMZ) Poly(Ɛ-Caprolactone Diol)-Based Polyurethane-Loaded CSNPs 229
8.2.6 The SiRNA-Loaded Galectin/CSNPs 231
8.2.7 For Breast Cancer Treatment 231
8.2.8 Lapatinib-Loaded Dextran Sulfate CSNPs 231
8.3 Conclusion 232
References 233

9 Applications of Modified Chitosan Nanoparticles in Water, Food, and Agriculture ... 239
9.1 Applications of Modified Chitosan Nanoparticles (CSNPs) in Water Treatment ... 239
9.1.1 Polyacrylamide-Grafted (PG) CSNPs for the Removal of Cu(II) ... 239
9.1.2 Chitosan-Alginate Nanoparticles (CANPs) for the Removal of Hg ... 240
9.1.3 Biomass of Cu and Co-Doped Alginate-Coated CSNPs for Removal of Heavy Metals ... 241
9.1.4 Applications of Modified Chitosan Nanoparticles CSNPs in Food Industry ... 241
9.1.5 Anthocyanins Loaded CSNPs for Food Products ... 241
9.1.6 Novel Nano-fibrillated Cellulose/Chitosan Nanoparticles for Food Packaging ... 242
9.1.7 Fish Gelatin/CSNPs for Food Packaging ... 242
9.1.8 Catechins and Catechins-Zn Complex Loaded β-CSNPs for Food Products ... 243
9.1.9 CSNPs with Summer Savory (*Satureja hortensis* L.) Essential Oil ... 243
9.1.10 CSNPs Loaded with 2,5-Dihydroxybenzoic Acid and Protocatechuic Acid ... 244
9.1.11 Tea Polyphenole (TP) Zn Complex-Loaded β-Chitosan Nanoparticles ... 244
9.2 Applications of Modified Chitosan Nanoparticles (CSNPs) in the Field of Agriculture ... 245
9.2.1 Chitosan-Based Polymers and Nanoparticles for Fruits and Seeds Coating ... 245
9.2.2 Growth Promotion and Improved Disease Resistance ... 246
9.2.3 Crop Productivity and Conservation of Agronomic Commodities ... 247
9.2.4 Recent Advances of Chitosan Nanoparticles Applications in Plants ... 248
9.2.5 Developing Nano-delivery Systems for Agriculture ... 250
9.3 Conclusion ... 251
References ... 252

10 Chitosan as Tools to Combat COVID-19 ... 259
10.1 Taxonomy of SARS-CoV-2 ... 259
10.2 Immunology and Immunopathology of COVID-19 ... 263
10.2.1 Innate Immunity to SARS-CoV-2 ... 263
10.2.2 Adaptive Immunity to SARS-CoV-2 ... 264

10.2.3 Immunopathology of COVID-19 266
10.3 Common Antiviral Approaches and Limitations 267
10.4 Antiviral Activities of Chitosan 269
10.4.1 Mechanism of Action of Chitosan 272
10.5 Direct Antiviral Effects of Chitosan 273
10.6 Immune-Boosting Effects of Chitosan 274
10.7 Chitosan as a Drug Delivery Vehicle for the Treatment of Viral Infections 276
10.8 Chitosan Potential Antiviral Effects for SARS-CoV-2 279
10.9 Chitosan Derivatives; A Novel Inhibitory Discovery Against Wild Type and Variants of SARS-CoV-2 281
10.10 Conclusion 287
References 287

11 Recent Challenges and Future Prospectus of Green Derivatized Chitosan Nanoparticles 297
11.1 Isolation, Preparation of Chitin, Chitosan, Their Functionalized Polymers and Bio-nanocomposites 297
11.2 Recent Challenges in Green Derivatization of Chitosan Nanoparticles (CSNPs) 298
11.3 Recent Strategies and Practical Aspects 298
11.4 Chitosan Nanoparticles Issues and Opportunities 299
11.5 Challenges to Be Addressed 300
11.6 Brief Outlook and Concluding Remarks 301
11.7 Open Questions 301
11.8 Future Prospects and Probabilities 301
References 302

About the Author

Khalid Mahmood Zia, Ph.D. is a renowned chemist and educator and currently serving as Professor and Chair at the Department of Chemistry, Government College University, Faisalabad, Pakistan. He earned his Ph.D. in Chemistry from the University of Agriculture, Faisalabad, in 2008. With a strong research focus on synthetic and natural polymers, nanocomposites, and biomaterials, he has made significant contributions to academia and industry.

Throughout his illustrious career, he has received numerous prestigious awards, including the HEC-Best Young Research Scholar Award (2012), Research Productivity Awards from PCST, Islamabad, the Australia-Award Pakistan (2017), and the Small Grant Scheme Award from the Australian Government (2018). He has authored over 180 impact factor research articles and six internationally published books, including *Algae Based Polymers, Blends, and Composites*, *Bionanocomposites: Green Synthesis and Applications*, and *Processing Technology for Bio-Based Polymers: Advanced Strategies and Practical Aspects*.

In addition to his academic achievements, he has held various administrative positions, including Director of the Office of Research, Innovation and Commercialization (ORIC), Director of the Quality Enhancement Cell (QEC), and Chairman of the Affiliation Committee at Government College University, Faisalabad. His impressive citation record exceeds 13,000, with an h-index of 55, further solidifying his reputation as a leading expert in his field.

His commitment to fostering strong ties between governments, industries, and academia has significantly impacted the scientific community, and his ongoing research endeavors continue to push the boundaries of polymer science and its applications.

Chapter 1
Chitosan Nanoparticles

1.1 Chitosan

1.1.1 Introduction

In recent years, the synthesis and application of green products are increasing tremendously to avoid the environment from hazardous. The green chemistry releases less waste, solvents, and chemicals which are harmful for human health. The natural biopolymers are environment-friendly, biodegradable, and biocompatible. The modification of synthetic polymers using natural products like chitin, chitosan, starch, etc., is on the top priority. The biopolymers show attractive properties due to which it widely used in different fields like the production of medicines, drug delivery, and food additives and preservatives [1].

Chitosan is a nitrogenous (amino-based) polysaccharide obtained by the deacetylation of chitin. Chitin presents in the cartilage of mollusks, cell wall of fungi, the exoskeleton of arthropods like shellfish, crabs, and scales of fish, and cuticles of insects (Fig. 1.1). The structure of cellulose, chitin, and chitosan resembles each other just the difference in the carbon number two at which –OH, –NH–CO–CH_3, –NH_2 present, respectively.

1.1.2 History

Henri Braconnot, a French professor of natural history, discovered chitin during research in mushrooms in 1811 after the discovery of a "material particularly resistant to usual chemicals" by A. Hachett, an English scientist in 1799, and in 1843, Lassaigne demonstrated the presence of nitrogen in chitin [2]. Henri Braconnot's name for chitin was fungine. In 1823, Odier found the same material in insects and plants and named it "chitine" [3]. Prof. C. Rouget observed that reaction of chitin

K. M. Zia, *Chitosan*, Smart Nanomaterials Technology,
https://doi.org/10.1007/978-981-96-2629-8_1

Fig. 1.1 Natural sources of chitin

with NaOH gives a product which soluble in acids unlike of chitin. Hoppe Seiler gave the name of this product as chitosan [3].

1.1.3 Physicochemical Properties

Chitosan is a natural, biodegradable, and biocompatible polysaccharide. After cellulose chitosan is the most abundant natural polymer, it is soluble in acidic media and insoluble in water due to amino groups. Degree of deacetylation (detachment of acetyl group during conversion of chitin to chitosan) affects the solubility of this polysaccharide. The cationic nature of chitosan is due to the introduction of positive charge in the polymer chain which makes it water soluble. The physicochemical properties of chitosan affect the functional properties. According to physical, chemical, and biological properties of chitosan, it uses in diverse fields [4]. Chitosan uses as a raw material for the synthesis of composites, nanoparticles, and hydrogels and for practical applications like tissue engineering, drug delivery, and wastewater treatment [3].

1.1.4 Molecular Weight

The effectiveness of chitosan and its derivatives, which are abundantly used in almost all fields of life, depends upon the crystallinity, degree of deacetylation, and molecular mass of chitosan. The reaction conditions like concentration, time, temperature, and repetition of alkaline steps change the molecular mass and degree of deacetylation of chitosan. At commercial level, the chitin is mainly extracted from marine species through acid and basic treatment. The acidic treatment involves the dissolution of calcium carbonate followed by basic treatment to dissolve proteins, and decolorization process is performed to obtain the colorless by-product [5].

1.1.5 Methods for Conversion of Chitin to Chitosan

Chitin is converted to chitosan by the removal of acetyl group at carbon number two (C-2) positions, and acetamide group is generated through a mechanism known as de-acetylation process. In de-polymerization reaction, molecular weight of final product changes because of removal of acetyl groups. The chitin (15–40% of crustacean shells) can be converted to chitosan by various methods, including chemical and enzymatic methods, and irradiation with ultrasound radiations.

1.1.5.1 Chemical Method

In chemical method, the chitin is homogenously and heterogeneously treated with acids and or alkali. In case of homogenous method, chitin is treated with concentrated base (alkali) for 3 h at 25 °C and then disperses in ice at 0 °C. Whereas in heterogeneous method, the chitin is dissolved in concentrated hot alkali and pH is neutralized using distilled water with repeated washings. In this process, de-acetylation process is difficult to produce, and therefore, thio-phenol is added as catalysts which reduce the degradation process and speed up the de-acetylation process. The effective de-acetylation (Fig. 1.2) process happens at 100 °C, and if concentration of alkali is four times more than the total amino acids present in polysaccharide. The release of toxic substances which cause environmental pollution and the usage of high energy are basic disadvantages [6]. The detailed schematic regarding extraction of chitin and chitosan from crustaceans is presented (Fig. 1.3).

1.1.5.2 Enzymatic Method

To overcome the drawbacks of the chemical method, an alternative method is used to convert the chitin into chitosan. In enzymatic deacetylation, the enzyme which is thermally stable, extracted from *Mucor rouxii* and *Absidia coerulea*, hydrolyze the

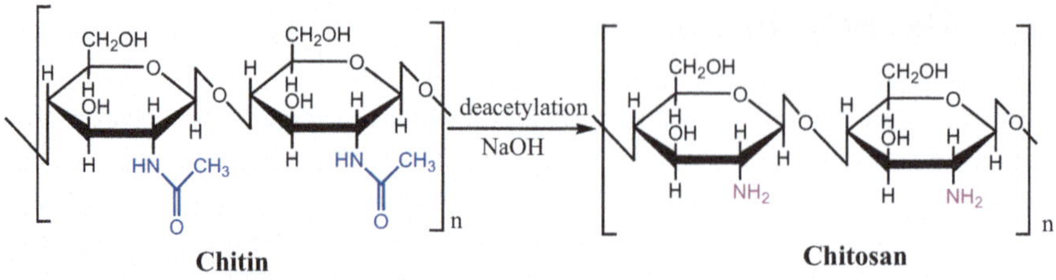

Fig. 1.2 Conversion of chitin to chitosan

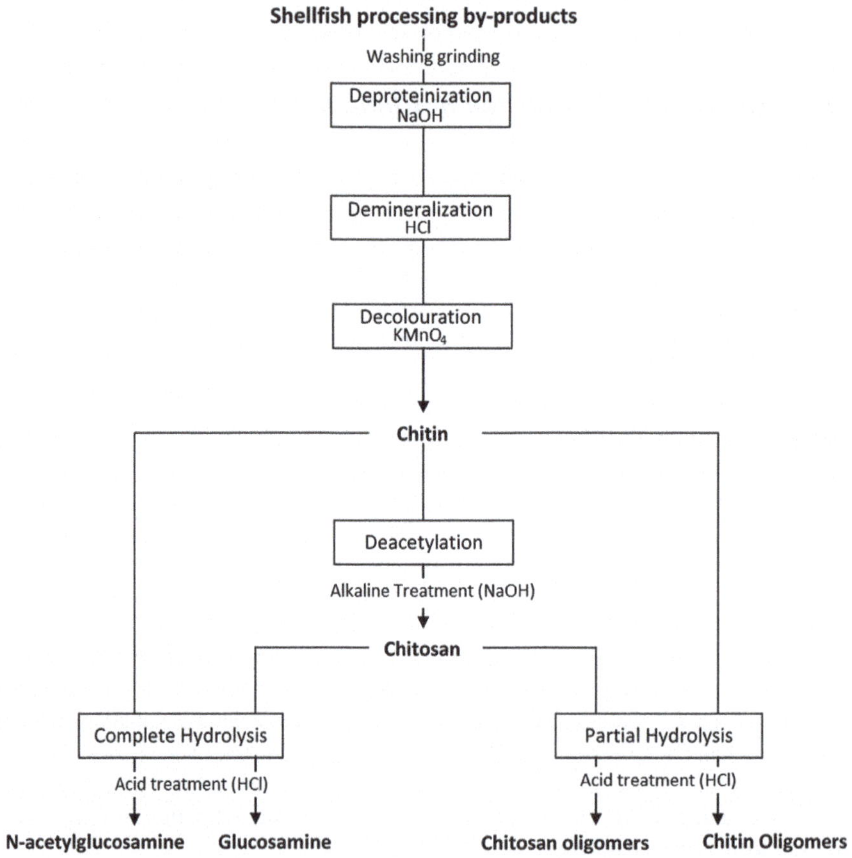

Fig. 1.3 Extraction of chitosan from crustaceans [6]

acetamide bonds. This method is carried out in both continuous and batch processes. Enzymatic treatment replaces the use of acid and base and nourishes the properties of product, and this method is environmentally friendly because it does not produce solvent and another waste material. The enzymes like cellulase, pepsins,

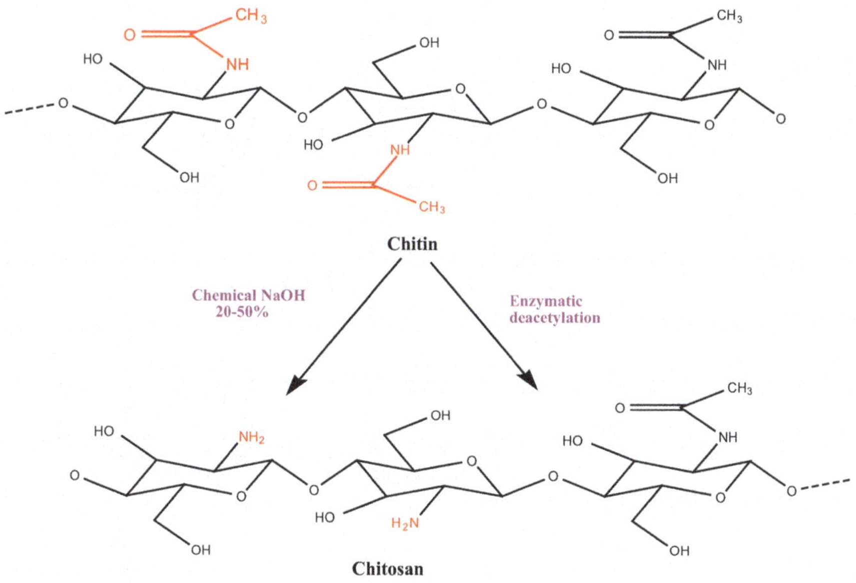

Fig. 1.4 Functional changes during conversion of chitin into chitosan

and amylase have the ability to breakdown the glycosidic linkage; i.e., the linkage presents between the chitin [7]. The functional changes during conversion of chitin to chitosan under chemical or enzymatic de-acetylation are presented in Fig. 1.4. Therefore, chitosan can be obtained through chitin by removal of acetyl group under enzymatic and chemical conditions (Fig. 1.4).

1.1.5.3 Irradiation with Ultrasound Radiations

The cost-effective and simple method is the de-acetylation of chitin under thermal conditions and gives a high yield of chitosan. Some physical methods like irradiation with ultrasound radiations of low frequency (20 kHz) also used for the manufacturing of chitosan [7].

1.1.6 Structure of Cellulose, Chitin, and Chitosan

Chitin and chitosan are derivatives of cellulose. If the hydroxyl group (–OH) at C-2 position of cellulose is replaced by acetamide (–$NHCOCH_3$) group, the resultant compound is chitin, whereas if the hydroxyl group (–OH) at C-2 position of cellulose is replaced by amine group (–NH_2), the resultant compound is chitosan (Fig. 1.5).

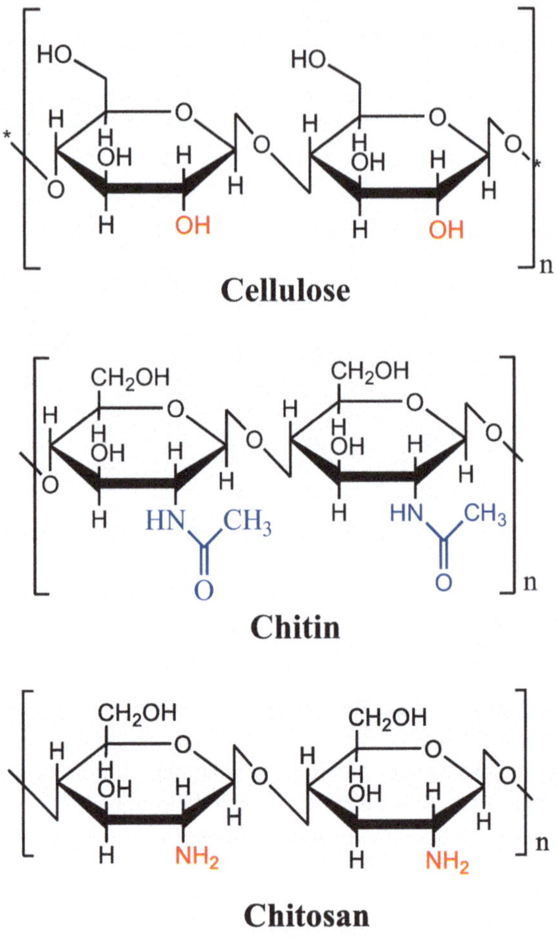

Fig. 1.5 Structures of cellulose, chitin, and chitosan

1.1.7 Derivatives of Chitosan

The properties like mechanical, chemical, and physical properties of chitosan-based materials can be enhanced by cross-linking, incorporating fillers, impregnation, and blending with different macromolecules. The structural modification is obtained by the presence of both the hydroxyl and amino group through different types of reactions. The cross-linking of amino and hydroxyl group of chitosan with different types of cross-linkers reduces adsorption power while grafting increases the adsorption capacity by adding more functional groups to the surface of chitosan. Chitosan has outstanding properties due to which it is abundantly used in almost all fields of life [8, 9].

Table 1.1 Derivatives of chitosan

S. No.	Derivative's name	Modification	Properties	References
1	Phosphorylated chitosan	By the addition of phosphate group at carbon 2 and 6	Tissue engineering	[5, 9]
2	Sulfated chitosan	By adding sulfite group at carbon 2 and 6	Biomedical applications	[2, 10]
3	Thiolated chitosan	By modification of amine group to 2-iminothiolane, thiobutylamidine, and cysteine	Drug delivery nasal and oral mucoadhesive	[4, 11]
4	Carboxymethyl chitosan	Carboxymethyl group at carbon 2 and 6	Drug delivery, biosensors, contact lenses	[3, 12]
5	Quaternized chitosan/N-alkyl	By modification of NH_2 to alkyl groups	More penetration of drug delivery	[5, 13]
6	N-acyl chitosan	By adding acetyl and formyl group	Increase solubility and penetration	[3, 14]
7	Hydrophobic chitosan	By introduction of –COOH group	Increases solubility	[2, 4, 15]

Structural variations of chitosan upgrade the physiochemical properties. The derivatives of chitosan obtained by varying the hydroxyl group and amino group at position six and two respectively are given (Table 1.1).

1.1.7.1 Phosphorylated and Sulfated Chitosan

The composites of chitosan are very well known because of their biocompatibility and biodegradability. Due to these properties, chitosan derivatives abundantly use in biomedical applications because of intrinsic antibacterial and its less foreign body reactions. The chitosan composites are prepared by replacing collagen, hydroxyapatite, alginate, and calcium phosphate [10]. These derivatives of chitosan mainly use in bone tissue engineering because bone is also composed of collagen and hydroxyapatite which are organic and inorganic in nature [16]. The schematic of phosphorylated and sulfated derivatives of chitosan is presented in Figs. 1.6 and 1.7 [9, 10].

1.2 Chitosan Nanoparticles (CSNPs)

As nanotechnology advances, nanostructures and nanophases have been applied to every field of science. An important building block of nanotechnology is nanoparticles. Nanoparticles are solid particles that measure in the nanometer range, ranging

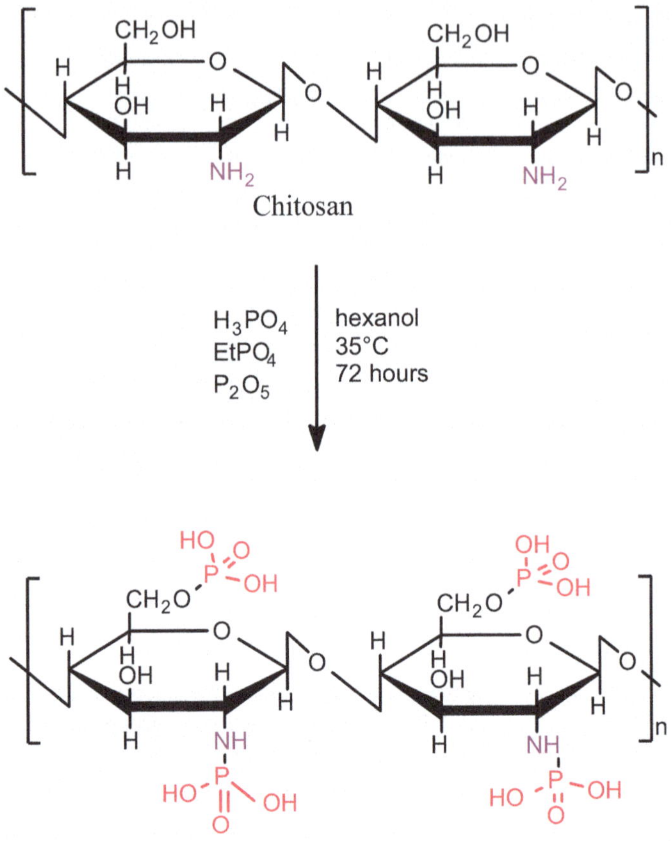

Fig. 1.6 Phosphorylated derivative of chitosan [9]

from 10 to 1000 nm. There are distinct physical and chemical properties that nanoparticles possess due to their large surface area and small size. The combination of polymers and nanotechnology has garnered significant interest over the last few years in different fields [17].

Natural and synthetic polymers can be synthesized into polymeric nanoparticles, and these particles are used for their stability and ease of surface modification. The use of polymer nanoparticles as carriers of bioactive molecules has garnered a great deal of interest in recent years. Biodegradable nanoparticles may serve as carriers of therapeutic molecules due to their biological properties, including biodegradability, biocompatibility, and bio-adhesiveness. Nanoparticles can also be fabricated from synthetic polymers such as polystyrene or from natural polymers such as protein and polysaccharide. The preparation of nanoparticles from synthetic polymers involves heat, organic solvents, or high shear forces which can negatively impact drug stability. On the other hand, nanoparticles prepared from natural polymers require only mild solvents and require little shear force to prepare [18]. A recent

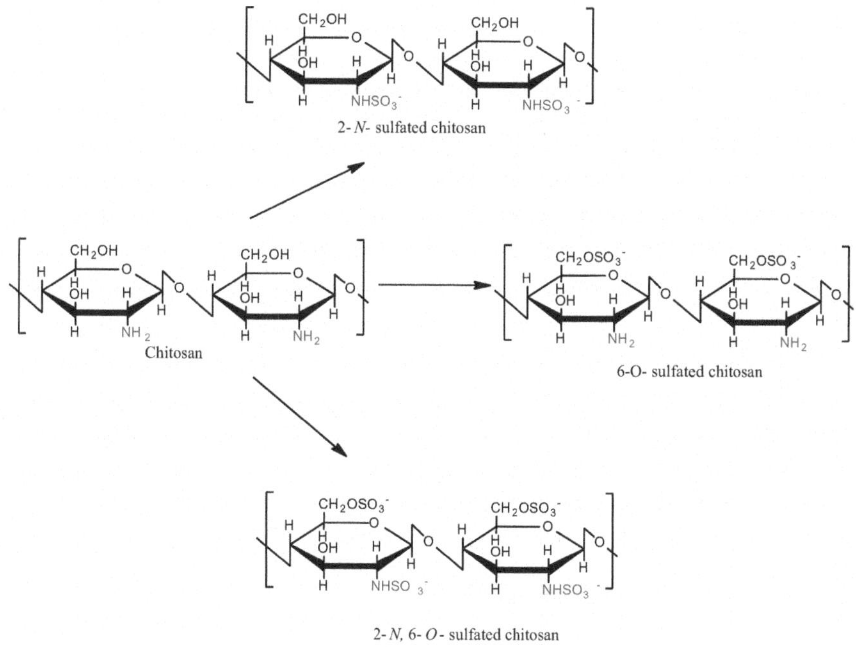

Fig. 1.7 Sulfated derivatives of chitosan [10]

study found that chitosan nanoparticles possess a number of biological properties, which have attracted considerable attention in the present scenario [19].

1.3 Applications and Limitations of Chitosan Nanoparticles

1.3.1 Applications of Chitosan Nanoparticles

Chitosan nanoparticles have great physical, chemical, biological, and antibacterial capabilities, making them a superior ecologically friendly material. It is bioactive and safe for people because chitosan is a natural material. Due to these special qualities, chitosan nanoparticles are used in a variety of applications including biomedical, biological and agriculture fields, wastewater treatment, textile industries, packaging of food, etc. [18–20].

1.3.1.1 Biomedical Applications

The positive charge on the chitosan is attracted by the negative charge present on most of the living tissues, hair, bone, and skin. Microbes also have negative charge on their external skin due to which chitosan has good binding power and helpful for biomedical applications as well as other living properties. Because of its significant biodegradability, antimicrobial, anti-inflammatory, wound healing properties, and biocompatibility, it performs extraordinary function in biomedical applications. Chitosan is easily molded into different sizes and porosities which are convenient for cell immobilization, drug delivery, and enzymes. Because of the above-mentioned properties chitosan nanoparticle widely used in different biomedical fields like (1) nanomedicines; (2) tissue engineering; (3) drug delivery [19].

Nanomedicines

Nanostructures and nanophases are being used in every field of science, especially nanomedicine, which has advanced with nanotechnology. The nanoparticles of chitosan can cross biochemical barriers, which is an advantage over chitosan. Chitosan is biocompatible and non-toxic due to which it is extensively used as a carrier for drug delivery systems in pharmaceutical industry. Several biomedical applications have made use of this property, including targeted delivery of drugs to the lungs, nasal, and colon implants [20].

Tissue Engineering

Tissue engineering is a process to replace or improve the biological functions using physiochemical and biochemical factors, combination of cells and material methods, and engineering. Different types of synthetic and natural materials are used and play an important role in tissue engineering of artificial organs. Chitosan derivatives play an important role in preparing different organs like bone, blood vessels, skin, and liver through tissue engineering [15].

Drug Delivery

The modification of chitosan by chemicals has a great influence on the delivery of various kinds of drugs in a target place. The nanoparticles synthesized from chitosan show an increase in loading efficiency and capacity. Norcantharidin, docetaxel, doxorubicin, and paclitaxel, the antitumor active compounds, were used as carriers of drugs. Low molecular weight chitosan is efficient for carrier as it has been approved as new carriers for nasal vaccine delivery in mice [21, 22].

1.3.1.2 Biological Applications

The biological applications are administered based on their special characteristics including (1) anti-inflammatory; (2) anticancer; and (3) antibacterial.

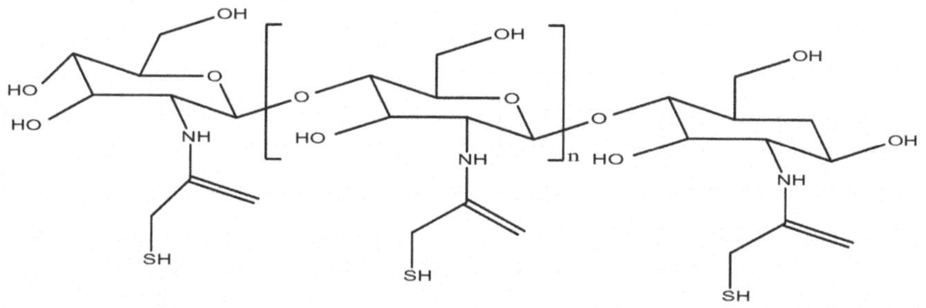

Fig. 1.8 Chemical structure of thiolated chitosan for drug delivery

Anti-inflammatory

The property of a substance or treatment which reduces the inflammation is known as anti-inflammatory property. For example, theophylline is an anti-inflammatory drug for allergic asthma but the usage of this drug has various side effects. Therefore, this drug is adsorbed on the modified chitosan nanoparticle which improves its anti-inflammatory effects (Fig. 1.8) [22, 23].

Antibacterial Activity

Chitosan has been shown to have antibacterial effects against a range of microorganisms, including bacteria, fungi, yeast, and viruses, with varying efficiencies of inhibition. The types of substituents like cationic species that can easily form cations, degree of deacetylation, molecular weight, and type of bacteria affect the antibacterial activity. For example, *Escherichia coli* (*E. coli*) which has a negative charge on its surface is electrostatically interacted by cationic chitosan derivatives. In numerous studies, it has been reported that the bacteria having different cell walls have shown a stronger antibacterial action against gram-positive bacteria than gram-negative bacteria [24, 25].

Anticancer Activity

The biocompatible chitosan-based drugs are used to overcome the side effects produced by traditional chemotherapeutic drugs. Chitosan-based drugs reduce the destruction of bone marrow and lymphoid. The introduction of phenyl and sulfate groups to chitosan is most widely used derivatives for cancer treatment (Fig. 1.9) [26, 27].

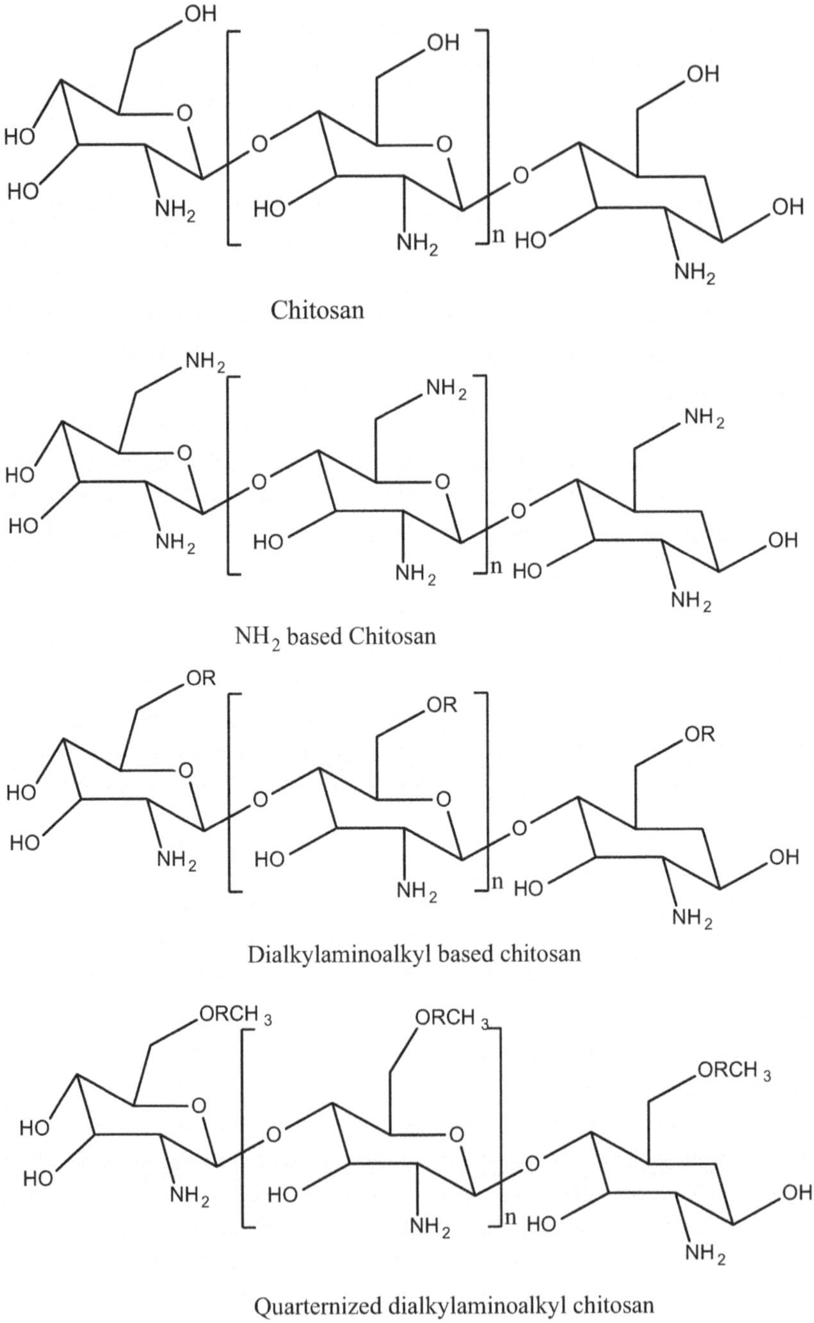

Fig. 1.9 Cationic derivatives of chitosan for cancer treatment [26, 27]

1.3.1.3 Agriculture Field

Since chitosan is a biopesticide, eco-friendly resulting enhances plants' natural capacity to protect themselves against fungus, it may also be used primarily as a natural seed treatment and plant development booster. Numerous pre- and postharvest diseases on different horticultural commodities have been halted by chitosan [28].

1.3.1.4 Wastewater Treatment

Different water assets have organic pollutants as well as effluents discharge from industries such as detergents, fertilizers, plasticizers, hydrocarbons, greases, biphenyls, pesticides, and pharmaceuticals. So, industrial wastewater has serious environmental threats. When industrial effluents are discharged into water bodies like lakes or rivers, it destroys water quality. To meet the best environmental standards, these inorganic and organic pollutants are efficiently removed from water resources. The chitin and chitosan being non-toxic and biodegradable nature can be used in wastewater treatment. The cationic properties of chitosan and chitin allow them to precipitate and coagulate in basic and neutral pH by binding contaminated medium with polymer. To remove wastewater pollutants, chitosan may act as an adsorbent material due to its hydroxyl (–OH) and amino ($-NH_2$) functional groups present in its backbone. Chitosan has the ability to modify in different chemical structures by it with various chemicals which allow its use in different fields. Chitosan has better adsorption capacity in neutral and in alkaline medium than in acidic media such as zeolites, oil palm ash bentonite, and polyurethane. On the other hand, the anionic suspended particles form aggregation with chitosan reduce the turbidity, chemical oxygen demand, and chlorides in wastewater. For the removal of heavy metals, dyes, organic pollutants (organochloride pesticides, oil impurities), chitosan acts as an adsorbing material [29, 30].

1.3.1.5 Fuel Cell

Fuel cells are environmental-friendly devices for power generation, energy conversion with no emission of power sources. Fuel cells being electrochemical energy storage system convert chemical energy into electrical energy by redox reaction. In fuel cells, electrodes (anode and cathode) are dipped in the electrolyte. Chitosan, being polycationic due to its amino group acts as polyelectrolyte complex within a variety of polyanions including collagen, alginate, DNA, synthetic polymers, and lingo-sulfonate via electrostatic interactions [8].

1.3.2 Limitations

Chitosan nanoparticles have been used in every fields of life; however, it has some limitations as it is soluble in low concentrations of inorganic acids like hydrochloric acid and acetic acid solution, which greatly affects its usage in various fields. The solubility of chitosan can be enhanced by chemical processes. Quaternary chitosan increases the solubility as well as retains the marvelous properties. In biomedical field, there is huge application of chitosan and chitosan nanoparticles. Quaternized chitosan nanoparticles have the great potential as in the future it will be used for the development of nanovaccines.

1.4 Preparation of Chitosan Nanoparticles (CSNPs)

For the synthesis of chitosan nanoparticles, different methods are used

- ionotropic gelation method;
- complex co-acervation method;
- co-precipitation method;
- microemulsion method;
- emulsion-based solvent diffusion method;
- reverse micellar method;
- polyelectrolyte complexes;
- emulsion-based solvent evaporation method.

1.4.1 Ionotropic Gelation Method

The chitosan nanoparticles (CSNPs) are widely prepared using this simple, mild method. The CS chains and polyanions like tripolyphosphate (TPP), which serve as cross-linker, form nanoparticles through electrostatic interactions. Depending on the mutual ratio between chitosan and polyphosphate ions, nanoparticles of different diameters are formed (Fig. 1.10). Chitosan nanoparticles are often prepared with tripolyphosphate (TPP), because it is non-toxic, multivalent, and capable of forming gels through ionic interactions.

In this method, 2 M acetic acid is slowly added at 25 °C under stirring at 9000 rpm to low molecular weight chitosan. During the reaction between chitosan and TPP, the amine moieties of the chitosan can cross-link with the phosphine residues of the TPP, resulting in nanoparticle formation. In this process, chitosan is transformed into a gel-like compound that cannot be dissolved in water. An acetic acid is a naturally occurring compound that can be used safely and effectively by the body. After adjusting the pH to 5.5 with 2 M NaOH solution, the chitosan solution is filtered through a cellulose acetate filter to remove any undissolved chitosan [31, 32].

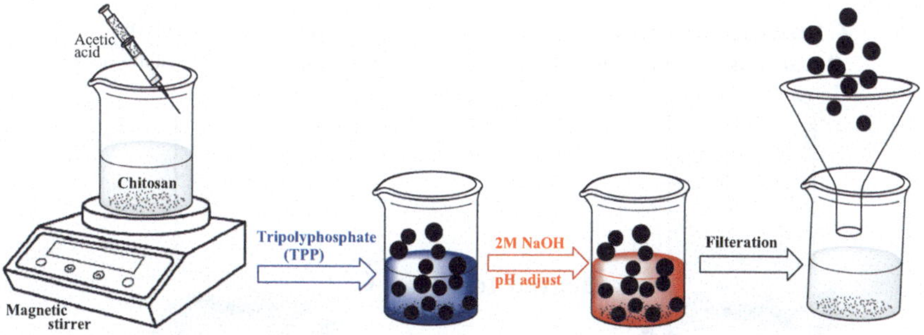

Fig. 1.10 Schematic diagram for ionotropic gelation method

1.4.2 Complex Co-acervation Method

Chitosan is dissolved in 1% acetic acid with gentle heating, and pH of the solution is adjusted to 5.5–5.7 with sodium hydroxide. A chitosan solution and a pDNA solution of 100 μg/mL in 50 mM sodium sulfate solution are preheated to 50–55 °C separately. An equal volume of both heated solutions is quickly mixed together and vortexed for 45 s (Fig. 1.11) [33].

1.4.3 Co-precipitation Method

In co-precipitation method, chitosan-based nanoparticles with a high degree of size uniformity are prepared by grafting D, L-lactic acid on chitosan to serve as a drug carrier for prolonged drug release. The lactic acid-grafted chitosan (LA-g-chitosan) is prepared by dehydrating the solvent cast thin film of chitosan containing lactic acids. The LA-g-chitosan nanoparticle is fabricated via a co-precipitation process by LA-g-chitosan in ammonium hydroxide to form co-acervate drops. Spherical and uniformly dispersed chitosan and lactic acid-modified chitosan (LA-g-chitosan) nanoparticles with a mean diameter of ca. 10 nm can be prepared using this method. Albumin encapsulation efficiency as high as 92% and 96% can be attained for chitosan and LA-g-chitosan nanoparticles, respectively [34].

1.4.4 Microemulsion Method

The method involved dissolving a surfactant in n-hexane, adding chitosan to acetic solution, then simultaneously adding glutaraldehyde while constantly stirring the mixture at room temperature [35]. The glutaraldehyde acts as a cross-linker during this process. A surfactant is used to create nanoparticles, and the system is stirred

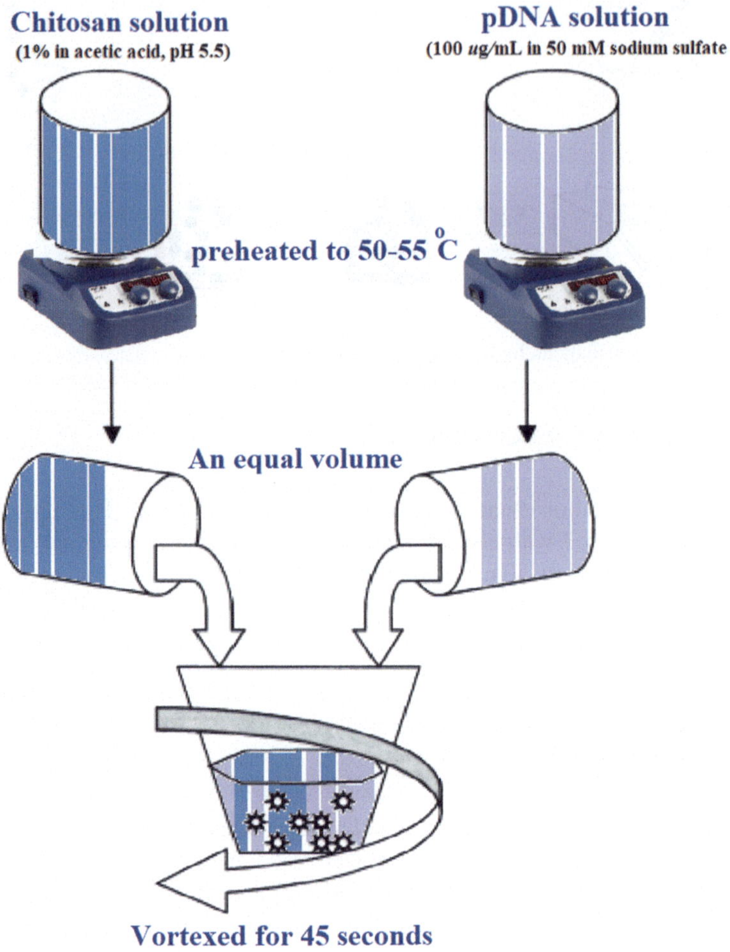

Fig. 1.11 Schematic presentation of the nanoparticle preparation by the complex co-acervation method [33]

overnight to allow the free amine group of chitosan to cross-link with glutaraldehyde. After evaporating the organic solvent, excess surfactant is precipitated with $CaCl_2$ and centrifuged to remove the precipitate. The characteristics considered as disadvantages of this method are the use of an antigenic agent, glutaraldehyde, and the incorporation of proteins to nanoparticles is not feasible as the covalent cross-linking may damage them.

1.4.5 Emulsion-Based Solvent Diffusion Method

The method begins by preparing an oil-in-water (O/W) emulsion by simultaneous inclusion of an organic phase such as methylene chloride or acetone, containing a hydrophilic drug [36]. Under mechanical stirring and high-pressure homogenization, an emulsion is obtained by injecting an organic phase into chitosan solution containing a stabilizing agent such as poloxamer. Polymer precipitation is caused by the diffusion of organic solvent into water, which forms nanoparticles as the emulsion is diluted with a large amount of water. This method could produce nanoparticles in the 300–500 nm range. In addition to the high shear forces used during nanoparticle preparation, this method also uses organic solvents [5].

1.4.6 Reverse Micellar Method

In this method, the aqueous solution of glutaraldehyde and chitosan is mixed with the organic solution containing an organic solvent, n-hexane, and lipophilic surfactant (Fig. 1.12). To avoid from turbidity of solution, the aqueous mixer was added with constant agitation. The concentration of water can be increased if large size of nanoparticles is required. The most useful surfactants are sodium bis (2-ethylhexyl) sulfosuccinate and cetyltrimethylammonium bromide CTAB. In order to form reverse micelles, chitosan is added to the organic solvent that contains the surfactant. The ultrafine nanoparticles having size less than 100 nm can be obtained through this method [32, 37]. The synthesis of nanoparticles completes in three different steps: (i) precipitation of $CaCl_2$ and surfactant; (ii) the elimination of unreacted materials through dialysis; and (iii) freeze and drying.

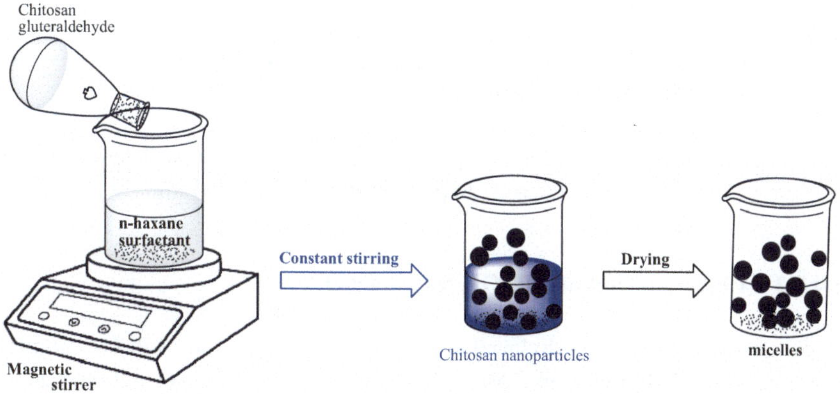

Fig. 1.12 Chitosan nanoparticles by reverse micellar method

1.4.7 Polyelectrolyte Complexes

The simplest method to prepare polyelectrolyte complexes (PECs) is co-acervation by mixing the acidic aqueous solution of chitosan (polycation) with an aqueous solution of polyanion. Because of its unique cationic character, chitosan can interact with anionic polymers to form polyelectrolyte complexes (PECs). These complexes, thanks to their easy preparation, biocompatibility, and pH sensitivity, have been widely studied as excipients for the formulation of active agents in multiple drug delivery systems. Mateescu et al. [38] summarized the aspects of drug delivery by PECs obtained by complexation of unmodified chitosan with natural polysaccharides or their derivatives, such as chondroitin sulfate, carboxymethyl starch, dextran sulfate, pectin, and alginate.

1.4.8 Emulsion-Based Solvent Evaporation Method

Precipitation and emulsion are formed by adding chitosan in aqueous phase, and nanoparticles are appeared as final product after the evaporation process and particle size range are 50–300 nm [39]. In this method, emulsions of chitosan are formed as a result of solvent evaporation. At start, chitosan solution is added in aqueous phase which result an emulsion. It is followed by the polymer–solvent evaporation and nanospheres are appeared as precipitated product. Then the chitosan is added in ethanol followed by the addition of pDNA-tris buffer along with high pouring rate of ethanol under continuous magnetic stirring. The reduced pressure is applied to remove the solvent resultantly nanoparticles which are formed.

Nanoparticles of chitosan for various purposes can be synthesized in variant size. A brief of such materials is given below.

1.4.8.1 Cyclosporine-A-Loaded PEGylated Chitosan-Modified Lipid-Based Nanoparticles

The efficiency of cyclosporine-A encapsulation on modified nanoparticles of chitosan is noted to be 69.22% and the average particle size may be 89.4 nm.

1.4.8.2 Chitosan-Modified PLGA Nanoparticles

In this method, 29 mg/ml ethyl acetate is used to dissolve 5 ml of PLGA, and this solution is added dropwise in 2.5%, w/v stabilizer PVA aqueous solution, and the cationic-polymer chitosan obtained after magnetic stirrer. For one hour, the emulsion is stirred at 1000 rpm and UltraTurrax T25 is used to homogenize the emulsion for 10 min at 13,500 rpm. The resultant product is used to prepare nanoparticles by diluting with Milli-Q water (50 ml) under constant stirring and the continuous stirring is continued to remove surplus ethyl acetate at room temperature overnight.

1.4.8.3 The Nanoparticles of Lectin/Chitosan

The nanoparticles of lectin/chitosan are hydrochlorothiazide (HTC) loaded nanoparticles latterly complexed with b-cyclodextrin (HCT0b-CD). The maximum recorded entrapment efficiencies are 81.8% and 91.1% loaded with HCT and HCT-b nanoparticles, respectively. The size ranges of such particles are 126–139 to 151–181 nm.

1.4.8.4 Nanospheres of Chitosan

In this method [39], 10 ml of ethyl acetate is used to dissolve 200 mg of PLGA at room temperature. An aqueous stabilizer solution of 100 mg poly(vinyl acetate) (PVA) is used to add to the above-prepared organic phase, and then further, 10 ml aqueous solution of 30 mg of chitosan is added into it. At room temperature, the emulsion is stirred for 3 h. Afterward, UltraTurrax T25 homogenizer is used to homogenize the emulsion for 10 min at 13,500 rpm. Resultantly, nanoprecipitation takes place by adding water under continuous stirring. A 40 °C temperature of water bath is sustained to remove organic solvent during stirring.

1.4.8.5 Cyclosporine-A-Loaded, PEGylated Chitosan-Modified Nanoparticles

The emulsification/solvent evaporation method is used to prepare nanoparticles. The solution A (a mixture of methylene chloride and acetone in 3:1) is used as solvent to dissolve soybean lecithin and cyclosporine-A. An aqueous solution of PEGylated chitosan and poloxamer is prepared as solution B. Under fast magnetic stirring, solution A is injected with slow speed into solution B and the medium is followed by ultrasonic dispersion by using a probe type ultra-sonicator (400 W, JY92-II). The ultrasonic-treatment is performed in 60 times at a 3 s interval in an ice bath after one second sonication. The solvent of emulsion (methylene chloride and acetone) is removed by reduced pressure by applying rotovap (RE-52C) at room temperature. The Heraeus Instrument is used to obtain precipitates by the centrifugation process at 12,000 × g at 4 °C for 30 min. The precipitates are dissolved in water to prepare an aqueous solution, and finally, sorbitol {10% (w/v)} is used to freeze-dried the prepared product.

1.4.8.6 Chitosan-Polylactic Acid (CS-PLA) Nanoparticle

In this method, a fine dispersion of chitosan-PLA nanoparticles is prepared by dissolving 120 mg of PLA in 10 ml of dichloromethane. The prepared solution is injected rapidly into 10 ml of 1% acetic acid solution containing 60 mg of CS and 200 mg of PEO. An emulsion is formed after the process of sonication for 60 min and strongly stirred to remove organic solvents. The nanoparticles are precipitated first by adding water and then cooled at 10 °C. Finally, rotary evaporator is used to evaporate the nanoparticle suspension [40].

1.5 Recent Advances in Preparations of CSNPs

1.5.1 Emulsification and Cross-linking

A new emulsification-cross-linking method has been developed for the preparation of chitosan microspheres [41]. In this method, the surfactant is dispersed in 2,3-dimethylbutane and CS is dissolved in a solution of acetic acid using this technique, and glutaraldehyde ($C_5H_8O_2$) is injected into the surfactant/C_6H_{14} combination while continuously stirring at ambient temperature (Fig. 1.13). This process is the first to be utilized to create chitosan nanofibers (CNF) and requires the synthesis of a W/O followed by the incorporation of a cross-linking agent that gives hardness to the synthesized droplets. There is cross-linking between aldehyde groups of C_5H_8O and amino groups (NH_2) of chitosan which is supplied after emulsification and, hence, after NP synthesis [42].

The researchers first created the chitosan nanoparticles (CSNPs), which are then employed to administer 5-fluorouracil. Some authors employed the approach for

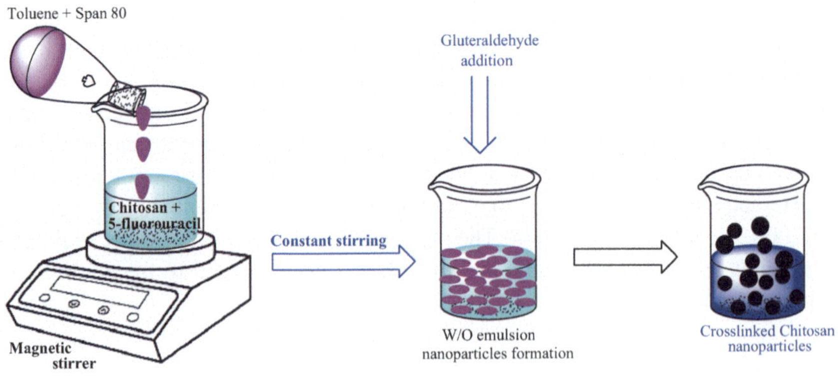

Fig. 1.13 Synthesis of cross-linked chitosan nanoparticles [42]

drug administration as well but changed the oil phase formation to petroleum ether and liquid paraffin [42, 43]. The degree of cross-linking and stirring time greatly affects the particle size of cross-linked chitosan nanoparticles [44].

Several disadvantages of this technique have been identified over time, including the requirement for time-consuming processes and the use of strong cross-linking molecules. Indeed, cross-linkers like glutaraldehyde have been proven to produce toxicity and endanger drug integrity. Furthermore, covalent cross-linking may destroy the nanoparticles; therefore, the protein or peptide incorporation is not feasible. As a result, the application of this approach to synthesize chitosan nanoparticles has been limited to a few works [41–43, 45].

1.5.2 Emulsion Droplet Coalescence

The modification of the cross-linking and emulsification processes mentioned above is initially published for the preparation of microparticles (Fig. 1.14). The approach is then modified by the same group of researchers to make chitosan nanoparticles (CSNP) coated with gadolinium as a method for cancer neutron-capture treatment. The CS is mixed in gadolinium aqueous solution, and a small portion (1 mL) is injected into 10 mL of paraffin liquid containing sorbitan sesquioleate (Span® 83). A high-speed homogenizer is used to agitate the mixture, resulting in a water–oil emulsion [46]. Similarly, another water-to-oil emulsion is made by combining 1.5 mL of sodium hydroxide with 10 mL of the same outer phase. After that, both emulsions are combined with a high-speed homogenizer, resulting in particle coalescence. This causes the CS particles to solidify due to the action of sodium hydroxide, which functions as a precipitating agent. Following that, ethanol, water, and toluene are used in a series of centrifugation and washing procedures.

The droplets of uniform composition are synthesized whenever two emulsions of similar outer phases are combined. The nanoparticles develop within the emulsion droplet. Decreased chitosan de-acetylation has been reported to increase the size of particles and limit nanoparticle capacity for drug attachment due to decreased ion–pair production and de-swelling capacity. It has been also discovered that 5-fluorouracil encapsulation effectiveness (about 70%) and release profile remain unchanged, which lasted 13 h on altering the chitosan content from 0.5 to 2.5%.

This method of synthesis [47] is feasible for the encapsulation of cyclosporine-A drug which is hydrophobic in nature. The ultimate characteristics of the vehicles are anticipated to be influenced by factors including chitosan molecular weight, reaction time, and homogenization rate. Moreover, the use of acetone as a solvent results in the formation of very small size droplets. In spite of the production of efficient vehicles, this procedure requires harsh conditions like high shear forces and organic solvents [47].

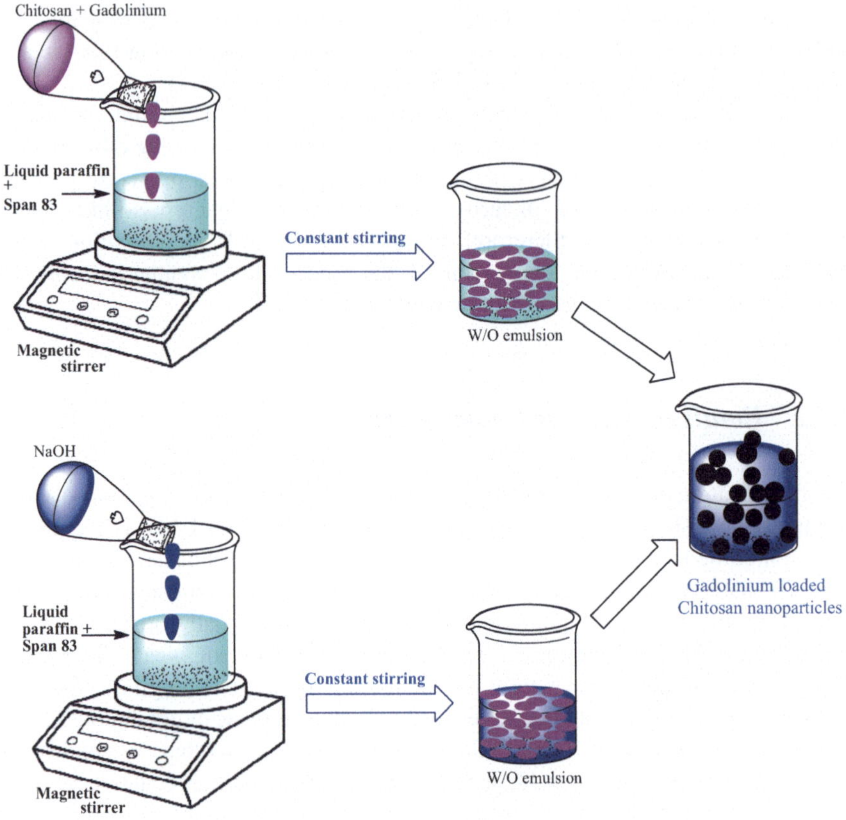

Fig. 1.14 Emulsion droplet coalescence [46]

1.5.3 Ionic Gelation Through Radical Polymerization

The approach is similar to ionic gelation, but this differs in that chitosan gelation happens simultaneously with the monomers of acrylic acid polymerization. The first stage is to mix an aqueous monomer solution of meth-acrylic or acrylic acid, including an aqueous solution of oppositely charged chitosan (CS), at 25 °C [46, 48]. Polyether or polyethylene glycol is sometimes added to the reaction solution, usually individually into the monomer solution or after combining with chitosan. The opposing charges of acrylic and CS or meth-acrylic acid cause an ionic contact and radical polymerization of both is triggered by the use of potassium per sulfate [48].

The reaction occurs in the presence of a nitrogen stream, as well as when the temperature is elevated to 60–70 °C. This reaction completes within around 6 h, following which the produced nanoparticle dispersion is allowed to stay overnight. Ultimately, the synthesized nanoparticles are washed using distill water to remove

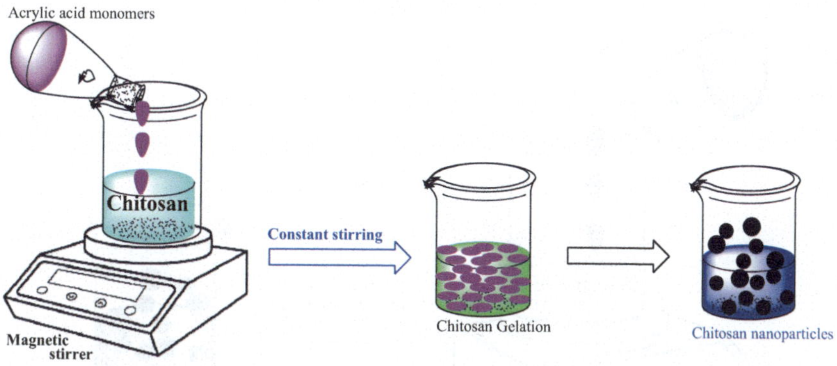

Fig. 1.15 Ionic gelation through radical polymerization [46, 48]

the unreacted moieties (Fig. 1.15). These nanoparticles have been utilized to give bovine serum silk, albumin, and insulin peptides orally. Increasing the MW of CS has been demonstrated to increase the MW of polyacrylic acid contained in the NP, demonstrating the presence of chitosan-induced template polymerization of acrylic acid.

Factors including the acrylic monomer/CS ratio and polymer concentration were discovered to have a significant impact on the physicochemical characterizations of NP. More precisely, it is stated that the zeta potential rises as the ratio of CS/acrylic acid increases and the size of nanoparticles is smaller for the minimum polymer ratio (1:1). The physiochemical properties of synthesized particles are greatly affected by different factors like polymer concentration and ratio of chitosan and acrylic moieties [37].

1.5.4 Desolvation

The desolvation method also known as phase separation or simple co-acervation, in which macromolecular aggregation occurs by partial desolvation of completely solvated molecules. The application of desolvating agents to synthesize chitosan-based particles was first studied for the creation of micron-sized carriers, but this approach is now commonly used to produce chitosan nanoparticles. Precipitating agents include non-solvents and sodium sulfate miscible with water and acetone (Fig. 1.16). This technique of generating chitosan nanoparticles is relatively easy and gentle since it includes the mixing of already formed solution of chitosan and dropwise addition of hydrophilic competing agent like sodium sulfate [49]. As the aqueous solution of chitosan comes into contact with salt, a gradual removal of the solvation water around the chitosan happens due to the salt's increased affinity for water. This reaction eventually results in polymer insolubilization and precipitation.

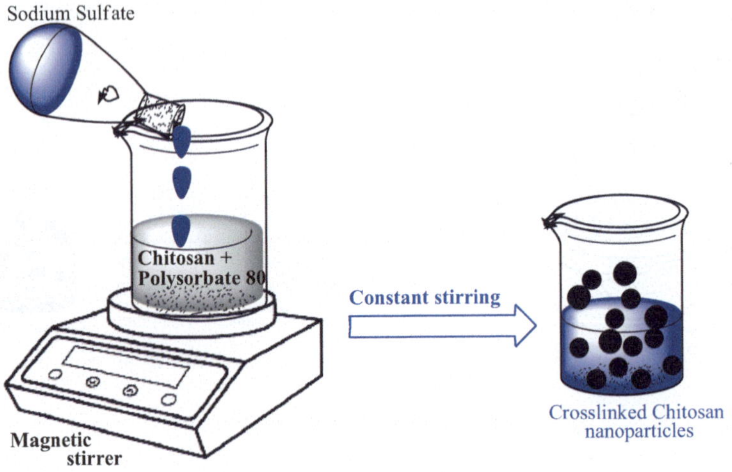

Fig. 1.16 Desolvation method with sodium sulfate [49]

The partial desolvation of chitosan occurs because water–salt interactions are more prominent than water–polymer interactions. This in turn increases the contact between chitosan molecules resulting in the formation of nanocarriers. The stabilization of the nanoparticle suspension occurs using a stabilizer polysorbate 80 which is frequently used in the preparation medium. Glutaraldehyde is used in the preparation process to harden the nanoparticles.

Factors such as concentration of chitosan, stirring rate, molecular weight of chitosan, concentration of desolvating agent may greatly affect the final properties of synthesized nanoparticles, and as a result, it is important to do parameter optimization. Furthermore, a relationship was discovered between the quantity of sulfate ions required and chitosan features such as molecular mass and de-acetylation degree [44].

1.6 Recent Development in Synthesis of Biocompatible Polymers Functionalized by Chitosan and Chitosan Nanoparticles

The studies on chitin and chitosan-based blends, composites, and nanocomposites have been comprehensively discussed [50]. Chitosan functionalized poly(vinyl alcohol) for prospects biomedical and industrial applications has been well documented [51]. Synthesis and molecular characterization of chitosan-based polyurethane elastomers using aromatic diisocyanate have been reported in the literature [52].

Influence of chitosan/1, 4-butanediol blends on the thermal and surface behavior of polycaprolactone diol-based polyurethanes have been reported [53]. Evaluation of cytotoxicity, hemocompatibility, and spectral studies of chitosan-assisted polyurethanes prepared with various diisocyanates have also been comprehensively studied [54]. Synthesis and molecular characterization of chitosan/starch blends-based polyurethanes and chitosan modified polyurethane bio-nanocomposites with biomedical potential have been well-documented [55, 56]. Effect of diisocyanate nature by mixture design approach with the mathematical modeling and experimental study of mechanical properties of chitosan-based polyurethanes have been reported in the literature [57]. Synthesis and molecular characterization of chitosan/alginate blends-based polyurethanes biocomposites and thermal degradation behavior and X-ray diffraction studies of chitosan-based polyurethane bio-nanocomposites using different diisocyanates have been comprehensively studied [58, 59]. Preparation of amylopectin and chitosan-based polyurethanes for sustained drug release studies and development of biocompatible aqueous polyurethane dispersions using chitosan and curcumin to improve physicochemical properties of textile surfaces have been administered and reported [60, 61].

1.7 Conclusion

The spark of the green technology and green products has illuminated the new paths of research in the recent years. It is ecologically and economically friendly technology. This chapter provides a detailed review of the emergence of chitosan as nanoparticles. It has elaborated the history of chitosan as a vibrant biological material. It has been elaborated in detail that chitosan is a natural biodegradable and biocompatible material which has been emerged as an excellent raw material for hydrogels, tissue engineering, drug delivery applications, agriculture, and wastewater treatment. Various synthetic modes of conversion of chitin into chitosan have been described including chemicals, enzymatic, and irradiation with ultrasound radiations; each method is unique and has pros and cons. Moreover, the details of the derivative of chitosan have also been included to provide an overview about the efficacy of the chitosan for various applications. The methods for the preparation of chitosan nanoparticles have also been listed. It has been described that chitosan nanoparticles possess a number of biological properties, which have attracted considerable attention in the present scenario. The chitosan nanoparticles have been discovered as an important reinforcement in various applications in particular biological and biomedical applications, covering a wide area of nanomedicine, tissue engineering, and drug delivery system. It has also been evaluated that the nanoparticles of chitosan are not devoid of limitations particularly associated with nanoparticles in general. It is soluble in low concentrations of inorganic acids like hydrochloric acid and acetic acid solution, which greatly affects its usage in various fields. However, this issue can be addressed by enhancing the solubility of chitosan by chemical process.

References

1. Asiri SM, Khan FA, Bozkurt A (2018) Synthesis of chitosan nanoparticles, chitosan-bulk, chitosan nanoparticles conjugated with glutaraldehyde with strong anti-cancer proliferative capabilities. Artif Cells Nanomed Biotechnol 46. https://doi.org/10.1080/21691401.2018.1533846
2. Jeuniaux C (1996) A brief survey of the early contribution of European scientists to chitin knowledge. In: Domard A, Jeuniaux C, Muzzarelli RAA, Roberts G (eds) Advances in chitin sciences. Jacques André Publ, Lyon, France, pp 1–9
3. Muzzarelli RAA, Muzzarelli C (2009) Chitin and chitosan hydrogels. In: Phillips GO, Williams PA (eds) Handbook of hydrocolloids. Woodhead Publishing Ltd., Cambridge, UK, pp 849–888
4. Kumari S, Kishor R (2020) Chitin and chitosan: origin, properties, and applications. In: Handbook of chitin and chitosan: volume 1: preparation and properties. https://doi.org/10.1016/B978-0-12-817970-3.00001-8
5. Boominathan T, Sivaramakrishna A (2021) Recent advances in the synthesis, properties, and applications of modified chitosan derivatives: challenges and opportunities. https://doi.org/10.1007/s41061-021-00331-z
6. Sharifi-Rad J, Quispe C, Butnariu M, Rotariu LS, Sytar O, Sestito S, Rapposelli S, Akram M, Iqbal M, Krishna A, Kumar NVA, Braga SS, Cardoso SM, Jafernik K, Ekiert H, Cruz-Martins N, Szopa A, Villagran M, Mardones L, Martorell M, Docea AO, Calina D (2021) Chitosan nanoparticles as a promising tool in nanomedicine with particular emphasis on oncological treatment. https://doi.org/10.1186/s12935-021-02025-4
7. Gomes LP, Paschoalin VMF, Del Aguila EM (2017) Chitosan nanoparticles: production, physicochemical characteristics and nutraceutical applications. https://doi.org/10.21577/1984-6835.20170022
8. Ma J, Sahai Y (2013) Chitosan biopolymer for fuel cell applications. https://doi.org/10.1016/j.carbpol.2012.10.015
9. Li QL, Chen ZQ, Darvell BW, Zeng Q, Li G, Ou GM, Wu MY (2006) Biomimetic synthesis of the composites of hydroxyapatite and chitosan-phosphorylated chitosan polyelectrolyte complex. Mater Lett 60. https://doi.org/10.1016/j.matlet.2006.03.046
10. Pires NR, Cunha PLR, MacIel JS, Angelim AL, Melo VMM, De Paula RCM, Feitosa JPA (2013) Sulfated chitosan as tear substitute with no antimicrobial activity. Carbohydr Polym 91. https://doi.org/10.1016/j.carbpol.2012.08.011
11. Sarti F, Bernkop-Schnürch A (2011) Chitosan and thiolated chitosan. In: Chitosan for biomaterials I, pp 93–110. https://doi.org/10.1007/12_2011_109
12. Mourya VK, Inamdara N, Ashutosh Tiwari N (2010) Carboxymethyl chitosan and its applications. Adv Mater Lett 1(1):11–33. https://doi.org/10.5185/amlett.2010.3108
13. Kim CH, Choi KS (2002) Synthesis and antibacterial activity of quaternized chitosan derivatives having different methylene spacers. J Ind Eng Chem 8(1):71–76
14. Lee KY, Ha WS, Park WH (1995) Blood compatibility and biodegradability of partially N-acylated chitosan derivatives. Biomaterials 16(16):1211–1216. https://doi.org/10.1016/0142-9612(95)98126-Y
15. Kim JH, Kim YS, Kim S, Park JH, Kim K, Choi K, Chung H, Jeong SY, Park RW, Kim IS, Kwon IC (2006) Hydrophobically modified glycol chitosan nanoparticles as carriers for paclitaxel. J Controlled Release. https://doi.org/10.1016/j.jconrel.2005.12.013
16. Tan W, Krishnaraj R, Desai TA (2001) Evaluation of nanostructured composite collagen—chitosan matrices for tissue engineering. Tissue Eng. https://doi.org/10.1089/107632701300062831
17. Agarwal M, Agarwal MK, Shrivastav N, Pandey S, Das R, Gaur P (2018) Preparation of chitosan nanoparticles and their in-vitro characterization. Int J Life-Sci Sci Res 4. https://doi.org/10.21276/ijlssr.2018.4.2.17
18. Yanat M, Schroën K (2021) Preparation methods and applications of chitosan nanoparticles; with an outlook toward reinforcement of biodegradable packaging. React Funct Polym 161:104849. https://doi.org/10.1016/j.reactfunctpolym.2021.104849

19. Yu J, Wang D, Geetha N, Khawar KM, Jogaiah S, Mujtaba M (2021) Current trends and challenges in the synthesis and applications of chitosan-based nanocomposites for plants: a review. Carbohyd Polym 261:117904. https://doi.org/10.1016/j.carbpol.2021.117904
20. Mohammed MA, Syeda JT, Wasan KM, Wasan EK (2017) An overview of chitosan nanoparticles and its application in non-parenteral drug delivery. Pharmaceutics 9(4):53. https://doi.org/10.3390/pharmaceutics9040053
21. Vinsova J, Vavrikova E (2012) Recent advances in drugs and prodrugs design of chitosan. Curr Pharm Des 14. https://doi.org/10.2174/138161208799316410
22. Sonia TA, Sharma CP (2011) Chitosan and its derivatives for drug delivery perspective. Adv Polym Sci 243. https://doi.org/10.1007/12_2011_117
23. Lee DW, Yun KS, Ban HS, Choe W, Lee SK, Lee KY (2009) Preparation and characterization of chitosan/polyguluronate nanoparticles for siRNA delivery. J Controlled Release 139. https://doi.org/10.1016/j.jconrel.2009.06.018
24. Geng X, Yang R, Huang J, Zhang X, Wang X (2013) Evaluation antibacterial activity of quaternary-based chitin/chitosan derivatives in vitro. J Food Sci 78. https://doi.org/10.1111/j.1750-3841.2012.02999.x
25. Huang J, Jiang H, Qiu M, Geng X, Yang R, Li J, Zhang C (2013) Antibacterial activity evaluation of quaternary chitin against *Escherichia coli* and *Staphylococcus aureus*. Int J Biol Macromol 52. https://doi.org/10.1016/j.ijbiomac.2012.10.017
26. Farley JH, Clear NP, Leroy B, Davis TLO, McPherson G (2006) Age, growth and preliminary estimates of maturity of bigeye tuna, *Thunnus obesus*, in the Australian region. Mar Freshw Res 57. https://doi.org/10.1071/MF05255
27. Lee J-K, Lim H-S, Kim J-H (2002) Cytotoxic activity of aminoderivatized cationic chitosan derivatives. Bioorg Med Chem Lett 12. https://doi.org/10.1016/S0960-894X(02)00530-9
28. Kashyap PL, Xiang X, Heiden P (2015) Chitosan nanoparticle based delivery systems for sustainable agriculture. https://doi.org/10.1016/j.ijbiomac.2015.02.039
29. Vidal RRL, Moraes JS (2019) Removal of organic pollutants from wastewater using chitosan: a literature review. https://doi.org/10.1007/s13762-018-2061-8
30. Picos-Corrales LA, Sarmiento-Sánchez JI, Ruelas-Leyva JP, Crini G, Hermosillo-Ochoa E, Gutierrez-Montes JA (2020) Environment-friendly approach toward the treatment of raw agricultural wastewater and river water via flocculation using chitosan and bean straw flour as bioflocculants. ACS Omega 5. https://doi.org/10.1021/acsomega.9b03419
31. Mohamad Sadeghi G (2017) Highly monodisperse chitosan nanoparticles prepared by a combined triple-method for potential use as drug carriers. Int J Nanomater Nanotechnol Nanomed. https://doi.org/10.17352/2455-3492.000013
32. Zhao LM, Shi LE, Zhang ZL, Chen JM, Shi DD, Yang J, Tang ZX (2011) Preparation and application of chitosan nanoparticles and nanofibers. https://doi.org/10.1590/S0104-66322011000300001
33. Bozkir A, Saka OM (2004) Chitosan nanoparticles for plasmid DNA delivery: effect of chitosan molecular structure on formulation and release characteristics. Drug Deliv J Deliv Target Therapeut Agents 11. https://doi.org/10.1080/10717540490280705
34. Bhattarai N, Ramay HR, Chou SH, Zhang M (2006) Chitosan and lactic acid-grafted chitosan nanoparticles as carriers for prolonged drug delivery. Int J Nanomed 1. https://doi.org/10.2147/nano.2006.1.2.181
35. Sailaja AK, Amareshwar P, Chakravarty P (2011) Different techniques used for the preparation of nanoparticles using natural polymers and their application. Int J Pharm Pharm Sci 3(2):45–50
36. Iqbal M, Zafar N, Fessi H, Elaissari A (2015) Double emulsion solvent evaporation techniques used for drug encapsulation. Int J Pharm 496(2):173–190. https://doi.org/10.1016/j.ijpharm.2015.10.057
37. Kahdestani SA, Shahriari MH, Abdouss M (2021) Synthesis and characterization of chitosan nanoparticles containing teicoplanin using sol–gel. Polym Bull 78. https://doi.org/10.1007/s00289-020-03134-2

38. Mateescu MA, Ispas-Szabo P, Assaad E (2015) Chitosan-based polyelectrolyte complexes as pharmaceutical excipients. Controlled drug delivery. The role of self-assembling multi-task excipients. Woodhead publishing series in biomedicine, vol 74. Elsevier, Kidlington, pp 127–161. https://doi.org/10.1016/B978-1-907568-45-9.00004-4
39. Mitra A, Dey B (2011) Chitosan microspheres in novel drug delivery systems. Indian J Pharm Sci 73(4):355. https://doi.org/10.4103/0250-474X.95607
40. Zohri M, Gazori T, Mirdamadi SS, Asadi A, Haririan I (2009) Polymeric nanoparticles: production, applications and advantage
41. Shi A, Guan Y, Zhang Y (2020) A new emulsification-crosslinking technique for preparation of physically crosslinked chitosan microspheres. J Bioact Compat Polym 35(4–5):289–300. https://doi.org/10.1177/0883911520933921
42. Grenha A (2012) Chitosan nanoparticles: a survey of preparation methods. J Drug Target 20(4):291–300. https://doi.org/10.3109/1061186X.2011.654121
43. Sun Y, Gu L, Gao Y, Gao F (2010) Preparation and characterization of 5-fluorouracil loaded chitosan microspheres by a two-step solidification method. Chem Pharm Bull 58(7):891–895. https://doi.org/10.1248/cpb.58.891
44. Divya K, Jisha MS (2018) Chitosan nanoparticles preparation and applications. https://doi.org/10.1007/s10311-017-0670-y
45. Gough JE, Scotchford CA, Downes S (2002) Cytotoxicity of glutaraldehyde crosslinked collagen/poly(vinyl alcohol) films is by the mechanism of apoptosis. J Biomed Mater Res 61(1):121–130. https://doi.org/10.1002/jbm.10145
46. Naskar S, Sharma S, Kuotsu K (2019) Chitosan-based nanoparticles: an overview of biomedical applications and its preparation. J Drug Deliv Sci Technol 49, 66–81. https://doi.org/10.1016/j.jddst.2018.10.022
47. Naskar S, Koutsu K, Sharma S (2019) Chitosan-based nanoparticles as drug delivery systems: a review on two decades of research. https://doi.org/10.1080/1061186X.2018.1512112
48. Usman A, Zia KM, Zuber M, Tabasum S, Rehman S, Zia F (2016) Chitin and chitosan based polyurethanes: a review of recent advances and prospective biomedical applications. Int J Biol Macromol 86:630–645. https://doi.org/10.1016/j.ijbiomac.2016.02.004
49. Verma D, Okhawilai M, Goh KL, Thakur VK, Senthilkumar N, Sharma M, Uyama H (2023) Sustainable functionalized chitosan based nano-composites for wound dressings applications: a review. Environ Res 116580. https://doi.org/10.1016/j.envres.2023.116580
50. Zuber M, Zia KM, Barikani M (2013) Chitin and chitosan based blends, composites and nanocomposites. In: Advances in natural polymers: composites and nanocomposites, pp 55–119. https://doi.org/10.1007/978-3-642-20940-6_3
51. Rafique A, Zia KM, Zuber M, Tabasum S, Rehman S (2016) Chitosan functionalized poly(vinyl alcohol) for prospects biomedical and industrial applications: a review. Int J Biol Macromol 87:141–154. https://doi.org/10.1016/j.ijbiomac.2016.02.035
52. Zia KM, Anjum S, Zuber M, Mujahid M, Jamil T (2014) Synthesis and molecular characterization of chitosan based polyurethane elastomers using aromatic diisocyanate. Int J Biol Macromol 66:26–32. https://doi.org/10.1016/j.ijbiomac.2014.01.073
53. Javaid MA, Zia KM, Ilyas HN, Yaqub N, Bhatti IA, Rehan M et al (2019) Influence of chitosan/1,4-butanediol blends on the thermal and surface behavior of polycaprolactone diol-based polyurethanes. Int J Biol Macromol 141:1022–1034. https://doi.org/10.1016/j.ijbiomac.2019.09.001
54. Javaid MA, Zia KM, Khera RA, Jabeen S, Mumtaz I, Younis MA et al (2019) Evaluation of cytotoxicity, hemocompatibility and spectral studies of chitosan assisted polyurethanes prepared with various diisocyanates. Int J Biol Macromol 129:116–126. https://doi.org/10.1016/j.ijbiomac.2019.01.084
55. Javaid MA, Zia KM, Zafar K, Khosa MK, Akram N, Ajmal M et al (2020) Synthesis and molecular characterization of chitosan/starch blends based polyurethanes. Int J Biol Macromol 146:243–252. https://doi.org/10.1016/j.ijbiomac.2019.12.234
56. Javaid MA, Khera RA, Zia KM, Saito K, Bhatti IA, Asghar M (2018) Synthesis and characterization of chitosan modified polyurethane bio-nanocomposites with biomedical potential. Int J Biol Macromol 115:375–384. https://doi.org/10.1016/j.ijbiomac.2018.04.013

57. Javaid MA, Younas M, Zafar I, Khera RA, Zia KM, Jabeen S (2019) Mathematical modeling and experimental study of mechanical properties of chitosan based polyurethanes: effect of diisocyanate nature by mixture design approach. Int J Biol Macromol 124:321–330. https://doi.org/10.1016/j.ijbiomac.2018.11.183
58. Rizwan M, Zia KM, Javaid MA, Zuber M, Aftab W, Rehman S (2021) Synthesis and molecular characterization of chitosan/alginate blends based polyurethanes biocomposites. Int J Biol Macromol 180:324–331. https://doi.org/10.1016/j.ijbiomac.2021.03.064
59. Javaid MA, Rizwan M, Khera RA, Zia KM, Saito K, Zuber M et al (2018) Thermal degradation behavior and X-ray diffraction studies of chitosan based polyurethane bio-nanocomposites using different diisocyanates. Int J Biol Macromol 117:762–772. https://doi.org/10.1016/j.ijbiomac.2018.05.209
60. Javaid MA, Jabeen S, Arshad N, Zia KM, Hussain MT, Ullah I et al (2023) Preparation of amylopectin and chitosan based polyurethanes for sustained drug release studies. Sustain Chem Pharm 33:101086. https://doi.org/10.1016/j.scp.2023.101086
61. Arshad N, Javaid MA, Zia KM, Hussain MT, Arshad MM, Tahir U (2023) Development of biocompatible aqueous polyurethane dispersions using chitosan and curcumin to improve physicochemical properties of textile surfaces. Int J Biol Macromol 251:126196. https://doi.org/10.1016/j.ijbiomac.2023.126196

Chapter 2
Preparation and Functionalization of the Chitosan Derivatives

2.1 Introduction

The purpose of this chapter is to demonstrate the procedures for preparing and functionalizing chitosan derivatives [1–3]. Chemical and enzymatic techniques of production are used, and chitosan is functionalized through cross-linking, grafting, carboxymethylation, phosphorylation, and etherification [4–7]. On chitosan derivatives, various research articles and review papers have been published [1–10], and the author has endeavored to summarize the comprehensive body of work in this book chapter.

2.2 Chitosan and Its Derivatives

Around the world, a diverse range of species, including crustaceans, arthropods, nematodes, mollusks, diatoms, filamentous fungal cell walls, and basidiomycetes, contains chitin, and an amino polysaccharide is available abundantly [11]. After cellulose, it is the second most prevalent polysaccharide. Despite its abundance, it has confined applicability due to its weak solubility and low reactivity. Chitosan is also a naturally occurring polysaccharide [12]. It is found in the cell walls of "Zygomycetes," a kind of fungus, as shown in Table 2.1. One of the most important components of these fungus cell wall is chitosan, aside from the fact that it is not found in any other kingdom of fungi. The amount of chitosan in different fungus species varies; Table 2.1 compares the amount of chitosan found in different fungus groups.

Table 2.2 shows that *Rhizopus oryzae* TISTR3189 has the greatest ability to produce chitosan among all strain groups [14]. Commercially, crab shell and shrimp chitin are deacetylated with a strong alkali at high temperatures for extended periods to obtain chitosan, which is a time-consuming and expensive process. Since it is

K. M. Zia, *Chitosan*, Smart Nanomaterials Technology,
https://doi.org/10.1007/978-981-96-2629-8_2

Table 2.1 Major components of fungal cell wall [13, 14]

Taxonomic group	Fibrillar polymers	Matrix polymers
Oomycetes	β (1,3), β (1,6)-glucan; cellulose	Glucan
Chytridiomycetes	Chitin; glucan	Glucan
Zygomycetes	Chitin; chitosan	Polyglucuronic acid; glucuronomannoproteins
Basidiomycetes	Chitin; β (1,3)-β (1,6) glucans	α (1,3)-glucan; xylomannoproteins
Ascomycetes/deuteromycetes	Chitin; β (1,3)-β (1,6) glucans	α (1,3)-glucan; glucuronomannoproteins

particularly demanded by industries as raw material, therefore chitin's deacetylation serves as the primary manufacturing process for chitosan.

Figure 2.1 demonstrates how various processed food wastes from shellfish are used to produce chitin and chitosan. Chitosan is built up of linear structured (1 → 4)-linked 2-amino-deoxy-D-glucan units [3]. Chitosan has a molecular weight range of 3800–20,000 Da depending on the degree of deacetylation. Molecular weight variations and degrees of deacetylation have an impact on the properties of chitosan [11]. It is increasingly being used in commercial and biomedical applications, especially in the synthesis of micro- and nanoparticles and composites. As a polymer, due to its biodegradability, biocompatibility, and nontoxicity, it is also used as a drug delivery system [12]. Chitosan does not dissolve in water and is only soluble in dilute acidic solutions of tartaric, citric, and acetic acids. It does not dissolve in sulfuric or phosphoric acids because chitosan sulfate and chitosan phosphate are formed when sulfuric and phosphoric acids react with it [13, 14]. In order to generate the ionizable soluble R–NH_3 form, the free amino groups of the chitosan's glucosamine units—which appear to have a pKa of 6.5—are protonated in a weak acidic solution [13, 14].

Synthesizing structurally modified chitosan for enhanced solubility and prospective applications has been increasingly popular in recent years. Derivatization does not modify the skeleton of chitosan, but it does provide it with new and better properties. By chemically modifying the hydroxyl or amino functional groups on

Table 2.2 Amount of chitosan produced by various fungi groups [14]

Species	Chitosan produced (mg g^{-1})[a]	Chitosan content (%)[a]
Rhizopus oryzae TISTR3189	138	14
Lentinus edodes no. 1	33	3.3
Pleurotus sajor-caju no. 2	12	1.2
Zygosaccharomyces rouxii TISTR5058	36	3.6
Candida albicans TISTR5239	44	4.4

[a] Cell dry weight

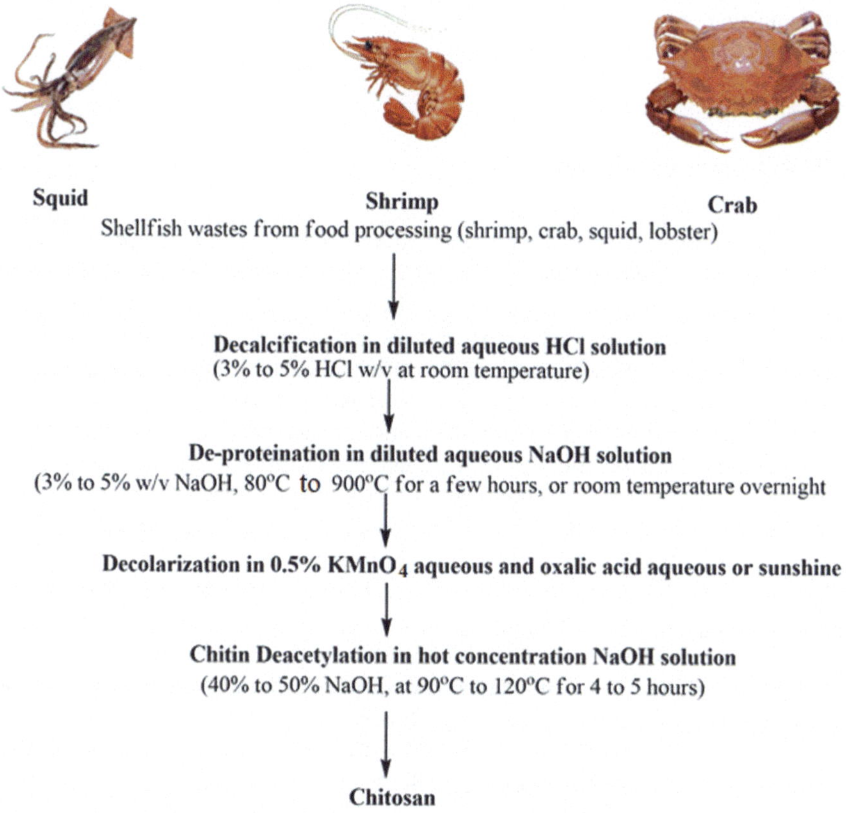

Fig. 2.1 Steps involved in the production of chitin and chitosan [11]

chitosan, several chitosan derivatives can be produced; however during the derivatization process, secondary functional groups, such as alkyl or carboxymethyl groups, are added to the chitosan structure, which considerably increased chitosan solubility without altering its cationic nature, making it soluble even at neutral and alkaline pH. Chitosan forms a polymer having polyampholytic properties when it is attached to a carboxylic group.

The grafted copolymerization method has been widely used among several modification methods. Grafting is the covalent bonding of a functional group to the chitosan backbone. Reactive groups can be grafted in two different ways: acetylated or deacetylated hydroxyl groups on the C_3 and C_6 carbons, as well as deacetylated free amino groups. These grafted, derivatized chitosans exhibit significantly improved chelating and complexation properties, as well as bacteriostatic, antibacterial, antioxidant, adsorption, and water solubility. Even though the grafted modified chitosans have certain intriguing properties including biodegradability and biocompatibility, mucoadhesive, high drug permeation, especially for hydrophilic macromolecules, protect acid-sensitive drugs, achieve colon targeting, enhanced drug release in the small intestine's basic conditions, and overcome chitosan's lower solubility in neutral pH [5–7, 15, 16].

2.3 Chitosan Isolation

The fungal biomass is treated with alkali (dilute NaOH) at high temperature (90–120 °C) to isolate chitosan from the fungal cell wall. The alkali treatment dissolved alkali soluble carbohydrates, lipids, proteins while remaining cell wall as *alkali insoluble material (AIM)*. In the next step, the dissolution of cell wall occurs in an acid media (i.e., 2–10% acetic acid, at 25–90 °C for 1–24 h). This step produced the other components of cell walls, which are *alkali and acid insoluble material (AAIM)*. In the last step, the fungal chitosan precipitation is accompanied by elevating the H 9–10 and chitosan is obtained by centrifugation. For the extraction of chitosan by HCl or acetic acid, the isolated product has very high phosphate impurities and very low chitosan yield. Recently, the trend has been changed, for acid treatment now the use of hot dilute sulfuric acid yields chitosan with good purity and very low impurities. Although at ambient temperature, chitosan is insoluble in dilute sulfuric acid, it is readily soluble in hydrochloric acid, lactic acid, citric acid, and acetic acid solutions. It is, however, soluble in boiling sulfuric acid. Chitosan's temperature-dependent solubility provided adaptability to other cell wall components such as polyphosphates and chitin. By treating the cell wall with hot dilute sulfuric acid, only chitosan is dissolved in the hot acid solution, while the other components of the cell wall are removed by filtration. In the final step, cooling of hot acid solution makes chitosan insoluble [13, 14].

2.4 Preparation of Chitosan Derivatives by Chemical Method

Glycosidic bond, C_2–NH_2, acetyl amino, C_3–OH, and C_6–OH are the functional groups of chitosan molecules. The acetyl amino-bond is the most stable of the reactive bonds, unlike the glycosidic bond, which can be easily broken. C_3–OH, on the other hand, is attached to the secondary hydroxyl group and cannot be freely rotated due to the substantial steric barrier. Due to various sorts of reaction designs, the overall chemical characteristics of C_2–NH_2 and C_6–OH give chitosan molecules an advantage in introducing distinct functional groups. Finally, the site of derivatization and functionalization depends on the desired applications of chitosan derivatives [12, 17]. Figure 2.2 illustrates the chemical transformation of chitosan [12].

By grafting small molecules or polymer chains onto the backbone structure of chitosan, or by quaternizing amino groups, chemical derivatization, chemical modification, and chemical functionalization of chitosan can be achieved. One of the

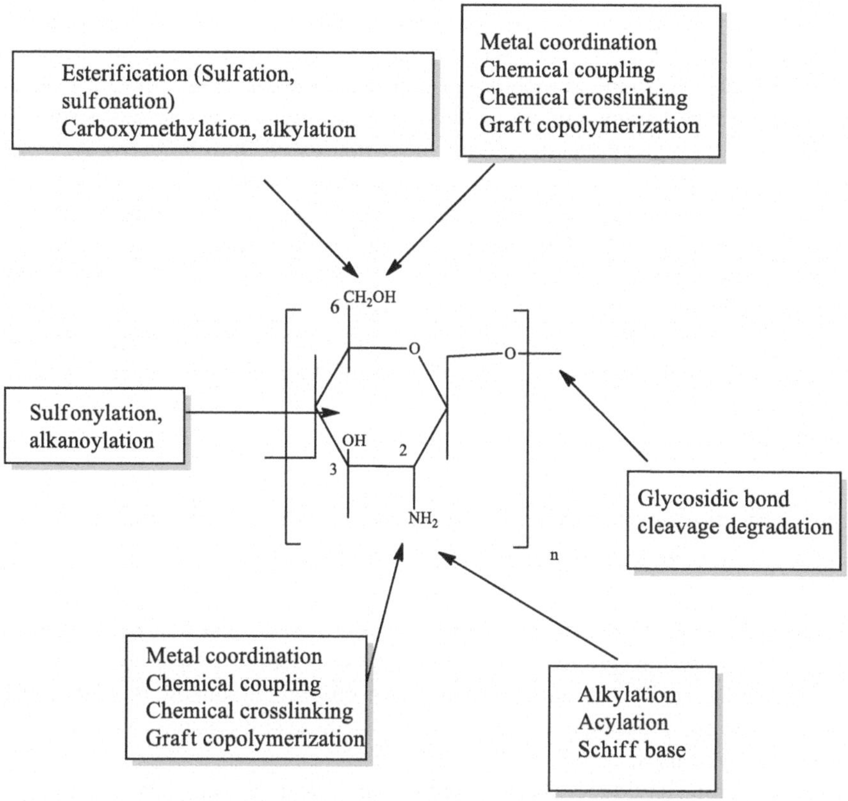

Fig. 2.2 Schematic representation of chitosan chemical reaction [12]

most frequent chemical methods for derivatizing chitosan is acylation. This method involves adding various organic acids, organic acid derivatives, and aliphatic or aromatic acyl groups to the chitosan molecular chain. The intramolecular or intermolecular hydrogen bonding of chitosan is destroyed by acylation, which diminishes the crystalline structure of chitosan while increasing its water solubility. The following is the order of active sites for acylation of chitosan: C_2–NH_2 > C_6–OH > C_3–OH. The acylation which occurs at C_2–NH_2 produces amide that is known as *N*-acylation. When C_6–OH is active for reaction, an ester is synthesized; however if protective functional group is present at C_2–NH_2, the obtained product is referred as *O*-acylation.

Alkylation is another chemical method for the production of chitosan derivatives. The introduction of alkyl group to chitosan enhances the water solubility of chitosan by significantly destroying the intermolecular hydrogen bonding in chitosan molecules. The long-chain alkyl groups lessened the solubility due to their hydrophobic nature, thus solubility of chitosan is tunable by modifying the chain length of alkyl group. The alkyl groups can be introduced at any of the reactive sites, i.e., C_2–NH_2, C_6–OH, and C_3–OH. *N*-alkylation occurs at C_2–NH_2 functional group, while alkylation at C_6–OH or C_3–OH produced *O*-alkylation. The *N*-alkylation is more favorable due to the availability of highly nucleophilic lone pairs. The addition of quaternary ammonium group, which is hydrophilic in nature and positively charged, is a well-known technique for the synthesis of chitosan derivatives. An introduction of quaternary ammonium group weakens hydrogen bonds, along with enhancing water solubility, and also adds chargeability to chitosan molecules. The quaternization occurs at C_2–NH_2 at site, while the chemical quaternization generally done by three methods, i.e., direct addition of quaternary ammonium substitution method, *N*-alkylation method and by epoxy derivatives (open loop) method. Besides these chemical methods for the synthesis of chemical derivatives, there are some other methods for the chemical functionalization of chitosan and it will be discussed later [11, 18–23].

2.4.1 *Synthesis of* N-*Schiff Base Type* N-*Secondary Amino Type Chitosan Crown Ethers Derivatives*

According to the particular molecular structure of the crown ethers, it has very good metal ion complexing selectivity. The chitosan crown ethers' properties depend on the structural variations of chitosan and crown ethers compounds. Regarding the concentration and separation of heavy-metal ions, these chitosan crown ethers offer a wide variety of uses.

The *N*-Schiff base-type chitosan crown ether and *N*-secondary amino-type chitosan crown were reported by Peng et al. using the formally synthesized crown ethers 4′-formyl-benzo-15-crown-5 and 4′-formyl-benzo-18-crown-5 [24]. The reaction occurred between 4′-formyl-benzo-15-crown-5 or 4,4′-formyl-benzo-18-crown-6 and chitosan to synthesis *N*-Schiff base-type cross-linked chitosan crown ethers. The first stage yielded *N*-Schiff base-type chitosan crown ether, and the obtained *N*-Schiff base-type chitosan crown ether derivative are turned to *N*-secondary amino-type chitosan crown by the reaction with sodium borohydride in the second step, as shown in Fig. 2.3 [24].

Figure 2.3 depicts the synthetic scheme for the production of chitosan crown ethers (I–III). The *N*-Schiff base-type benzo-21-crown-7-chitosan is prepared by dissolving 2.5 g of chitosan powder in 70 ml of 10% acetic acid solution and diluting with methanol. In 20 ml methanol, 5.8 g 4′-formyl-benzo-21-crown-7 are dissolved. This solution is then gradually added to the chitosan solution, with continuous stirring under nitrogen for 2 h. As the reaction proceeded, the solution became yellow. A light yellow solid is produced after the solution has been kept in the oven for 36 h at 80 °C. Soxhlet extraction is used to remove any unreacted 4′-formyl-benzo-21-crown-7 by repeatedly washing the product in methanol. Chitosan crown ether (I), which is formed, is a grayish product that is yielded up to 85%. The *N*-secondary amino-type benzo-21-crown-7-chitosan **(II)** is synthesized by carrying out the steps

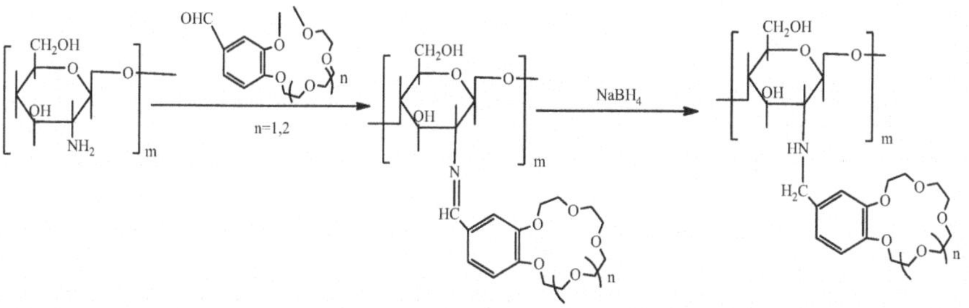

Fig. 2.3 Chitosan derivative synthesis reaction routes: *N*-Schiff base-type chitosan crown ethers ($n = 1$ for crown ether-I; $n = 2$ for crown ether-II) and *N*-secondary amino-type-chitosan crown ethers ($n = 1$ for crown ether-III; $n = 2$ for crown ether-IV) [24]

of **(I)**: 1.0 g of sodium borohydride is dissolved in 20 ml of methanol and then gently added to the resultant sticky solution of **(I)**. After the mixture is refluxed for 24 h, a faintly yellow solid is produced, which is extracted using a Soxhlet method and washed with methanol to get rid of any unreacted 4′-formyl-benzo-21-crown-7 and sodium borohydride. The yellow-colored dried chitosan crown ether **(II)** 89% is produced. The same process described in **(I)** is used to synthesize *N*-Schiff base-type benzo-21-crown-7 cross-linked chitosan **(III)**. Cross-linked chitosan powder is swollen in methanol overnight and combined with 4′-formyl-benzo-21-crown-7 solution instead of chitosan powder. After washing and drying, the yield of *N*-Schiff base-type benzo-21-crown-7 cross-linked chitosan (III) is around 70% [23].

Tang et al., on the other hand, synthesized three chitosan crown ethers by reacting 4′-formyl-benzo-21-crown-7 with chitosan and cross-linked chitosan as illustrated in Fig. 2.4 [23].

Preparation of Chitosan Crown Ethers: Peng et al. reported the chitosan crown ether (I–IV) routes in Fig. 2.3 [24]. The proposed technique is utilized to synthesize *N*-Schiff base-type 4′-formyl-benzo-15-crown-5-chitosan **(I)** and *N*-Schiff base-type 4′-formyl-benzo-15-crown-6 chitosan **(II)**. The chitosan powder (1.5 g) is dissolved in 100 ml of 10% w/v acetic acid solution. This solution is then diluted with ethanol. In 20 ml ethanol, 4′-formyl-benzo-15-crown-5 and 4′-formyl-benzo-15-crown-6 are dissolved separately. Under nitrogen, these solutions are dropped into a chitosan solution. The solution is refluxed and agitated for 24 h, yielding a thick solution that is kept at 80 °C in an oven for a longer period of time until 4 g solid mass of light brown-colored chitosan crown ethers is produced. The products are washed by the Soxhlet extraction method by using ethanol for the removal of any unreacted benzo-15-crown-5 and benzo-15-crown-6. The yields of benzo-15-crown-5-chitosan **(I)** and benzo-15-crown-6-chitosan **(II)** are 89 and 80%, respectively.

N-secondary amino-type benzo-15-crown-5-chitosan **(III)** is synthesized using step **(I)**. After obtaining the sticky solution, 1.0 g of sodium borohydride is dissolved in 20 ml of ethanol and slowly added to the above solution. After 6 h of refluxing the acquired mixture, a dried, pale yellow solid is obtained. The product is washed with ethanol using Soxhlet method to remove unreacted sodium borohydride, and up

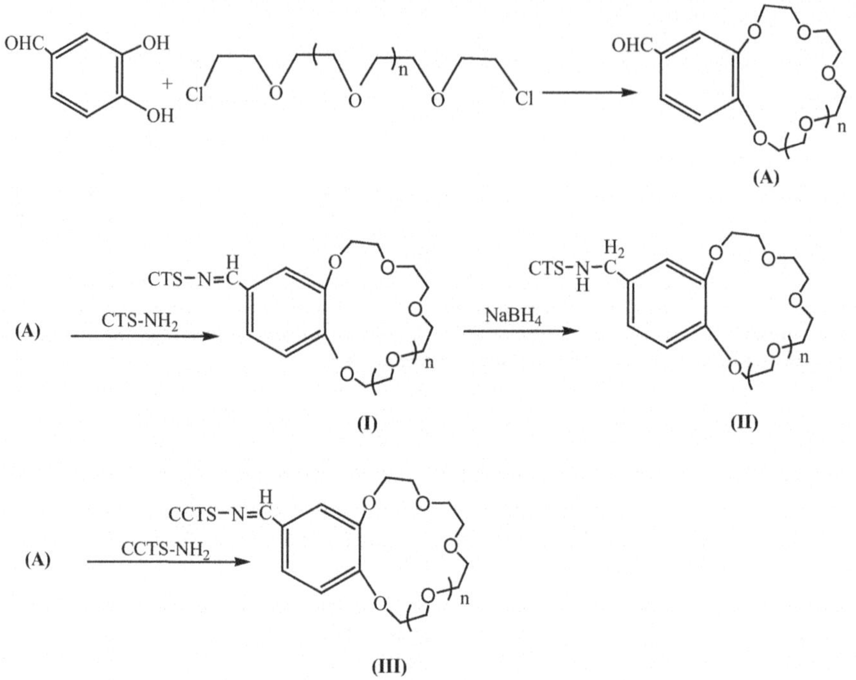

Fig. 2.4 Synthetic scheme for chitosan-crown ethers (I–III), $n = 3$ [23]

to 90% of the dried product **(III)** is produced. The *N*-secondary amino-type benzo-18-crown-6-chitosan **(IV)** was produced using the same procedure used for **(III)** from 4′-formyl-benzo-8-crown-6, chitosan, and sodium borohydride. The yield of the product is 82% [24].

It is possible to conclude that Tang et al. [23] and Peng et al. [24] utilized the identical approach with the exception that Peng et al. [24] used ethanol as a diluent in the subsequent method. The addition of methanol or ethanol has no impact on the product yield.

2.4.2 Synthesis of Schiff Base Type N-Heterocyclic Chitosan Derivatives

Electrophiles are more likely to attack and react with the amine group at C_2 in chitosan's 2-amino-2-deoxy-D-glucopyranose units than with the hydroxyl groups at C_3 and C_6. The reductive amination process is thus the most trustworthy method for adding alkyl or aryl functional groups to the amine group of chitosan. When chitosan is treated to aliphatic or aromatic aldehydes, it can produce biopolymer Schiff base intermediates. When these intermediates are treated with sodium borohydride, it can be reduced to *N*-alkyl or *N*-aryl chitosan derivatives. The synthesis of *N*-heterocyclic chitosan derivatives using a reductive amination procedure and five heterocyclic aldehydes with various levels of substitution is described below. Table 2.3 shows the chemical structure, degree of acetylation, degree of deacetylation, and degree of substitution of chitosan derivatives which are all determined using ^{1}H-NMR [7].

As illustrated in Fig. 2.5, Kim et al. [8] proposed a method for the synthesis of chitosan derivatives, with the following synthesis procedure. The 300 ml, 1% v/v acetic acid solution is used to dissolve the 18 mmol of chitosan. Dropwise additions of one of the heterocyclic aldehydes listed below are added to the chitosan solution at room temperature: benzo-[d]-[1,3]-dioxole-5-carbaldehyde, 4-oxo-4-H-chromene-3-carbaldehyde, 5-methylfuran-2-carbaldehyde, furan-2-carbaldehyde, and 3-pyridine carboxyaldehyde. The pH of the reaction mixture is adjusted to 4.5 by adding 1 M NaOH solution and stirring the reaction mixture for an hour. Sodium borohydride solution 10% w/v (1.5 to aldehyde) is added to this mixture and agitated again for 1.5 h. The *N*-heterocyclic chitosan derivative precipitated when the pH of the solution is adjusted to 10. The unreacted aldehydes are then extracted by Soxhlet using diethyl ether/ethanol (1:1 v/v) for two days after the precipitates are rinsed with distilled water to neutralize them. The desired compounds are then produced after oven drying at 60 °C [7, 8].

2.4.3 Synthesis of Chitosan Derivatives with Quaternary Ammonium Salt

Chitosan's amine group may be easily modified by alkyl groups, and in order to obtain water-soluble cationic polyelectrolytes and novel chitosan derivatives containing quaternary ammonium salt, the amino groups in chitosan interact with aldehydes via Schiff base intermediates. The resulting *N*-alkyl chitosan derivatives are then quaternized using methyl iodide. Table 2.4 lists the quaternized *N*-alkyl chitosan derivatives with the degree of substitution.

Figure 2.6 depicts the synthetic scheme for *N*-alkyl chitosan derivatives, and as compared to chitosan, these *N*-alkyl quaternized chitosan derivatives exhibit extraordinarily high levels of antibacterial activity; for instance, it exhibits considerably stronger antibacterial activity against *Escherichia coli.* The length of the alkyl substitution chain is another factor that affects the antibacterial action and the hydrophobic characteristics of these compounds are enhanced by lengthening the alkyl substitution chain [8, 25].

***N*-Alkyl Chitosan Derivatives Synthesis**: Adopting the strategy portrayed in Fig. 2.6, Kim et al. [8] synthesized quaternary ammonium salt derivatives of chitosan. The 1% v/v acetic acid is used to prepare the 1% w/v chitosan solution. Aqueous aldehyde solutions (formaldehyde, n-butyraldehyde, n-octylaldehyde, n-dodecylaldehyde) of various concentrations are prepared and mixed with a chitosan solution at room temperature under continuous stirring for 30 minutes, after which

Table 2.3 Chemical structure, degree of acetylation, degree of deacetylation, and degree of substitution of chitosan derivatives determined using ^{1}H-NMR [7]

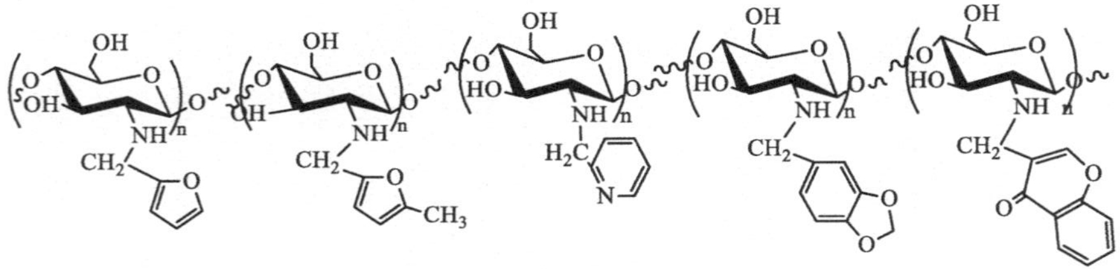

1 **2** **3** **4** **5**

Compound	Degree of deacetylation DDA[a]	Degree of substitution DS[b]	Degree of acetylation DA[c]
Chitosan	0.89	…	0.11
1	0.55	0.34	0.11
2	0.61	0.30	0.09
3	0.46	0.43	0.11
4	0.55	0.35	0.10
5	0.56	0.33	0.11

[a] DDA is calculated from the integral ratio between the proton on C-2 and the glucose unit proton

[b] DS is calculated from the ratio of areas of the proton in the heterocyclic moieties and the proton of pyranose unit

[c] DA is the temporary value at the present stage because the peak of CH_3 in NHAc is overlapped with the peak of NH_2 in chitosan or NH_2 and NH groups in the synthesized derivatives in the ^{1}H-NMR spectra, so the equation that is applicable for the calculation of DA is as follows: $DA = 1 - (DDA + DS)$

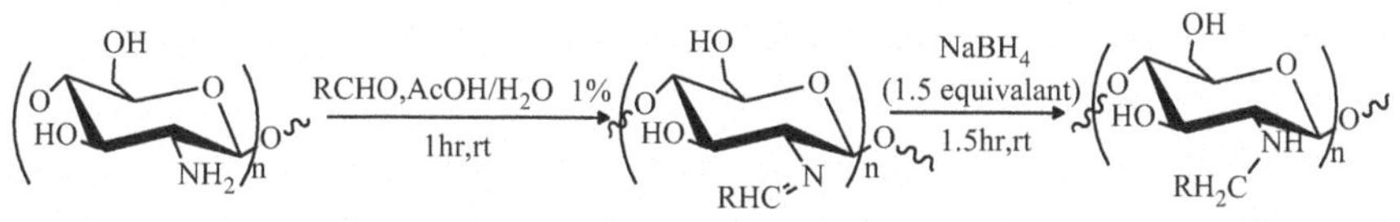

Fig. 2.5 Synthesis route for *N*-heterocyclic chitosan derivatives [7]

Table 2.4 Degree of the substitution of chitosan derivative through elemental analysis

Samples		Degree of *N*-alkylation		Degree of quaternization	
(Mol. of aldehyde/glucosamine unit)		C/N ratio	Titration	C/N ratio	Potentiometry
N-trimethyl chitosan	(fivefold excess)	–	–	0.63	0.58
N-butyl chitosan	(fivefold excess)	0.81	0.77	0.58	0.56
N,*N*-dibutyl chitosan	(11-fold excess)	0.50	0.48	0.28	0.25
N-octyl chitosan	(15-fold excess)	0.90	0.86	0.24	0.20
N-dodecyl chitosan	(15-fold excess)	0.90	0.85	0.20	0.18

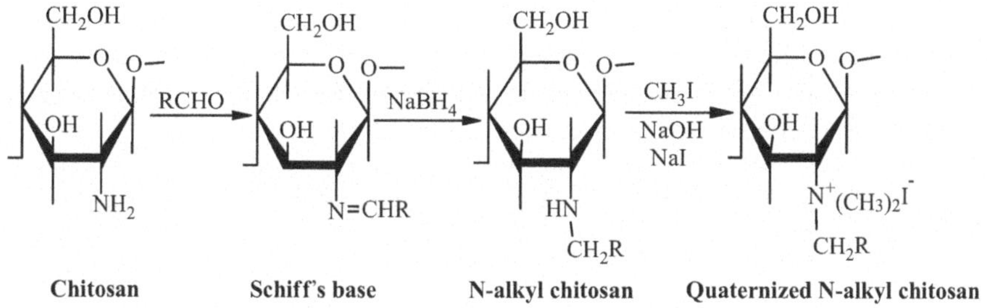

Fig. 2.6 Scheme of synthesis for quaternized *N*-alkyl chitosan derivatives [8]

the solution's pH is adjusted to 4.5. This solution is then combined with a 10% w/v sodium borohydride solution, and the mixture is agitated for an hour. By raising the pH of the solution to 10, the *N*-alkyl chitosan precipitated out. To remove the excess base, the precipitates are thoroughly washed with distilled water. Unreacted aldehydes and inorganic by-products are then removed over the period of 2–7 days using the Soxhlet method of the several *N*-alkyl chitosan derivatives, *N*-trimethyl chitosan, *N*-butyl chitosan, *N*-octyl chitosan, and *N*-dodecyl chitosan are filtered out and dried for 48 h at 40 °C in a vacuum.

Synthesis of Quaternized *N*-Alkyl Chitosan Derivatives: The 250 ml of *N*-methyl-2-pyrolidone is dissolved in 5 g of each synthesized *N*-alkyl chitosan derivative for 12 h at room temperature, and NaOH (1.4 M) and CH_3I are added in each dispersion. Additionally, NaI is added to each reaction medium to adjust the concentration to 0.2 M with continuous stirring at 50 °C for 12 h. The solution is then precipitated with acetone and collected. The quaternized *N*-alkyl chitosan derivatives are produced after drying and are characterized by FTIR and NMR [8].

2.4.4 Synthesis of Sugar-Bearing Chitosan Derivatives

The synthetic procedure Park et al. reported [10] for the sugar-bearing derivatization of chitosan is depicted in Fig. 2.7. Prior to *N*-acetylation, chitosan and gluconic acid are covalently bonded and the entire methodology is as follows: 1-ethyl-3-(3-dimethylaminopropyl) carbodiimide hydrochloride (EDC) were used as an active ester intermediate during the chemical conjugation of chitosan with gluconic acid. In 100 ml of 0.1 N HCl, gluconic acid (0.1 or 0.2 equiv/[–NH2]) and chitosan (1 g) are dissolved. Prior to adding the EDC (1.5 equiv/[gluconic acid]) and *N*-hydroxysuccinimide (NHS), the pH of the mixture is adjusted to 5.0 by adding 1 N HCl and 1 N NaOH.

The resulting mixture stirred for 24 h at room temperature. After stirring, 1 N NaOH is used to raise the pH of the solution to 9.0. This solution is dialyzed for 5 days against an excess of distilled water. Lyophilization of the solution produced sugar-based chitosan (SBC). *N*-acetylation of SBC is accomplished by dissolving 0.5 g SBC in 50 ml of 0.5% acetic acid solution. The solution is then dissolved in 40 ml of 1,2-propanediol and degassed at room temperature for one day without stirring. The acetylating solution is comprised of 10 ml of 1,2-propanediol and the necessary quantity of acetic anhydride, which is gently added and agitated at room temperature for 5 h. Using the 1.0 N, NaOH, the pH of the resultant solution is adjusted to 9.0. As

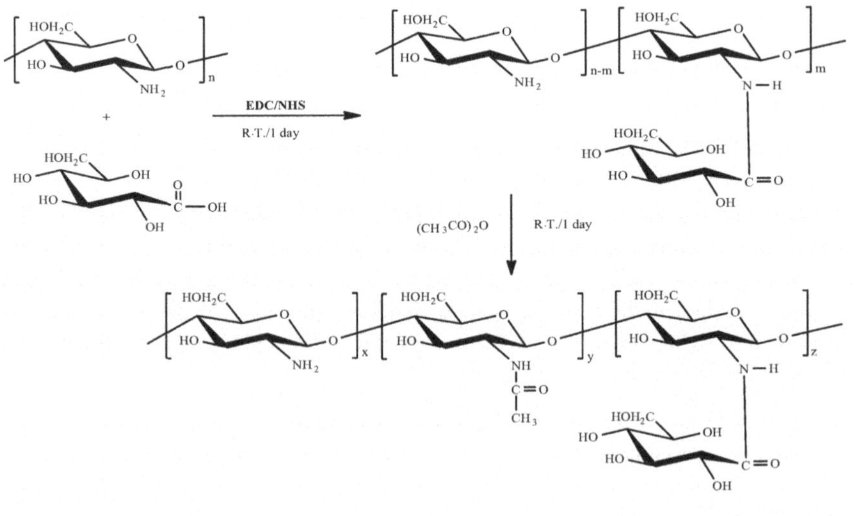

Fig. 2.7 Synthesis route of sugar-bearing chitosan derivatives with different degrees of acetylation [10]

stated earlier, the *N*-acetylated SBCs are produced through dialyzing against distilled water and lyophilization. Finally, the characterization is performed using ^{1}H-NMR [10].

2.4.5 *Preparation of Chitosan Derivatives from Gladius of Squid* **Sepioteuthis lessoniana**

The crushed interior shell of *Sepioteuthis lessoniana* is deproteinized and demineralized by Subhapradha et al. to extract chitin, which is subsequently deacetylated with 40% NaOH to get chitosan. In order to create gelatinous chitosan, 50 ml of DMF-SO_3 is added to a 50 ml solution of chitosan in a DMF-formic acid combination while the mixture is being continually stirred. After 2.5 h of processing at a temperature of 40–60 °C, ethanol is added to the reaction mixture in order to precipitate the product. Under pressure, the combination of product is filtered. After being washed with ethanol, the precipitates are once again dissolved in distilled water. With 2 M NaOH, the pH of the solution is adjusted to 7–8. Using a 12,000 Da MW cut-off dialyzer membrane, the solution is dialyzed for 48 h against distilled water. Sulfated chitosan is produced by lyophilizing and concentrating the product.

To produce the phosphorylated chitosan, 2 g of chitosan powder and 30 g of urea are dissolved in 50 ml of DMF and 5.2 ml of orthophosphoric acid. At 150 °C, the mixture is stirred for an hour. The reaction mixture precipitated and rinsed with methanol after cooling. The mixture's pH is adjusted to 10–11 after the residues are dissolved in distilled water. For 48 h, the solution is dialyzed with distilled water and a 12,000 Da MW cut-off dialysis membrane. The final step is to lyophilize the product to produce phosphorylated chitosan [9].

2.4.6 *Synthesis of Novel Quaternary Chitosan Derivatives via* N-*Chloroacyl-6-*O-*Triphenylmethyl Chitosans*

Holappa et al. reported the quaternization process by using *N*-chloroacyl-6-*O*-triphenylmethyl-chitosan as the starting material. *N*-phthaloyl-chitosan, *N*-phthaloyl-6-*O*-triphenylmethyl-chitosan, and 6-*O*-triphenylmethyl-chitosan are utilized to synthesize this compound. *N*-chloroacyl-6-*O*-triphenylmethyl-chitosan has a complete *N*-chloroacetylation of degree of substitution (0.85), whereas *N*-Chlorobutryl-6-*O*-triphenylmethyl-chitosan has a partial degree of substitution (0.67). Though 15% of the *N*-acetyl groups in the starting material have not yet undergone any reactions, the degree of substitution 0.85 is still regarded as complete substitution. Below is discussed precursor synthesis, derivatization techniques, and synthetic routes. The various synthetic methods used to synthesize quaternary chitosan derivatives are shown in Fig. 2.8 [26].

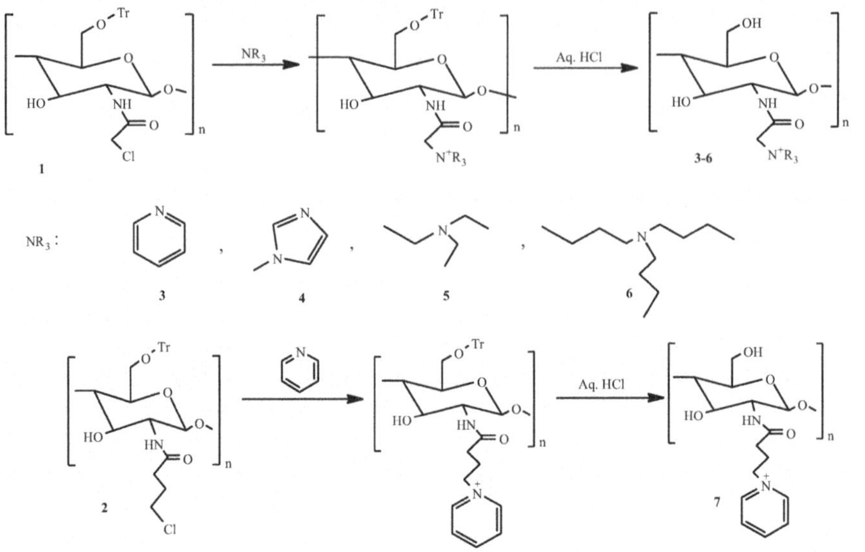

Fig. 2.8 Synthetic schemes for several quaternary chitosan derivatives using *N*-chloroacyl-6-*O*-triphenylmethyl-chitosan [26]

***N*-Phthaloyl-Chitosan (1)**: For the preparation of *N*-Phthaloyl-Chitosan, in 200 ml of DMF containing 5% v/v water, 10 g of chitosan is disseminated and swirled overnight. The chitosan solution is then mixed with 27.6 g phthalic anhydride and agitated for 8 h at 120 °C. The reaction mixture is then immersed in cold water, filtered, and washed with methanol. The product yield is 92%. The degree of substitution is 0.81 based on the C/N ratio of elemental analysis, suggesting that about 95% of free amino groups substituted from starting material, that is confirmed by FTIR [27].

***N*-Phthaloyl-6-*O*-Triphenylmethyl-Chitosan (2)**: In order to prepare *N*-Phthaloyl-6-*O*-triphenylmethyl-Chitosan, a 215 ml solution of pyridine containing 14.31 g of *N*-phthaloyl-chitosan is mixed with 43.2 g of triphenyl chloromethane. This is stirred at 90 °C continuously for 24 h. The product is washed with ethanol and diethyl ether after the solvent has evaporated. The yield is about 100%, and the degree of substitution for 6-*O*-triphenylmethylation is about 0.83, according to both FTIR and the C/N ratio of the elemental analysis [27].

6-*O*-Triphenylmethyl-Chitosan (3): For the preparation of 6-*O*-Triphenylmethyl-Chitosan, a mixture of 295 ml of water, 26.37 g of *N*-phthaloyl-6-*O*-triphenylmethyl-chitosan, and 148 ml of hydrazine monohydrate is stirred at 100 °C for 15 h. Following the evaporation of the solvents, the product is washed with water, ethanol, and diethyl ether, 98% of the product is obtained. The elemental analyses, FTIR, and ^{13}C CP/MAS NMR spectra all provided proof that the phthalimide moiety has been completely eliminated [27].

***N*-Chloroacyl-6-*O*-Triphenylmethyl-Chitosan (4)**: For this purpose, 6-*O*-triphenylmethyl-chitosan is acylated using 4-chlorobutryl chloride **(4b)** and chloroacetyl chloride **(4a)**. The 45 ml of *N*-methylpyrrolidone (NMP) is used to dissolve 1 g of 6-*O*-triphenylmethyl-chitosan, and the mixture is then cooled to room temperature under argon for the acylation. The chloroacyl chloride is added dropwise to the aforementioned solution after being dissolved in 5 ml of NMP. After 15 min in an ice bath, the reaction mixture swirled for 3 h at room temperature. Diethyl ether is used to precipitate out the resulting polymer, while methanol and diethyl ether are used to wash it. For the 6-*O*-triphenylmethyl-chitosan, the amounts of chloroacetyl chloride and 4-chlorobutryl chloride that are used for free amino are 85 μl, 169 μl, and 338 μl for chloroacetyl chloride and 120 μl, 240 μl, and 480 μl for 4-chlorobutryl chloride, respectively. The yield of products **(4a)** is about 83%, 81%, and 99% while products **(4b)** yields 70%, 85%, and 45%, respectively. The acquired relative yields are calculated using the degree of substitutions determined from *N*-methylpiperazine end products [28].

***N*-(1-Carboxymethyl-2-Pyridinium)-Chitosan-Chloride (5)**: The preparation of *N*-(1-Carboxymethyl-2-pyridinium)-Chitosan-Chloride and *N*-Chloroacyl-6-*O*-triphenylmethyl-Chitosan (4) involves dissolving 300 mg of the compound in 10 ml of pyridine. The solution is then agitated at 60°C for 72 hours under an argon atmosphere to ensure proper reaction conditions. The product is washed with methanol, and diethyl ether after the solvent has been evaporated. The reported yield of the product is 51%.

At room temperature, the 6-*O*-triphenylmethyl protecting group is removed for three hours stirring with 170 mg of the preceding reaction product and 20 ml of 1 M HCl. After being evaporated, the reaction mixture is washed with methanol and diethyl ether. From proton ^{1}H-NMR, the degree of substitution is determined [26].

***N*-[1-Carboxymethyl-2-(1-Methylimidazolium)]-Chitosan-Chloride (6)**: In order to prepare *N*-[1-carboxymethyl-2-(1-methylimidazolium)]-chitosan-chloride, 10 ml of 1-methylimidazole is stirred at 60 °C under argon for 72 h with 295 mg of *N*-Chloroacyl-6-*O*-triphenylmethylchitosan **(4)**. Following the entire evaporation of the solvent, it is washed with methanol and diethyl ether. A product yield of about 34% is obtained. While stirring for three hours at room temperature with 105 mg of the preceding reaction product and 10 ml of 1 M HCl, the 6-*O*-triphenylmethyl protecting group is removed. After being rinsed with methanol and diethyl ether, the reaction mixture evaporated. From ^{1}H-NMR, the degree of substitution is determined [26].

***N*-(1-Carboxymethyl-2-Triethylammonium)-Chitosan-Chloride (5)**: For the preparation of *N*-(1-carboxymethyl-2-triethylammonium)-chitosan-chloride, the 400 mg of *N*-chloroacyl-6-*O*-triphenylmethyl-chitosan and 200 mg of KI is added to 5.05 ml of triethylamine, 20 ml of *N*-methylpyrrolidone, then the mixture is stirred for 72 h at 60 °C under argon. The product is then precipitated out with diethyl ether when the reaction mixture is cooled in cold water. Methanol and diethyl ether are used to wash the product, nevertheless. About 33% is the yielded of the product. With

180 mg of the preceding reaction product and 18 ml of 1 M HCl at room temperature, the 6-*O*-triphenylmethyl protecting group is removed for 3 h of stirring. The finished product is washed with methanol, and diethyl ether after the reaction mixture is evaporated until it completely dries. The degree of substitution determined by ^{1}H-NMR and the product yield reported is about 45% [26].

***N*-(1-Carboxymethyl-2-Tributylammonium)-Chitosan-Chloride (6)**: For this preparation, the 20 mg of *N*-methylpyrrolidone is combined with 400 mg of *N*-chloroacyl-6-*O*-triphenylmethyl-chitosan **(4)**, 239 mg of KI, and 8.58 mg of tributylamine after being continuously stirred for 72 h at 60 °C under argon. After stirring and cooling the reaction mixture with cold water, diethyl ether is used to precipitate the product, and then the methanol and diethyl ether are used to wash it, resultantly, only 20% of the yield has been reported.

In order to remove the 6-*O*-triphenylmethyl protecting group, 12 ml of 1 M HCl is mixed with 120 mg of the preceding reaction product for 3 h stirring at room temperature. The mixture of the reaction is dried by evaporation and then washed with methanol and diethyl ether. Using ^{1}H-NMR, the degree of substitution is estimated, and the product yield reported is about 83% [26].

***N*-(1-Carboxybutyl-4-Pyridinium)-Chitosan-Chloride (7)**: For the preparation of *N*-(1-carboxybutyl-4-pyridinium)-chitosan-chloride, a 220 mg of *N*-chlorobutyryl-6-*O*-triphenylmethyl-chitosan is dissolved in 8 ml of pyridine and then exposed to argon for 72 h at 60 °C. Following the solvent's evaporation, methanol and diethyl ether are used to wash the product. The product's yield as reported is about 57%.

The 140 mg of the preceding product is stirred with 14 ml of 1 M HCl at room temperature for three hours in order to remove the 6-*O*-triphenylmethyl protecting group. Methanol and diethyl ether are used to wash the end product once the reaction mixture has dried. By using ^{1}H-NMR, the degree of substitution is estimated and the product yield of about 56% is been reported [26].

2.4.7 Enzymatic Method

Chitosan is derivatized and functionalized via chemical processes, which demand for extremely severe conditions, reactive materials, and a significant number of cross-linking agents. When chemically modified polymer derivatives are used in biomedical applications and foodstuff, it can have deleterious impacts on human health and is not environmentally friendly. Chemical cross-linkers may also have detrimental outcomes (due to leftover cross-linkers or secondary reactions that result in undesirable by-products). The synthesis and functionalization of natural polymers via enzymatic techniques have been researched as alternatives to environment-unfriendly, hazardous, and non-specific chemical processes due to rising environmental safety concerns [11, 29].

It may be advantageous to derivatize or functionalize natural polymers using enzymatic methods in a variety of ways. Although published techniques of chitosan derivatization and enzymatic functionalization are less frequent than those from traditional chemical procedures, the trend has been changing. Enzyme-based methods have the power to completely remove the safety and health problems associated with harsh reactive chemicals. Because biocatalysts are extremely selective, it may be used to eliminate the processes necessary for both full protection and deprotection. The macromolecular structure may also be suitably adjusted by the enzyme specificity properties for improved control of polymer performance [30–32].

2.4.8 Enzymatic Synthesis of Chitosan Derivatives

The composition of biopolymers determines which enzymatic tools should be used to functionalize and derivatize chitosan (related to target enzyme activities and functional groups). Peroxidases (POD) and oxidase enzymes like polyphenoloxidase (PPO), which also encompasses laccases and tyrosinases, constitute the major groups of enzymes. The other enzymes, such as lipases, phosphorylases, and transglutaminases, can be used as secondary catalysts. Enzymatically aided reactions may also be used to cross-link functionalized biomaterials in order to enhance their useful properties, such as stability [11].

Laccases

Laccases are oxidoreductases, blue multicopper oxidases enzymes which are most popular among all oxidase enzyme in many biotechnological applications.

The *o*- and *p*-diphenols are oxidized by laccases via a free radical mechanism, which produces the corresponding quinones in the presence of molecular oxygen, which is reduced to water during this oxidation. Additionally, laccases are capable of converting monophenols directly into the equivalent semiquinones without the requirement for hydroxylation, which is required for the tyrosinase-catalyzed melanization process to begin. Figure 2.9 depicts the ability of laccases to oxidize lignin, methoxyphenols, and aminophenols [11].

The oxidation of chitosan by laccases affects its antibacterial and antioxidant properties. Using *Tinea versicolor* laccases, the previous study demonstrated how to graft phenolic chemicals such as catechol, caffeic acid, tannic acid, gallic acid, 2,5-dihydroxy benzoic acid, and quercetin to chitosan at pH 4.5–6.5. However ethyl ferulate and ferulic acid by using *Myceliophthora thermophila* laccases at pH 7.5. Laccase from *Trametes* sp. oxidizes a polyphenolic extract from *Hamamelis virginiana* in the presence of succinate buffer at pH 4.5 in order to covalently cross-link gelatin and chitosan. This process is shown in Fig. 2.10. The polyphenolic extract is allegedly incorporated into the gelatin/chitosan network via the Michael addition technique [11].

Tyrosinases

The typical substrate for tyrosinase is tyrosine, but this enzyme, copper monooxygenase (also known as monophenol monooxygenase) may trigger two distinct types of

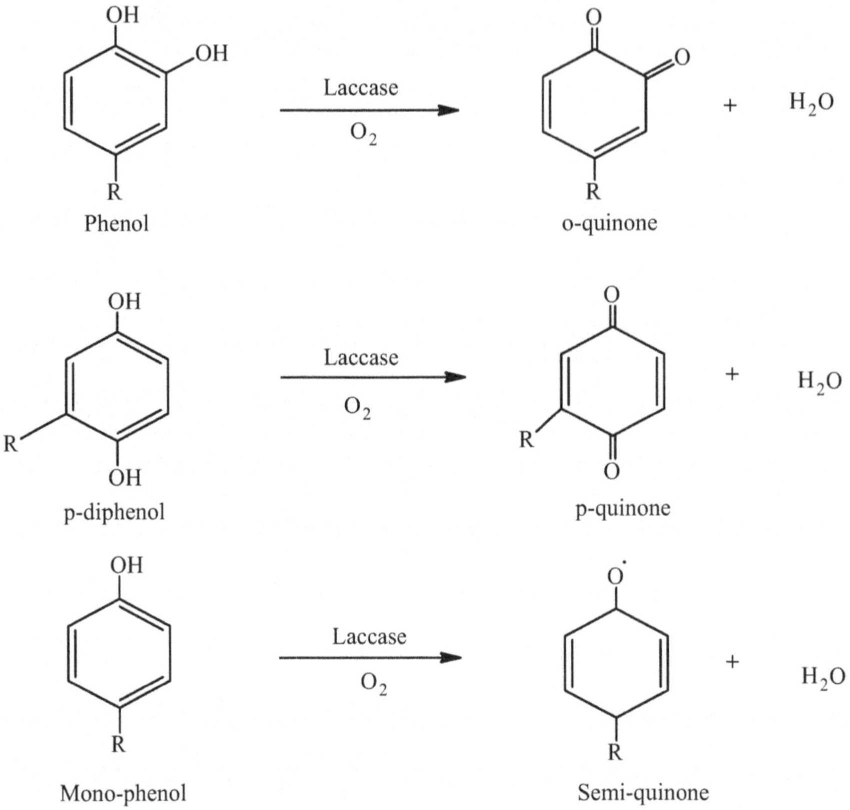

Fig. 2.9 Catalyzed processes using laccases [11]

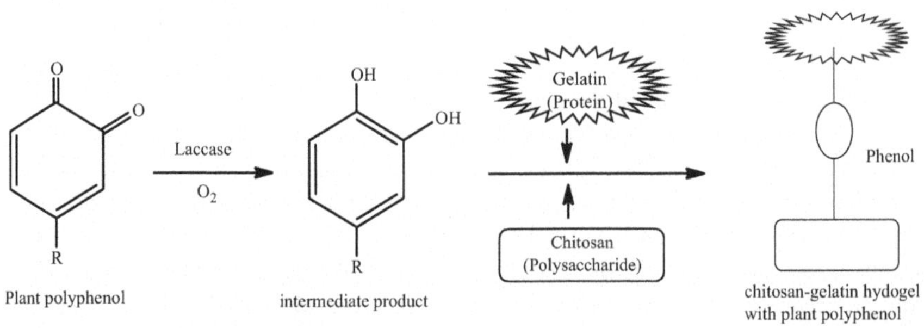

Fig. 2.10 Addition of a plant phenolic extract from *Hamamelis virginiana* and a laccase-based enzyme to a gelatin/chitosan conjugate [11]

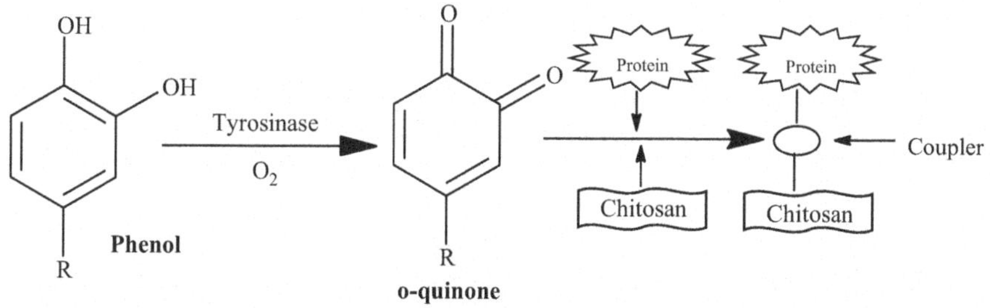

Fig. 2.11 Oxidation of phenolics by tyrosinase and their coupling with an active chitosan conjugate [11]

oxygen-dependent activities. First, monophenols are oxidized by the enzymes cresolase or monophenolase, and second, *o*-diphenols are converted into highly reactive *o*-quinones by either catecholase or monophenolase. Due to both types of tyrosinase activities, it can have broad substrate specifications. Tyroninase catalyzes the hydroxylation reaction, which changes monophenols into *o*-diphenols. The analogous *o*-quinones of these *o*-diphenols are then produced by oxidizing them by reducing oxygen to water. These *o*-quinones undergo simultaneous Schiff base and Michael addition reactions with primary amines (NH_2).

Research on the tyrosinase-catalyzed process revealed that the functional properties of chitosan, particularly its antioxidant and antibacterial properties have been altered. Numerous studies have been published regarding the enzymatic derivatization of chitosan via tyrosinase for the grafting-oxidation reaction of phenol derivatives, including phenolic acids like caffeic acid and chlorogenic acid, phenolic esters like dodecyl gallate and octyl gallate, and flavonoids like catechin, epicatechin, fisetin, hesperidin, quercetin, and rutin as shown in Fig. 2.11.

Additionally, many researchers reported that phenols such as cresol, aniline, 4-hydroxystyrene, and arbutin may be grafted onto chitosan at pH 7. For the filtering of wastewater by mushroom tyrosinase, the oxidation of 1- and 2-naphthol to their corresponding quinones and subsequent grafting of these quinones to chitosan at 40 °C and pH 8 have also been described. According to the biochemical method, Fig. 2.12 depicts the tyrosinase-catalyzed grafting-oxidation reaction for the production of protein or peptide-polysaccharide bioconjugates. The 3,4-dihydroxyphenylalanine is hydroxylated in the first step, and then it undergoes a second oxidation to form its corresponding *o*-quinones. In the subsequent step, these active *o*-quinones are coupled with the amino group of chitosan via the two separate methods of Schiff based and Michael Mantan adduct formation.

Peroxidases

Peroxidases, which are iron (III) protoporphyrin IX enzymes, are widely dispersed in all living things, including plants, animals, and microbes. In the presence of hydrogen peroxide, peroxidases also converted phenolic compounds into their corresponding quinones. In order to functionalize commercial polymers like chitosan,

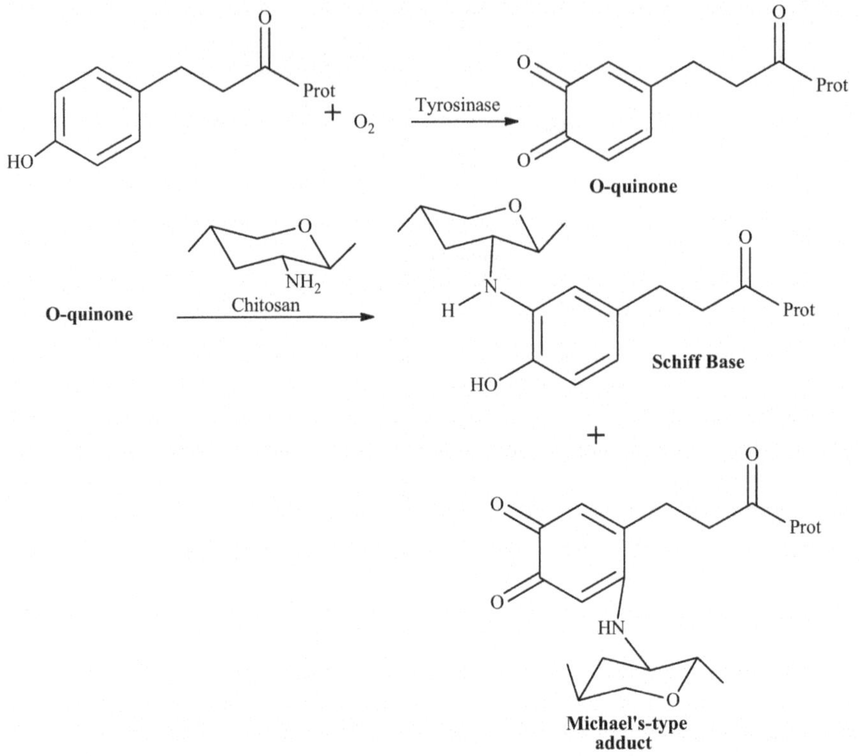

Fig. 2.12 Tyrosinase oxidation procedure for binding a peptide to chitosan

these enzymes provided a novel approach. Because chitosan is soluble at pH values below 6.0, the peroxidases for chitosan are best because it has much greater substrate specificities and an appropriate pH range of 5.0–6.0. Figure 2.13 illustrates how horseradish peroxidase works effectively in the presence of H_2O_2 and grafted phenolic compounds to chitosan successfully at pH 4.5 in a phosphate buffer containing dodecyl gallate. According to reported research, flavonoids such as chrysin, rutin, hesperidin, quercetin, and naringin have been effectively grafted on chitosan after being oxidized to their respective active intermediates by the chloroperoxidase derived from marine fungus. At pH 3, a solution with 20% isopropyl alcohol and 80% (60 mM) acetate buffer is used to conduct this enzymatic reaction.

The studies on peroxidases enzymes provided information about the formation of hydrogels at neutral pH due to the chemical modification of chitosan derivatives. This procedure is divided into two stages: first, the chitosan derivatives are produced chemically by mixing chitosan with substrates that contain α-hydroxy acids while for the second step, in the presence of horseradish peroxidase and phosphate buffer pH 7.4, injectable, biodegradable chitosan hydrogels are produced, as shown in Fig. 2.14 [11].

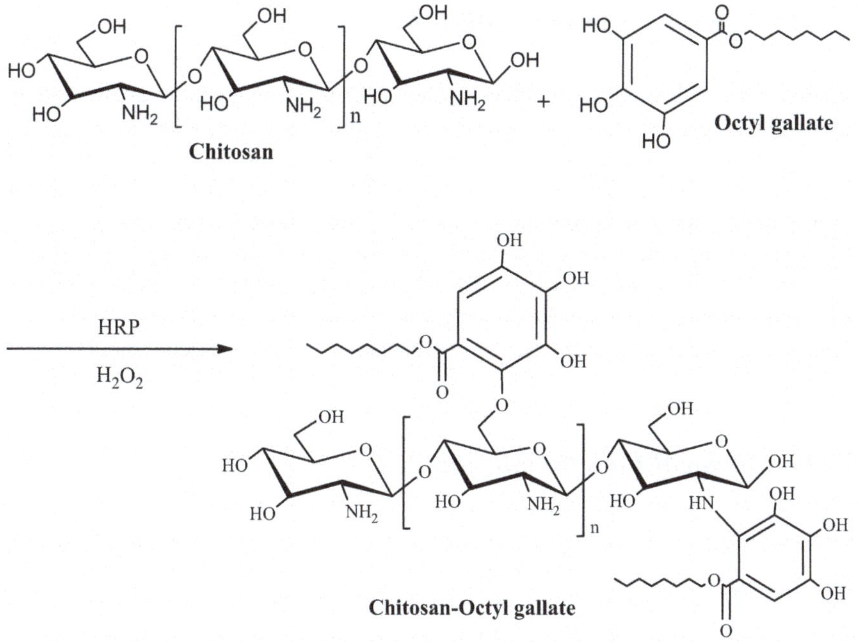

Fig. 2.13 Peroxidase-catalyzed grafting of octyl gallate to chitosan [11]

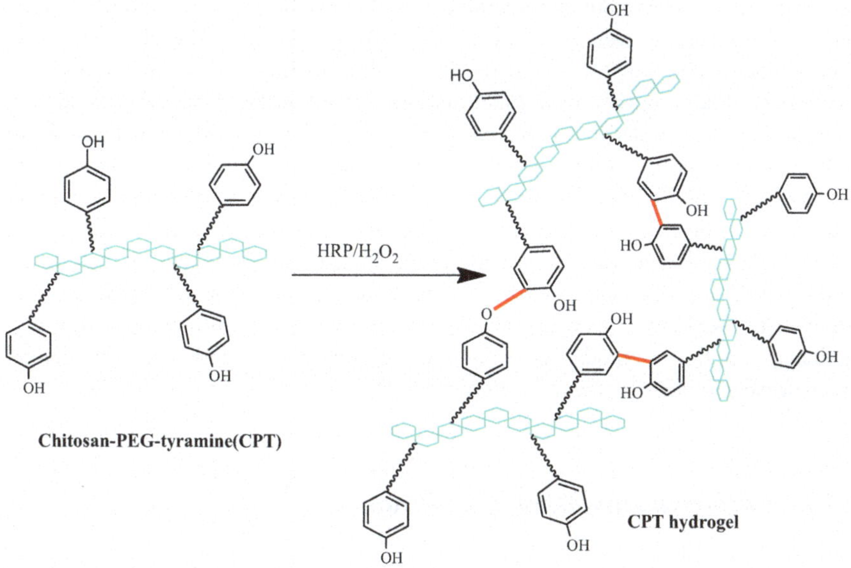

Fig. 2.14 In situ gel production of chitosan-PEG-tyramine (CPT) conjugates in the presence of horseradish peroxidase and H_2O_2 [11]

2.5 Functionalization of Chitosan

Chitosan has a wide range of applications; however, those applications are limited by factors like its poor solubility at neutral pH, its intra- and intermolecular hydrogen bonding in network structure, its low antioxidant potential owing to its inability to donate hydrogen. In spite of this, it has poor workability, poor cell attachment because of excessive hydrophilicity, rigid, and brittle physical properties. Additionally, it has strong antioxidant capabilities for breaking chains. In order to enhance chitosan's capacity as an antioxidant and its biological and physiochemical properties by incorporating active molecules into the chitosan backbone, a series of researches employing either chemical or enzyme approaches have been carried out [11].

2.5.1 Chemical Functionalization

Polysaccharides can be used in a variety of areas because of their chemical functionalization. For the specific end-user applications in a variety of domains including the biomedical, biotechnological, and pharmaceutical industries, the chemical processes offer a wide choice of derivatives with better features. Alkylation, acylation, hydroxyalkylation, carboxyalkylation, quaternization, sulphatation, thiolation, phosphorylation, and grafted copolymerization are only a few of the functionalization that has been made. There are innumerable possible applications for modified variants, and research in this area has accelerated significantly. The chitosan backbone is chemically functionalized by the addition of tiny functional groups. The compounds that comprise these functional groups might be alkyl, carboxymethyl, or phenolic compounds like catechin, caffeic acid, ferulic acid, or gallic acid to make chitosan more soluble in alkaline and neutral pHs without compromising the material's antioxidant qualities, cationic characteristics, potential to inhibit pathogens, ability to adhere to cells, or binding properties to cationic dyes. The most widely used techniques for synthesizing chitosan functionalization include the redox pair of ascorbic acid (Vitamin C) and hydrogen peroxide (H_2O_2) in an inert atmosphere, methyl methacrylate (MMA) and potassium persulfate, carbodiimide-mediated coupling method, living cationic polymerization, and radiation method (microwave and electron beam irradiation) [11].

2.5.2 Carboxymethylation of Chitosan

Carboxymethylated chitosan is synthesized through carboxymethylation, a process in which some of the chitosan's –OH groups are substituted out for $–CH_2COOH$ groups. Reactive functional ligands with –COOH and $–NH_2$ groups still have the

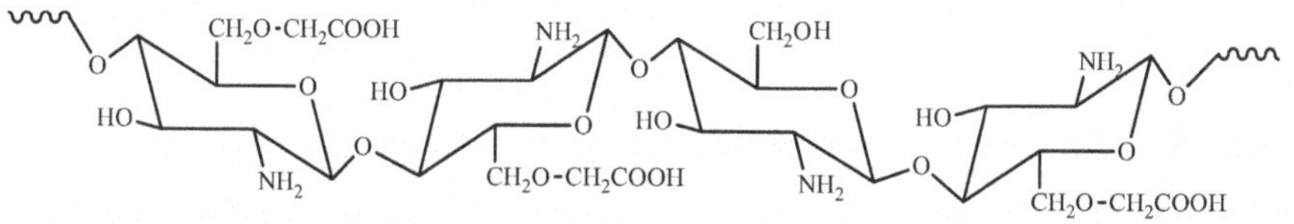

Fig. 2.15 Structure of carboxymethyl chitosan [33]

ability to go through chemical changes in order to enhance their physical properties of dye binding and metal chelation [33].

Preparation: The approach described by Farag and Mohamed for carboxymethylating chitosan is as follows: 100 ml of 20% w/v NaOH and 5 g of chitosan are swirled for 15 min. After that, 15 g of monochloroacetic acid is added dropwise to the reaction mixture, and stirring is again continued at a temperature of 40 ± 2 °C for two hours. After stirring, 10% v/v acetic acid is added to the reaction mixture to neutralize it, and then more than 70% methanol is added. The product (carboxymethyl chitosan) is filtered, then methanol is used to wash it. The product is vacuum-dried for 8 h at 55 °C to produce dry carboxymethyl chitosan (6.5 g approx.). Subsequently, the degree of substitution is estimated, and it is determined to be around 0.75. Figure 2.15 shows the structure of carboxymethyl chitosan [33, 34].

2.5.3 *Phosphorylation of Chitosan*

Bi-phosphonates and phosphonic acids, which are organic phosphorus compounds, are crucial because these are more efficient in complexing metals than carboxylic groups. The presence of an amino group, which produces NH_2–CH_2–PO_3^{2-} from the chitosan molecule, strengthens the PO_3^{2-}'s metal-binding capabilities by increasing the strong donor effects for a mono-dentate ligand. Heras et al. claimed that the *N*-methylene phosphonic chitosan produced in a single step has consistent modifications and did not appreciably change the chitosan's properties, such as its capacity to form films [35]. The *N*-methylene phosphonic chitosan's chemical composition is depicted in Fig. 2.16 [35].

Preparation: Heras et al. reported the method for the preparation of *N*-methylene phosphonic chitosan following the method reported in the established literature. Following the procedure, the chitosan 2% w/v solution is prepared in glacial acetic acid 1% v/v solution. One part of the chitosan solution and one part of the phosphorous acid are dissolved in the water drop by drop while being stirred continuously for an hour. This reaction mixture is heated to 70 °C and refluxed for an hour before a quantity (36.5% w/w) of formaldehyde is added drop by drop. The reaction mixture is thereafter heated for six hours. The resultant light yellow solution is dialyzed against distilled water with a cut-off value of 2500 Da until the pH of the water reaches to 6.8 [35].

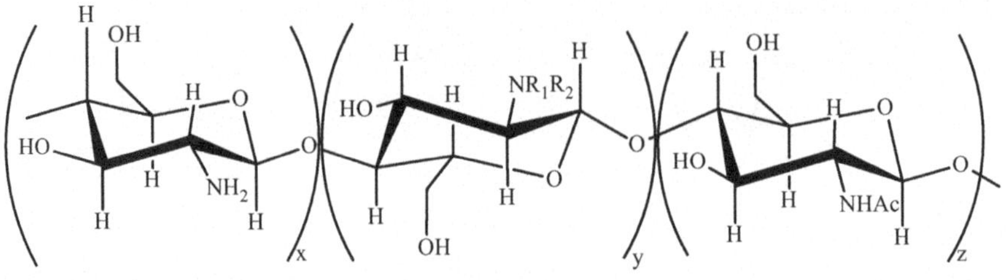

Fig. 2.16 Chemical composition of *N*-methylene phosphonic chitosan [35]

2.5.4 Graft Copolymerization of Chitosan

During the grafted copolymerization process, side groups are incorporated into chitosan's main chain, giving it unique properties. These modified chitosan can change the surface of a fabric or cellulosic surface and also improve chitosan's antibacterial characteristics. These grafted-copolymerized chitosan surfaces can be used to improve the anticoagulant properties of tissue engineering materials. The development of sustained drug release and other biopharmaceuticals is greatly facilitated by the grafted-copolymerized chitosan [36, 37].

Preparation: The novel chitosan-*O*-poly(ethylene glycol) grafted copolymers are produced by Gorochovceva and Makuška using the established process. With continuous stirring for 16 h at 60 °C, Methyl-Polyethylene-Glycol Iodide (MPEGI) (13.4 g), *N*-Phthaloyl Chitosan (0.9 g), and Ag_2O (1.55 g) are dissolved in 60 ml DMF. After adding 40 ml of hydrazine monohydrate and 80 ml of water, the reaction mixture is then heated at 90 °C with magnetic stirring for 15 h. The hydrazine monohydrate is subsequently extracted from the mixture by rotary evaporation, yielding a viscous solution. This mixture is diluted with water, vacuum-dried, and 8.2 g PEGylated-*O*-Chitosan is produced. The synthesis procedure is shown in Fig. 2.17 [37].

2.5.5 Esterification of Chitosan

The esterification of chitosan happens when chitosan molecules react with oxygen-containing inorganic acids, oxygen-containing mineral acids, or their anhydrides. The most preferred acids for esterifying chitosan are phosphoric, sulfuric, and chlorosulfuric acids. Sulfated chitosan can be used in place of heparin and sulfated heparin in a variety of biological applications, such as antiviral, anticoagulant, specific protein binding, and promoting osteogenic differentiation. The mechanism

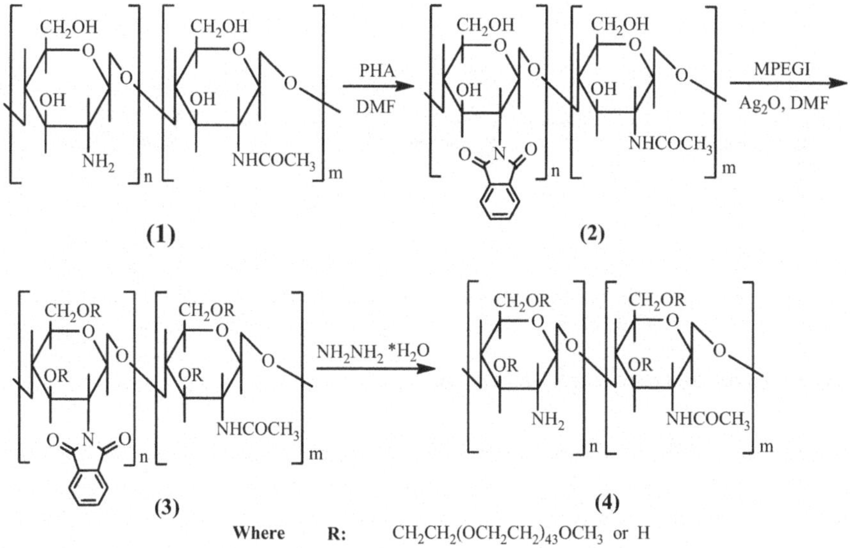

Fig. 2.17 Process for synthesizing PEGylated-*O*-Chitosan [37]

of sulfated chitosan was found same as heparin. As a high-strength fiber, the esterified chitosan derivatives can be utilized [18, 36].

Preparation: The established proposed the chemical process for producing multi-derivative chitosan hydroxypropyl chitosan with maleic acid sodium (HPCs-g-MAS) (MAS) by esterifying chitosan with propylene epoxide and then grafting and copolymerizing. The hydroxypropyl chitosan (HPC) is prepared from propylene epoxide and chitosan under alkali conditions. For alkalization, 3 g of chitosan is mixed with 30 g (% w/v) of a NaOH solution at − 18 °C. A mixture of 30 ml of isopropyl alcohol and alkali-treated chitosan is stirred at 40°C for one hour. The 30 ml of propylene epoxide is added to the reaction mixture along with constant stirring, and it is refluxed at 60 °C for two hours. The reaction mixture's pH is adjusted to 7.0 by adding 1% v/v HCl before filtering it. The filtrate is then thoroughly washed with 95% v/v ethyl alcohol. The product is vacuum-dried for 48 h at 60 °C. This obtained product HPCs and then used for the synthesis of grafted copolymerization. In this procedure, 15 ml of deionized water, 0.2 g of HPCs, and a predetermined amount of the sodium salt of maleic acid are placed in a round-bottom flask and heated in a water bath at 70°C. For the removal of dissolved oxygen, nitrogen gas bubbled out for 30 min. Ammonium persulfate (0.1 mmol) dissolved in 10 ml water and slowly added in reaction mixture for initial grafting of polymers. However, the reaction products precipitated in acetone before being filtered, and then it is again washed in acetone before being dried at 60 °C under vacuum. Homopolymers are extracted using the Soxhlet technique by refluxing with ethyl alcohol for 24 h, followed by 48 h of vacuum-sealed drying at 60 °C [38].

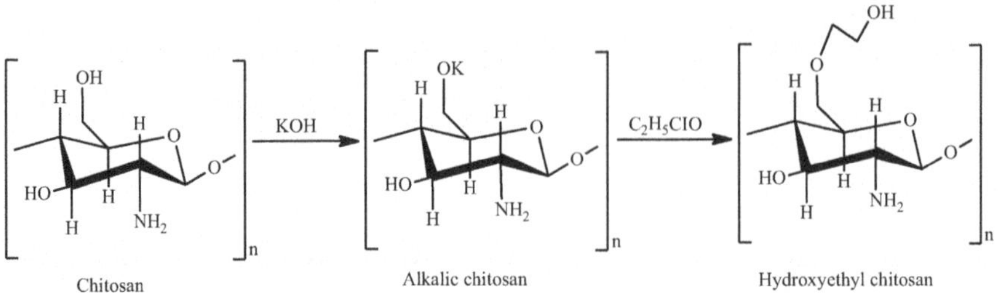

Fig. 2.18 Schematic for the synthesis of hydroxyethyl chitosan (HECs) from chitosan [39]

2.5.6 *Etherification of Chitosan*

Through the synthesis of the appropriate alkylating agent (alkyne derivatives), chitosan's –OH groups undergo etherification. Deacetylation processes transform these produced alkyne derivatives into chitosan ether derivatives. The ether derivatives of chitosan are non-cytotoxic, do not influence on growth of fibroblasts, and also do not have significant irritation. However, these derivatives cause delayed inflammatory response and delayed hypersensitivity. Hydroxyethyl chitosan (HEC) is a good biocompatible and biodegradable material that may be used in the medical field. The HECs have bacteriostatic, bactericidal effective against *E. coli*, and hygroscopic moisturizing effects. It is utilized as a natural textile softener and finishing agent, as well as a preservative in cosmetics [36, 39].

Preparation: Shao et al. synthesized and reported the process to produce hydroxyethyl chitosan (HEC) by following the steps [39]. The 100 g chitosan is stirred for 2 h at 65 °C in 1 L of 0.2 M NaOH. The precipitates are moistened, neutralized with distilled water, and then again alkalized with a 50% solution of KOH. This alkalized reaction mixture cooled at – 20 °C for 24 h. Then reaction mixture is mixed with 95% ethyl alcohol along with the addition of chloroethanol (C_2H_5ClO) with continuous stirring about 30 h at 45 °C, so that it completely dissolves in water. This mixture is filtered through cellulose acetate (1.2 μm) membrane filter. The product obtained after dialysis and freeze-drying under vacuum. The scheme of reaction is illustrated in Fig. 2.18 [39].

2.5.7 *Cross-linking of Chitosan*

There are lot of methods used for the cross-linking of chitosan, i.e., chemical cross-linking with glutaraldehyde and formaldehyde, with Mo(IV) polyoxyanions, blocked diisocyanate. The hydroxyl groups containing polymers needed more drastic conditions, i.e., high temperature, low pH, etc., for establishing cross-linking with aldehydes; however, amino groups containing polymers can easily cross-linked with the same reagents with mild conditions in which Schiff bases formed. The chemical cross-linking of chitosan with glutaraldehyde considerably changes a number of the micro- and macroproperties, including ion permeability, water absorption, mechanical, and chemical characteristics. Additionally, there is a significant chance that chemical cross-linking will increase the wetting strength of membranes and hydrogels [40, 41].

Preparation: Li et al. synthesized chitosan and glutaraldehyde cross-links using the functional strategy. For this purpose, 2 g of chitosan is dissolved in a 50 ml solution of acetic acid (5%) beforehand, following this 15 ml of glutaraldehyde (25%) is added for 24 h at room temperature; resultantly, the mixture is continually mixed to produce the water gel [40]. By washing the cross-linked complex more than eight times with double-distilled water, the unreacted, non-cross-linked glutaraldehyde is separated. The cross-linked chitosan glutaraldehyde lyophilized is preserved at −70 °C. Figure 2.19 depicts the structural differences between chitosan and chitosan-cross-linked with glutaraldehyde [40].

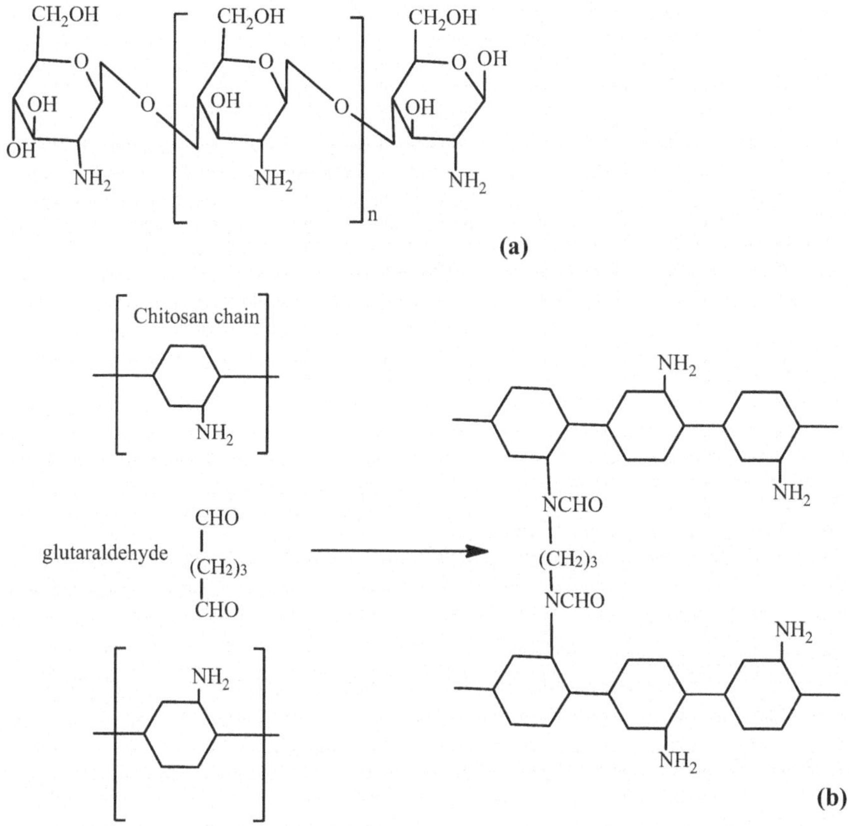

Fig. 2.19 Structure of chitosan and chitosan-cross-linked with glutaraldehyde [42]

In addition to the above-mentioned derivatives of chitosan, surface characteristics and XRD studies of chitin/1,4-butane diol blended polyurethane elastomers (PUEs) and UV-irradiated chitin-based PUEs have also been reported [43, 44]. Polysaccharide-based bionanocomposites, properties, and applications have been comprehensively discussed [45].

2.6 Summary

The purpose of this chapter is to demonstrate the procedures for preparing and functionalizing chitosan derivatives. The authors have endeavored to summarize the comprehensive body of work on the chemical and enzymatic techniques for the production of chitosan derivatives. Chitosan is a naturally occurring polysaccharide. It's found in the cell walls of "Zygomycets," a kind of fungi. The fungal biomass is treated with alkali (dilute NaOH) at high temperature (90–120 °C) to isolate chitosan from the fungal cell wall. The alkali treatment dissolved alkali soluble carbohydrates, lipids, proteins while remaining the cell wall as alkali insoluble material. In the last step, the fungal chitosan precipitation is accompanied by elevating the pH 9–10 and chitosan is obtained by centrifugation. The whole process is very time-consuming and sufficient chitosan does not produce. Since it is particularly demanded by industries as raw material, therefore chitin's deacetylation serves as the primary manufacturing process for chitosan. Synthesizing structurally modified chitosan for enhanced solubility and prospective applications has been increasingly popular in recent years. Derivatization does not alter the skeleton of chitosan, but it does provide it with new and better properties. Small functional groups, such as alkyl or carboxymethyl groups, are introduced into the structure during the derivatization process. Grafting is the covalent bonding of a functional group to the chitosan backbone. There are two types of reactive groups that can be grafted: deacetylated free amino groups and acetylated hydroxyl groups on C_3 and C_6 carbons. Chitosan molecules have four distinct functional groups. The bonds between the functional groups are C_2–NH_2, acetyl amino, C_3–OH, and C_6–OH. Chitosan molecules have distinct chemical characteristics that give them an advantage in introducing distinct functional groups. The site of derivatization and functionalization depends on the desired applications of chitosan derivatives. By grafting small molecules or polymer chains onto the backbone structure of chitosan, chemical derivatization, chemical modification, and chemical functionalization can be achieved. Acylation diminishes the crystalline structure while increasing its water solubility. The acylation which occurs at C_2–NH_2 produces amide that is known as *N*-acylation. Alkylation is another chemical method for the production of chitosan derivatives. Alkyl groups can be introduced at any of the reactive sites, i.e., C_2–NH_2, C_6–OH, and C_3–OH.

Chemical Methods: An introduction of quaternary ammonium group weakens hydrogen bonds, along with enhancing water solubility, and also adds chargeability. The *N*-Schiff base-type chitosan crown ethers were reported by Peng et al. and were synthesized by reacting 4′-formyl-benzo-15-crown-5 or 4,4′-formyl-benzo-18-crown-6 with chitosan and sodium borohydride. Tang et al., on the other hand, synthesized three different types of crown ether from the same reaction. Chitosan crown ethers have very good metal ion complexing selectivity. Regarding the concentration and separation of heavy-metal ions, it offers a wide variety of uses. Electrophiles are more likely to attack and react with the amine group at C_2 in chitosan's 2-amino-2-deoxy-D-glucopyranose units than with the hydroxyl groups. The reductive amination process is the most trustworthy method for adding alkyl or aryl functional groups to an amine. The synthesis of *N*-alkyl, *N*-aryl, and 5 heterocyclic aldehydes with various levels of substitution is proposed and reported by Kim et al. Chitosan's amino groups interact with aldehydes via Schiff base intermediates. Kim et al. reported the synthesis of quaternary ammonium salt derivatives of chitosan. The *N*-alkyl chitosans are quaternized using methyl iodide to produce new antibacterial derivatives. The hydrophobic characteristics of these compounds are enhanced by lengthening the alkyl substitution chain. The quaternization process described by Holappa et al. used *N*-chloroacyl-6-*O*-triphenylmethyl-chitosan as the starting material, and this study reported the precursor synthesis, derivatization techniques, and synthetic routes. Park et al. reported the synthetic procedure for the sugar-bearing derivatization of chitosan. The entire methodology used 1-ethyl-3-(3 dimethylamino-propyl) carbodiimide hydrochloride as an active ester intermediate. The crushed interior shell of *S. lessoniana* is deproteinized and demineralized by Subhapradha et al. to extract chitin, which is subsequently deacetylated with 40% NaOH to get chitosan and then converted into sulfated chitosan and phosphorylated chitosan from the extracted chitosan.

Enzymatic Methods: The synthesis and functionalization of natural polymers via enzymatic techniques have been researched as alternatives to environment-unfriendly, hazardous, and non-specific chemical processes. Enzyme-based methods have the power to completely remove the safety and health problems associated with harsh reactive chemicals. The composition of biopolymers determines which enzymatic tools should be used to functionalize and derivatize chitosan. Peroxidases (POD) and oxidase enzymes like polyphenoloxidase (PPO) are the major groups of enzymes. Using *T. versicolor* laccases, researchers have shown how to graft phenolic chemicals such as catechol, caffeic acid, tannic acid, and gallic acid to chitosan at pH 4.5–6.5. Ethyl ferulate and ferulic acid can be added by using *M. thermophila* laccase at pH 7.5 or higher. Numerous studies have been published regarding the enzymatic derivatization of chitosan via tyrosinase for the grafting-oxidation reaction of phenol derivatives, including phenolic acids like caffeic acid and chlorogenic acid. Phenols such as cresol, aniline, 4-hydroxystyrene, and arbutin may be grafted onto chitosan at pH 7 and 40 °C by tyrosinases. Horseradish peroxidase worked effectively in the presence of H_2O_2 and grafts phenolic compounds to chitosan successfully. The studies provided information about the formation of hydrogels at neutral pH due to the chemical modification of chitosans by peroxidase.

Functionalization of Chitosan: The chitosan backbone is chemically functionalized by the addition of tiny functional groups. These functional groups might be alkyl, carboxymethyl, or phenolic compounds like catechin, caffeic acid, ferulic acid, or gallic acid. There are innumerable possible applications for modified variants, and research in this area has accelerated significantly. Carboxymethylated chitosan is an organic compound with a unique property that allows it to undergo chemical changes in order to enhance its physical properties, such as dye binding and metal chelation. Farag and Mohamed reported the approach for preparing carboxymethylating chitosan. Bi-phosphonates and phosphonic acids are crucial because these are more efficient in complexing metals than carboxylic groups. Heras et al. claimed that the *N*-methylene phosphonic chitosan produced in a single step has consistent modifications and did not appreciably change its properties. The novel chitosan-*O*-poly(ethylene glycol) grafted copolymers were produced by Gorochovceva and Makuška. The development of sustained drug release and other biopharmaceuticals is greatly facilitated by the grafted-copolymerized chitosan. The most preferred acids for esterifying chitosan are phosphoric, sulfuric, and chlorosulfuric acids. As a high-strength fiber, it can be utilized by grafting copolymerizing with propylene epoxide and maleic acid sodium. Xie et al. proposed the chemical process for producing multiderivative hydroxyl-propyl chitosan with maleic acids sodium (HPCs-g-MAS). Shao et al. synthesized and published the process to produce hydroxyl ethyl chitosan (HECs). The HEC is a good biocompatible and biodegradable material and has bacteriostatic, bactericidal effective against *E. coli*, and hygroscopic moisturizing effects. Li et al. synthesized chitosan and glutaraldehyde cross-links and claimed that chemical cross-linking of chitosan with glutaraldehyde considerably changes a number of the micro- and macroproperties, including ion permeability, water absorption, mechanical, and chemical characteristics.

References

1. Praveen P, Rao V (2014) Synthesis and thermal studies of chitin/AgCl nanocomposite. Procedia Mater Sci 5. https://doi.org/10.1016/j.mspro.2014.07.411
2. Prakash B, Jothirajan MA, Umapathy S, Amala V (2013) Synthesis and characterization of biodegradable ultrasonicated films made from chitosan/Al_2O_3 polymer nanocomposites. Phys Procedia. https://doi.org/10.1016/j.phpro.2013.10.014
3. Chen X, Ma Z (2014) Multiplexed electrochemical immunoassay of biomarkers using chitosan nanocomposites. Biosens Bioelectron 55. https://doi.org/10.1016/j.bios.2013.12.037
4. Xie DF, Martino VP, Sangwan P, Way C, Cash GA, Pollet E, Dean KM, Halley PJ, Avérous L (2013) Elaboration and properties of plasticised chitosan-based exfoliated nano-biocomposites. Polymer (Guildf) 54. https://doi.org/10.1016/j.polymer.2013.05.017
5. Ahmed TA, Aljaeid BM (2016) Preparation, characterization, and potential application of chitosan, chitosan derivatives, and chitosan metal nanoparticles in pharmaceutical drug delivery. https://doi.org/10.2147/DDDT.S99651
6. Alves NM, Mano JF (2008) Chitosan derivatives obtained by chemical modifications for biomedical and environmental applications. https://doi.org/10.1016/j.ijbiomac.2008.09.007
7. El Badawy M (2008) Chemical modification of chitosan: synthesis and biological activity of new heterocyclic chitosan derivatives. Polym Int 57. https://doi.org/10.1002/pi.2333

8. Kim CH, Choi JW, Chun HJ, Choi KS (1997) Synthesis of chitosan derivatives with quaternary ammonium salt and their antibacterial activity. Polym Bull 38. https://doi.org/10.1007/s002890050064
9. Subhapradha N, Ramasamy P, Srinivasan A, Shanmugam A, Shanmugam V (2013) Preparation of chitosan derivatives from gladius of squid *Sepioteuthis lessoniana* (Lesson, 1830) and antimicrobial potential against human pathogens. J Biol Sci 13. https://doi.org/10.3923/jbs.2013.257.263
10. Park JH, Cho YW, Chung H, Kwon IC, Jeong SY (2003) Synthesis and characterization of sugar-bearing chitosan derivatives: aqueous solubility and biodegradability. Biomacromolecules 4. https://doi.org/10.1021/bm034094r
11. Aljawish A, Chevalot I, Jasniewski J, Scher J, Muniglia L (2015) Enzymatic synthesis of chitosan derivatives and their potential applications. https://doi.org/10.1016/j.molcatb.2014.10.014
12. Riva R, Ragelle H, Des Rieux A, Duhem N, Jérôme C, Préat V (2011) Chitosan and chitosan derivatives in drug delivery and tissue engineering. https://doi.org/10.1007/12_2011_137
13. Pochanavanich P, Suntornsuk W (2002) Fungal chitosan production and its characterization. Lett Appl Microbiol 35. https://doi.org/10.1046/j.1472-765X.2002.01118.x
14. Gholizadeh Aghdam M (2010) Extraction of chitosan from fungal cell wall by sulfuric acid studying the effect of deacetylation degree and temperature on recovery chitosan. Ind Biotechnol
15. Xie W, Xu P, Liu Q (2001) Antioxidant activity of water-soluble chitosan derivatives. Bioorg Med Chem Lett 11. https://doi.org/10.1016/S0960-894X(01)00285-2
16. Sashiwa H, Kawasaki N, Nakayama A, Muraki E, Yamamoto N, Aiba SI (2002) Chemical modification of chitosan. 14:1 synthesis of water-soluble chitosan derivatives by simple acetylation. https://doi.org/10.1021/bm0200480
17. Argüelles-Monal WM, Lizardi-Mendoza J, Fernández-Quiroz D, Recillas-Mota MT, Montiel-Herrera M (2018) Chitosan derivatives: introducing new functionalities with a controlled molecular architecture for innovative materials. https://doi.org/10.3390/POLYM10030342
18. Wang W, Meng Q, Li Q, Liu J, Zhou M, Jin Z, Zhao K (2020) Chitosan derivatives and their application in biomedicine. https://doi.org/10.3390/ijms21020487
19. Mourya VK, Inamdar NN (2008) Chitosan-modifications and applications: opportunities galore. https://doi.org/10.1016/j.reactfunctpolym.2008.03.002
20. Cai J, Dang Q, Liu C, Fan B, Yan J, Xu Y, Li J (2015) Preparation and characterization of *N*-benzoyl-*O*-acetyl-chitosan. Int J Biol Macromol 77. https://doi.org/10.1016/j.ijbiomac.2015.03.007
21. Shibano M, Nishida S, Saito Y, Kamitakahara H, Takano T (2014) Facile synthesis of acyl chitosan isothiocyanates and their application to porphyrin-appended chitosan derivative. Carbohydr Polym 113. https://doi.org/10.1016/j.carbpol.2014.05.099
22. Huang X, Xu C, Li Y, Cheng H, Wang X, Sun R (2019) Quaternized chitosan-stabilized copper sulfide nanoparticles for cancer therapy. Mater Sci Eng C 96. https://doi.org/10.1016/j.msec.2018.10.062
23. Tang XH, Tan SY, Wang YT (2002) Study of the synthesis of chitosan derivatives containing benzo-21-crown-7 and their adsorption properties for metal ions. J Appl Polym Sci 83. https://doi.org/10.1002/app.2316
24. Peng C, Wang Y, Tan S, Cheng G (1998) Preparation of chitosan derivatives. Synthesis of *N*-Schiff base type and *N*-secondary amino type chitosan-crown ethers. Polym J 30. https://doi.org/10.1295/polymj.30.843
25. Jia Z, Shen D, Xu W (2001) Synthesis and antibacterial activities of quaternary ammonium salt of chitosan. Carbohydr Res 333. https://doi.org/10.1016/S0008-6215(01)00112-4
26. Hollapa J, Nevalainen T, Soininen P, Másson M, Järvinen T (2006) Synthesis of novel quaternary chitosan derivatives via *N*-chloroacyl-6-*O*-triphenylmethylchitosans. Biomacromolecules 7. https://doi.org/10.1021/bm050707j

27. Holappa J, Nevalainen T, Savolainen J, Soininen P, Elomaa M, Safin R, Suvanto S, Pakkanen T, Másson M, Loftsson T, Järvinen T (2004) Synthesis and characterization of chitosan *N*-betainates having various degrees of substitution. Macromolecules 37. https://doi.org/10.1021/ma0358780
28. Holappa J, Nevalainen T, Soininen P, Elomaa M, Safin R, Másson M, Järvinen T (2005) *N*-chloroacyl-6-*O*-triphenylmethylchitosans: useful intermediates for synthetic modifications of chitosan. Biomacromolecules 6. https://doi.org/10.1021/bm049372g
29. Lin YH, Liang HF, Chung CK, Chen MC, Sung HW (2005) Physically crosslinked alginate/*N*,*O*-carboxymethyl chitosan hydrogels with calcium for oral delivery of protein drugs. Biomaterials 26. https://doi.org/10.1016/j.biomaterials.2004.06.011
30. Gübitz GM, Paulo AC (2003) New substrates for reliable enzymes: enzymatic modification of polymers. https://doi.org/10.1016/j.copbio.2003.09.010
31. Bertoni F, Barbani N, Giusti P, Ciardelli G (2006) Transglutaminase reactivity with gelatine: perspective applications in tissue engineering. Biotechnol Lett. https://doi.org/10.1007/s10529-006-9046-2
32. Cunha AG, Gandini A (2010) Turning polysaccharides into hydrophobic materials: a critical review. Part 2. Hemicelluloses, chitin/chitosan, starch, pectin and alginates. https://doi.org/10.1007/s10570-010-9435-5
33. Farag RK, Mohamed RR (2013) Synthesis and characterization of carboxymethyl chitosan nanogels for swelling studies and antimicrobial activity. Molecules 18. https://doi.org/10.3390/molecules18010190
34. Luo Y, Teng Z, Wang X, Wang Q (2013) Development of carboxymethyl chitosan hydrogel beads in alcohol-aqueous binary solvent for nutrient delivery applications. Food Hydrocoll 31. https://doi.org/10.1016/j.foodhyd.2012.11.011
35. Heras A, Rodríguez NM, Ramos VM, Agulló E (2001) *N*-methylene phosphonic chitosan: a novel soluble derivative. Carbohydr Polym 44. https://doi.org/10.1016/S0144-8617(00)00195-8
36. Zhao D, Yu S, Sun B, Gao S, Guo S, Zhao K (2018) Biomedical applications of chitosan and its derivative nanoparticles. https://doi.org/10.3390/polym10040462
37. Gorochovceva N, Makuška R (2004) Synthesis and study of water-soluble chitosan-*O*-poly(ethylene glycol) graft copolymers. Eur Polym J 40. https://doi.org/10.1016/j.eurpolymj.2003.12.005
38. Xie W, Xu P, Wang W, Liu Q (2002) Preparation and antibacterial activity of a water-soluble chitosan derivative. Carbohydr Polym 50. https://doi.org/10.1016/S0144-8617(01)00370-8
39. Shao K, Han B, Gao J, Song F, Yang Y, Liu W (2015) Synthesis and characterization of a hydroxyethyl derivative of chitosan and evaluation of its biosafety. J Ocean Univ China 14. https://doi.org/10.1007/s11802-015-2544-x
40. Li B, Shan C-L, Zhou Q, Fang Y, Wang Y-L, Xu F, Han L-R, Ibrahim M, Guo L-B, Xie G-L (2013) Synthesis, characterization, and antibacterial activity of cross-linked chitosan-glutaraldehyde. Mar Drugs 11:1534–1552
41. Beppu MM, Vieira RS, Aimoli CG, Santana CC (2007) Crosslinking of chitosan membranes using glutaraldehyde: effect on ion permeability and water absorption. J Memb Sci 301. https://doi.org/10.1016/j.memsci.2007.06.015
42. Mirzaei BE, Ramazani A, Shafiee M, Danaei M (2013) Studies on glutaraldehyde crosslinked chitosan hydrogel properties for drug delivery systems. Int J Polym Mater Polym Biomater 62. https://doi.org/10.1080/00914037.2013.769165
43. Zia KM, Barikani M, Zuber M, Bhatti IA, Barmar M (2009) Surface characteristics of polyurethane elastomers based on chitin/1,4-butane diol blends. Int J Biol Macromol 44(2):182–185. https://doi.org/10.1016/j.ijbiomac.2008.12.004
44. Zia KM, Barikani M, Bhatti IA, Zuber M, Barmar M (2009) XRD studies of UV-irradiated chitin based polyurethane elastomers. Carbohyd Polym 77(1):54–58. https://doi.org/10.1016/j.carbpol.2008.12.002
45. Zafar R, Zia KM, Tabasum S, Jabeen F, Noreen A, Zuber M (2016) Polysaccharide based bionanocomposites, properties and applications: a review. Int J Biol Macromol 92:1012–1024. https://doi.org/10.1016/j.ijbiomac.2016.07.102

Chapter 3
Green Derivatization of Chitosan Nanoparticles

3.1 Chitosan Derivatives: New Functionalities

Within the polymer family, chitosan has wide spread attention not only in variation of the degree of polymerization but also in extent of acetylation and in molecular weight. Along the polymer chain, chitosan has unique distribution of acetylated and deacetylated units. The chitosan specification depends upon isolation process and its source whether it is natural or not. Thus, it is obligatory to know about the intrinsic properties of chitosan, as well as their effect on the properties of chitosan derivatives. Chitosan has widely used as pollutant adsorbent due to its high adsorption potential and low cost. For the removal of heavy metal ions, chitosan has a number of hydroxyl groups, that has the capacity to bind with these metal ions. Besides of all above advantages, chitosan also has some flaws such as poor mechanical strength, lower heat stability, and pH sensitivity. All these parameters limit their applications in various industrial fields. Thus, in order to overcome above-mentioned drawbacks, cross-linking and graft modification of chitosan can be used which enhances its performance. Chitosan in powder form isn't suitable due to its porosity and specific surface area. Chitosan nanoparticles (CSNPs) and its hydrogels can be synthesized by physical modification. Composites with enhanced adsorption properties can also be synthesized by the combination of chitosan with other materials such as zeolite and graphene etc. [1].

Biological and physicochemical properties of chitosan can be improved by its derivatization. In addition to the synthesis process, the type of the chemicals utilized to functionalize the chitosan derivatives has a significant impact on their ultimate properties. A 1,2,3-triazole is used to create an organoselenium-chitosan derivative [2]. Chitosan has unique properties such as biodegradability, biocompatibility and non-toxicity. Besides all of these key qualities, its mucoadhesive property also plays an important role in the development of many industries such as pharmacy, biomedicine, food industry, nanotechnology, and biotechnology [3]. The development of unique features in natural polymers is an emerging topic in materials science.

K. M. Zia, *Chitosan*, Smart Nanomaterials Technology,
https://doi.org/10.1007/978-981-96-2629-8_3

Modification of polysaccharides can be done by numerous ways. Chitosan is vulnerable to modification due to the presence of free amino group at C-2 position and hydroxyl group at C-3 and C-6 position [4]. Many researchers highlighted the need of chitosan derivatives with enhanced properties, such as increased stability especially at basic and neutral pH values. Chitosan derivatives with these enhanced properties can be used for therapeutic chemical delivery [5].

3.2 Chemical Derivation of Chitosan

Number of qualities, such as solubility, hydrophilic character, affinity for bioactive compounds, and gelling capabilities, can be enhanced by the chemical derivatization of chitosan [6]. Other techniques, in which the reactions can be regulated stoichiometrically, are favored when particular novel functionalities are pursued. Polymer science has developed a number of methodologies to synthesize new polymer-based hydrogels, tissue engineering scaffolds, mineral chelation, drug delivery systems, and toxic substances. Chitosan qualities such as its solubility in organic solvents or in acidic solutions can be improved or changed by adding lipophilic or hydrophilic molecules [7]. Necessary and particular elements of chitosan have been covered in previous researches [8, 9].

The synthesis of chitosan derivatives with controlled molecular architecture has been achieved not only through modifications of existing methods but also by introducing new approaches. For the development of novel materials, it is prerequisite to understand the relationship between characteristics and chemical structure. Redundant depolymerization of polysaccharides, unchecked increase of substituent groups in the main chain, and irregular distribution are the results of non-selective chemical derivatization of chitosan. Amino group must be shielded due to their high reactivity, so that it may be able to hydroxyl group to participate in functionalization reaction. Several techniques have been proposed; however, the N-phthaloylation of chitosan has been the most popular thus far [10, 11]. The N-phthaloyl chitosan with the derivative of phthalic anhydride is used to protect the amino group for additional chemical alterations. Thus, chitosan is protected from undesired reactions. The N-phthaloylation would be carried out in the mixture of DMF and water to avoid O-phthaloyl substitution [12].

Hydrazine monohydrate is used to remove the phthaloyl protection at the end of chemical modification to resynthesize the free amino group. Although to save chitosan's amine moieties N-phthaloylation is considered to be one of the most effective strategy, but it has two drawbacks: chitosan's solubility in aqueous solutions is affected by its N-phthaloylation. It can only be dissolved in aprotic polar organic solvents, which has been linked to the derivative's increased crystallinity when compared to pristine chitosan [11]. Obviously, the precursor's solubility in various organic solvents may be helpful for performing the O-substitution reaction in the latter. The chitosan chain depolymerizes by unblocking the reaction by using hydrazine monohydrate, causing its mechanical characteristics to deteriorate

[13]. The production of Schiff bases with aldehydes is another method for safeguarding amines. N-aryl aldehydes have been identified as the ideal alternative, as their Schiff bases disrupt interchain packing, making the hydroxyl group more accessible for reactions. The N-benzylidene chitosan, N-m-toluylidene chitosan, and N-salicylidene chitosan are the most effective among them [14, 15]. At acidic pH values, the deprotection is simple, yielding O-substituted chitosan derivatives with little chain breakdown. However, the protection of the amine is far from perfect under certain situations, which is the fundamental downside of this technique [16]. The preservation of amino groups has also been achieved by dissolving chitosan in methane sulfonic acid [17]. Due to the intense acidic conditions employed to dissolve it, the polymer chain of the chitosan degrades significantly [18].

Kolb and Sharpless created the term "click chemistry" to describe "a collection of near-perfect" reactions in which two functional groups only react with each other. Reactions in click chemistry are considered to be high yielding reactions and carried out at low temperature and wide range of pH. These types of reactions are insensitive to oxygen and water. Highly regioselective products are need to made that don't require any sophisticated purification. One distinguishing feature is that it follows modular reactions because it mimics the biological processes in nature [19]. In chitosan derivatization cycloadditions are one of the most commonly waged reaction. Another important reaction of click chemistry is Huisgen's reaction (Fig. 3.1) [19].

It is a cycloaddition reaction between azides and alkynes that synthesizes two region-isomer triazoles. The reaction is carried out with or without the use of catalyst. The Cu(I) is used as catalyst. The reaction occurs immediately in aqueous medium at medium temperature. The Diels–Alder reaction is also the type of cycloaddition reaction between dienophile and diene (Fig. 3.2). The reaction has thermodynamically reversible property that depends upon the reactants with a completely stable reaction product [20].

Chitosan derivatives that produced by click chemistry have unique features such as photochromic behavior, tunable thermos-sensitive capabilities, and pH-sensitive macromolecular networks [21]. The Diels–Alder cycloaddition offers the advantage of enabling materials to alter their physicochemical properties based on temperature,

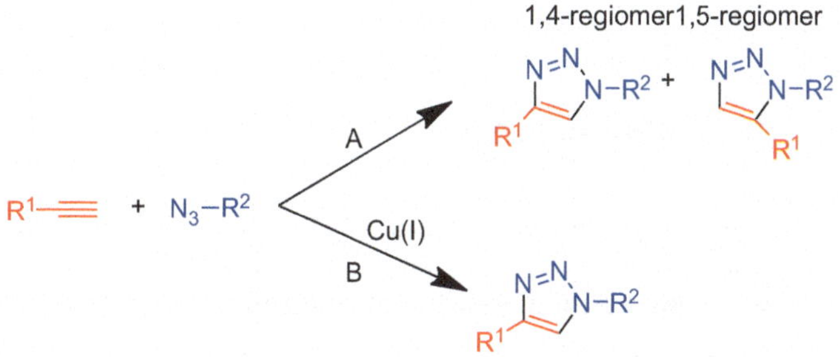

Fig. 3.1 Huisgen cycloaddition reactions in the absence (A) or presence (B) of Cu(I) catalyst

Fig. 3.2 Diels–Alder reaction between a diene and dienophile

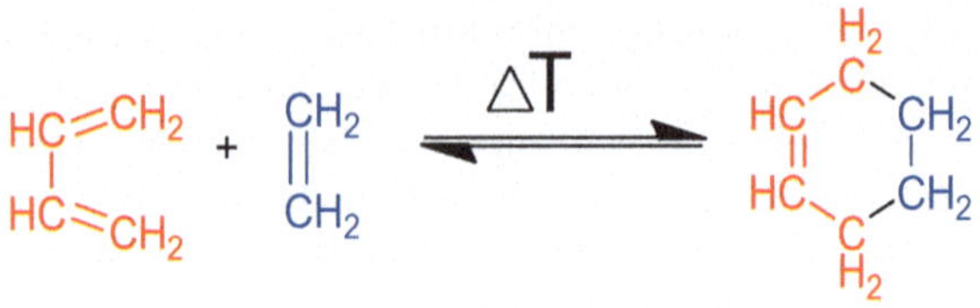

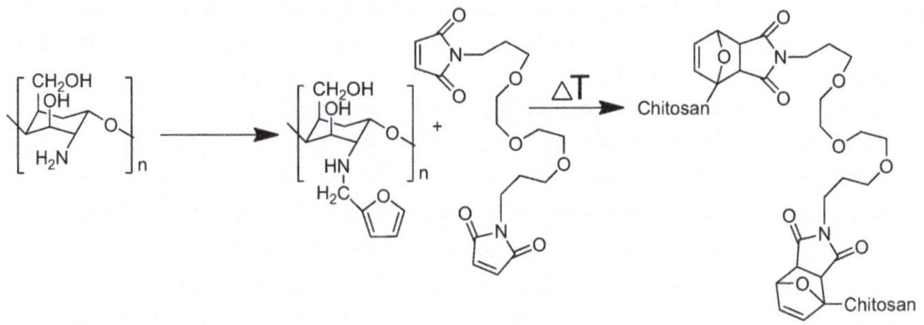

Fig. 3.3 Synthetic scheme for the preparation of N-(furfural) chitosan by Schiff base formation/reductive amination process and Diels–Alder cycloaddition with bis-maleimide [23]

a feature absent in Huisgen's cycloaddition. The furan's structure gives it a distinct dienic property, making it ideal for the formation of this type of reaction [22]. This property of furan chitosan allows researchers to look into the possibility of reaction between maleimides (dienophile) and furan chitosan derivative. This method is utilized to create a new chitosan hydrogel with fascinating drug carrier properties that could be exploited to create new biotechnological and biomedical materials [23] (Fig. 3.3).

The number of studies using click chemistry to alter chitosan is rapidly increasing [19–23]. Click chemistry's application to the synthesis of regioselectively controlled chitosan derivatives, on the other hand, is still in its early phases, and we should expect it to gain even more traction in the future years. Thus, it is doubtlessly ascribable to the reaction's simplicity as well as the good options provided by these techniques for introducing chemical alterations with high molecular architecture control.

3.3 Trimethyl Chitosan (TMC)

Recently N,N,N-trimethyl chitosan chloride (TMC), the quaternary derivative of chitosan has appeared as an attractive, pliable, and functional polymer. This positive-charged quaternary chitosan derivative is soluble over a wide pH range and has improved mucoadhesive, permeability, drug delivery, and DNA transport capabilities. Derivatization or grafting of TMC can also be done to personalize other properties like cell recognition. The TMC and its derivatives have better antibacterial properties than chitosan, with the protonated amino groups being responsible for this. The repulsive interactions between protonated chitosan amine groups are larger than quaternary groups in TMC (–N^+ $(CH_3)_3$). These interactions allow the TMC to be more flexible and fairly interact with cell membrane of bacteria [24].

3.3.1 Synthesis of Trimethyl Chitosan

Methyl iodide alkylation is enhanced at high temperatures in a strong alkaline environment, where it neutralizes acid by-products, preventing acidification and protecting unreacted primary amino groups from protonation, thereby enabling the quaternization (methylation) of amino groups in chitosan. (Fig. 3.4). The degree of quaternization (DQ) can be adjusted by increasing the number of reaction stages, extending the reaction duration, regulating reaction steps, or using chitosan with different deacetylation grades. High degrees of quaternization (DQ) facilitate O-methylation at C-3 and C-6 hydroxyl groups of chitosan, as confirmed by spectroscopic characterization. In general, O-methylation resulted in less soluble molecules. As a consequence, researchers are looking for trimethyl chitosan (TMC) polymers with a high DQ but low degree of O-methylation. Domard et al. [25] described the synthesis of N,N,N-trimethyl chitosan utilizing a suspension of 5 g chitosan in 250 ml N-methyl-2 pyrrolidinone reacting with CH_3I and NaOH for 3 h at 36 °C (chitosan: CH_3I: NaOH in molar ratio 1:15:2). This method, however, resulted in significant chitosan depolymerization. To prevent polymer breakdown and regulate the different quaternization parameters, the literature reported the changes in the reactant ratio (chitosan: CH_3I: NaOH in a molar ratio of 1:15:3.45). A ^{1}H-NMR investigation, on the other hand, shows that such a technique would predominantly create dimethylated polymers with just 10–15% quaternization. In terms of solvent/reagent addition sequence, researchers adjusted the technique, reporting one-step and two-step synthesis. In a one-step synthesis, chitosan is quaternized in a single step by reacting with CH_3I and NaI in NMP, followed by alkalization with aq. NaOH. In a two-step synthesis, chitosan quaternization occurs through sequential treatment with NaI in aq. NaOH, followed by reaction with CH_3I/NMP. The resultant product is washed with ethanol and ether before being methylated once again, although with less CH_3I this time. Dimethylation is much reduced when the basic reaction is repeated. The degrees of quaternization of TMC polymers are lower when dimethyl-amino-pyridine is employed as the base than when sodium hydroxide is utilized. Degradation of the polymer is also reduced. The observed difference may be attributed to the relatively weaker alkaline properties of dimethyl-amino-pyridine (DMAP) compared to sodium hydroxide (NaOH). While the DQ is kept to a bare minimum (12.5–34.4%), combining the two bases has no effect on polymer degradation. In an effort to improve the DQ by expanding the number of steps, O-methylation is detected. Dimethylation is also decreased when the number of reaction steps is increased. On the other hand, a longer response time increased both DQ and demethylation [24–26].

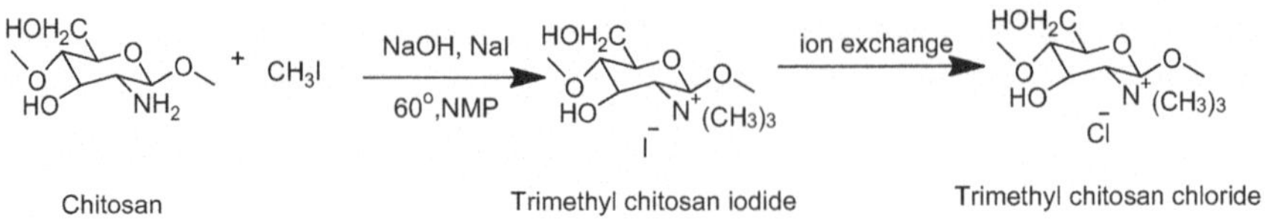

Fig. 3.4 Synthesis of N,N,N-trimethyl chitosan [4, 24]

3.4 Carboxymethyl Chitosan

Trimethylation of chitosan amino groups is the simplest quaternization process [27]. Furthermore, carboxymethylation as a hydrophilic alteration can increase chitosan solubility, yielding carboxymethyl chitosan (CMC), which has several biological uses including wound healing, bioimaging, tissue engineering, and drug/gene delivery [28]. Carboxymethylation occurs at the C-6 hydroxyl and NH_2 positions, producing water-soluble N, O-carboxymethyl chitosan compounds characterized by the presence of primary or secondary amine groups [29]. Trimethyl carboxymethyl chitosan derivatives are produced to increase antibacterial activities [30]. The CMC may also link metal ions to make self-polishing anticorrosive pigments, which include CMC copper ion, a powerful antiseptic that attacks algae [31]. The CMC's antibacterial activity can also be boosted by including different nanoparticles into its composite materials [32]. Due to its higher number of chelating groups and enhanced chain flexibility, the CMC supramolecular hydrogels have better water retention, hydrophilicity, biodegradability, bioactivity, antioxidant capacity, antimicrobial activities, and stronger metal ion adsorption capacities than chitosan hydrogels [33]. The CMC compounds are applied to avoid postsurgical adhesion which resultantly affect a patient's long-term health by causing impaired organ function, intestinal obstruction, diminished fertility, reoperation difficulties, and possibly pain, illness, and death [34]. Because it can regulate the rate of medicine release, the CMCs are used as pharmaceutics delivery systems. Controlled release systems enable the steady administration of hydrophilic and hydrophobic therapeutic agents at predetermined dosages over prolonged intervals, utilizing tailored delivery devices [35].

3.4.1 Synthesis of Carboxymethyl Chitosan

Carboxymethyl cellulose (CMC) production involves alkalization and etherification in a temperature-controlled (25 °C) double-walled glass reactor. Pure chitosan is disseminated in isopropanol first, and then 40% NaOH aqueous solution is added dropwise over a 15-min period during magnetic stirring at room temperature. Following alkalization, monochloroacetic acid (MCA) dissolved in isopropanol (1:1) is allowed to react, which is stirred at room temperature for the required time (6.4–10.6 h). Then filtration takes place, and the solid result is suspended in methanol prior to getting neutralized with glacial acetic acid. After filtering the slurry, the solid is carefully washed in 80% ethanol to remove undesirable by-products before being dried at room temperature. The material is dissolved in distilled water for purification of carboxymethyl chitosan. Positive filtering through 0.45 m membranes is performed on the resultant solution before it is added to NaCl.

The last step is drying; CMC is precipitated by mixing 100% ethanol and washed thoroughly with the combination of water and ethanol. Ethanol content is increased to form different combinations (70, 80, and 90%) and dried at room temperature. The reaction yield is calculated using the weights of the parent chitosan and the resulting product [36].

3.5 Thiolated Chitosan

Chitosan is a polysaccharide composed up of N-acetylglucosamine copolymers and glucosamine. The ease with which chitosan may be chemically changed and salts produced with acids is due to the primary amino group. Because the amino groups are protonated, chitosan is soluble at acidic pH, but at alkaline and at neutral pH it is insoluble [37]. Thiolation of chitosan involves covalent attachment of thiol-bearing moieties to its polymeric backbone, yielding three distinct derivatives, i.e., cysteine-grafted chitosan, thioglycolic acid-grafted chitosan, and 4-Thio-butyl-amidine-grafted chitosan (chitosan-TBA) [38]. In comparison to unmodified chitosan, these thiolated chitosan exhibits a number of advantages, including greatly increased mucoadhesive and permeation-enhancing characteristics [39]. Thiolated chitosan is ideal excipients for controlled drug release dosage forms due to their excellent cohesive characteristics. Furthermore, at physiological pH values, thiolated chitosan solutions exhibit in situ gelling characteristics.

3.5.1 Synthesis of Thiolated Chitosan

The basic aim for thiol group immobilization is the large amino group at the two position of chitosan's glucosamine subunits. By creating amide or amidine bonds, sulfhydryl-carrying molecules may be covalently attached to this primary amino group.

Cysteine and thioglycolic acid ligands form amide bonds with chitosan's primary amino group through a water-soluble carbodiimide-mediated coupling reaction. The carboxylic acid group (–COOH) of the ligands reacts with chitosan's amino group ($-NH_2$) to form a stable amide linkage. Performing the synthesis at a pH below 5 prevents the formation of disulfide bonds caused by air oxidation. For the synthesis of amidine bonds, 2-iminothiolane is exploited as coupling reagent. It has the benefit

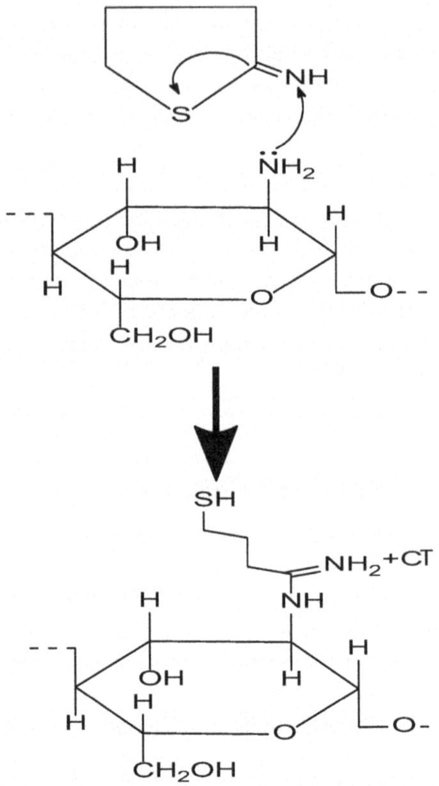

Fig. 3.5 Synthesis of 2-iminothiolane-based chitosan derivative [38]

of being a one-step coupling reaction. Furthermore, the thiol group of the reagent is secured from oxidation due to the chemical symmetry of the reagent. Figure 3.5 depicts the chemical interaction of chitosan with 2-iminothiolane [38].

3.5.2 *Properties of Thiolated Chitosan*

Thiolation of chitosan introduces reactive thiol groups that form disulfide bonds with cysteine-rich regions of mucosal glycoproteins, enhancing mucoadhesive properties through covalent cross-linking and increased polymer-mucus interactions [40]. Studies have shown that modifying chitosan with thioglycolic acid significantly enhances its mucoadhesive properties. Specifically, chitosan-thioglycolic acid conjugates demonstrated a 5–tenfold increase in mucoadhesion compared to unmodified chitosan, making them a promising material for mucosal drug delivery applications. The immobilization of thiol groups can significantly boost the permeation-enhancing effect of chitosan. Aside from their mucoadhesive and permeation-enhancing characteristics, thiolated chitosan have good cohesive qualities. Because of the decreased thiol functionalities on the chitosan backbone, thiolated chitosan can create inter and intramolecular disulfide bonds as well as disulfide interactions with mucus glycoproteins. This cross-linking of polymeric chains is responsible for the extraordinary stability of drug carrier systems based on thiolated chitosan.

3.6 Chitosan Enzyme Inhibitors

Because of its non-toxicity, outstanding cohesive characteristics, and polycationic nature, chitosan has shown to be an efficient emulsifier in a number of drug delivery methods. The polymer has been utilized in various pharmaceutical applications, including direct tablet compression, controlled release solid dosage form development, and enhancement of drug dissolution profiles [41]. Chitosan has no inhibitory effects for enzymes. Polycarbophil, for example, is found to be a powerful inhibitor of trypsin and carboxypeptidase B, whereas chitosan actually increased the activity of both enzymes. Chemically treated chitosan, on the other hand, has enzyme inhibitory characteristics. Though chitosan offers all of these benefits for non-invasive (poly)peptide delivery, its effectiveness as a carrier matrix is typically inadequate since embedded medications are susceptible to adverse pre-systemic proteolytic enzyme degradation.

Luminally secreted proteases include gastric proteases, i.e., pepsins (released through gastric pits into the stomach lumen) and pancreatic proteases, i.e., chymotrypsin, elastase, trypsin, carboxypeptidase A and B (released into the small intestine lumen). Pepsins, on the other hand, pose no harm to peptide and protein medications taken orally. Coating compounds like polymethacrylates or cellulose acetate phthalate ensure stability in acidic environments. The formulations appear to be particularly useful for administering therapeutic peptides and proteins orally. A coating material, on the other hand, cannot have a protective effect against pancreatic proteases since the therapeutic (poly)peptide must be released in the gut, which the coating would prohibit. Pancreatic proteases are enzymes that can endure the harsh conditions of the digestive system and remain active throughout the gastrointestinal tract, from the stomach to the colon. These peptidases are among the enzymes that have received the most attention. The bulk of these enzymes is metallo-peptidases, which implies that it needs metal ions like zinc, to maintain their conformation and/or function. Peptides having a molecular mass of less than 3 kDa are cleaved by membrane-bound peptidases [42]. In peptide delivery systems, chitosan has the potential to be an efficient tool. Unlike other polymers with similar advantages for peptide administration, such as poly(acrylates), chitosan has no enzyme inhibiting characteristics. In recent times, though, it has been proved that by making slight chemical adjustments to the polymer, this disadvantage may be addressed [43].

3.7 Chitosan-Ethylenediaminetetraacetic Acid (Chitosan-EDTA)

The EDTA has been effectively used to enhance chitosan-based adsorbents' performance [44]. The modification involves covalent bonding between chitosan's $-NH_2$ groups and EDTA's esterifying carboxylic groups and electrostatic interactions between protonated $-NH_2$ groups of chitosan and –COO or –OH groups of EDTA, increasing cross-linking density [45]. It is worth to mention here that surface modification of chitosan composites enhances adsorbate absorption. Furthermore, the interaction between chitosan's protonated $-NH_2$ moieties and EDTA's –COO or –OH moieties could increase cross-linking density, resulting in improved acid resistance and mechanical qualities. More functional groups responsible for attracting the adsorbate could be introduced by EDTA. Surface modification of chitosan as a composite matrix increases adsorbate absorption, as previously demonstrated using various fillers [46]. The benefits involved including the improved acid resistance and mechanical properties, leading functional groups for enhanced adsorbate attraction and increased adsorption capacity. The application involves heavy metal removal, e.g., Pb(II) and Cd(II) adsorption improved using chitosan-EDTA polymer combination and copper (Cu) heavy metal adsorbent modification using cross-linking approach [47, 48]. The advantages of EDTA as a cross-linker include its cost-effectiveness and lower toxicity compared to alternatives like EDC. Key findings highlight that the –COO and $-NH_2$ groups from chitosan and EDTA act as active sites for bond formation and adsorption. The key findings highlighting –COO and $-NH_2$ groups from chitosan and EDTA serve as active site. The polymer chitosan-EDTA has been extensively modified as a metal absorbent material. Juang et al. [48] used a cross-link approach to successfully modify the Copper (Cu) heavy-metal adsorbent made from chitosan and EDTA. In the bond formation and adsorption process, the presence of –COO, as well as $-NH_2$ from chitosan and EDTA, is an active site [49].

3.7.1 Preparation Method of Chitosan-EDTA

Chitosan-EDTA is made by cross-linking ethylenediaminetetraacetic acid (EDTA) with various compositions. By varying the weight of chitosan such as 2.25, 2.75, and 1.75 g, it is mixed in 50 mL of 2% acetic acid (b/v). The next step is the addition of 50 mL EDTA solution with varied composition ratios of 0.25, 0.5, and 0.75 g. Until the solution is solid, it is dripped into a 2 M NaOH solution. The hydrogel is then rinsed several times. The hydrogels are stored in a vacuum container and have already been applied to soil conditioner [49].

3.8 Half-Acetylated Chitosan

Half N-acetylated chitosan, a biocompatible and biodegradable polymer, has been developed as a water-soluble drug carrier. Its solubility, molecular weight, and chemical modification have been investigated [50]. The preparation methods involve chitosan depolymerization (i.e., treatment with $NaBO_3$ in aqueous acetic acid), molecular weight determination (i.e., size-exclusion chromatography with low-angle laser light-scattering photometry), and N-acetylation (i.e., reaction with acetic anhydride in aqueous acetic acid). The results revealed that the water solubility increased with decreasing molecular weight and half N-acetylated chitosan showed improved solubility in water and organic solvents. It is simple and efficient method developed, for preparing highly soluble chitosan with the confirmation of biocompatibility and biodegradation of half N-acetylated chitosan [50]. It is pertinent to mention here that water solubility enables homogeneous chemical modifications and biocompatible and biodegradable properties suitable for drug delivery applications [50].

3.8.1 Synthesis of Half-Acetylated Chitosan

For the preparation of half-acetylated chitosan, initially 20 g of chitosan is mixed with 500 ml of 2.8% acetic acid solution. Then, at room temperature, mixture of 4.5 ml anhydride and 50 ml ethanol is added by continuous stirring. By the addition of strong KOH solution, the mixture is precipitated after 1 h. The half N-acetylated becomes dried when the temperature reaches at 50 °C. The half N-acetylated chitosan has higher antioxidant activity, and it depends upon the concentration as well as molecular weight [51].

3.9 Acrylated Chitosan

The Michael type reaction is used to conjugate poly(ethylene glycol)diacrylate (PEGDA) to chitosan, resulting in a novel mucoadhesive polymer. Chitosan with a high molecular weight modification PEGDA outperformed both non-modified and thiolated chitosan in terms of mucoadhesive qualities. At molecular level, there are additional interactions between mucin glycoproteins and modified chitosan which is revealed by the rheology test of mucin/polymer mixture. This novel polymer has the potential to be a valuable polymeric carrier matrix for delivery systems, allowing the vehicle to spend more time on the mucosa surface. When compared to both non-modified alginate and thiolated alginate, this polymer showed a considerable increase in adhesion strength [52].

3.9.1 Synthesis of Acrylated Chitosan

The acrylated chitosan (chitosan-PEGAc) is synthesized using the process described by Ma et al. [53]. The required amount of chitosan, distilled water, and acetic acid are charged to a 500 mL round-bottom flask equipped with a magnetic stir bar. The flask is fitted with a reflux condenser and purged with N_2 before being placed in a 40 °C water bath and stirred for 60 min. After that, the required polyethylene glycol diacrylate (PEGDA) is added to the mixed solution, which is agitated for 24 h. The solution is filtered, rotary evaporated, then placed into acetone to remove the unreacted PEGDA and get the white precipitate. The precipitate is washed three times with acetone before being dried under vacuum at 30 °C overnight to get the chitosan derivative PEGDA-CS in Fig. 3.6 [53].

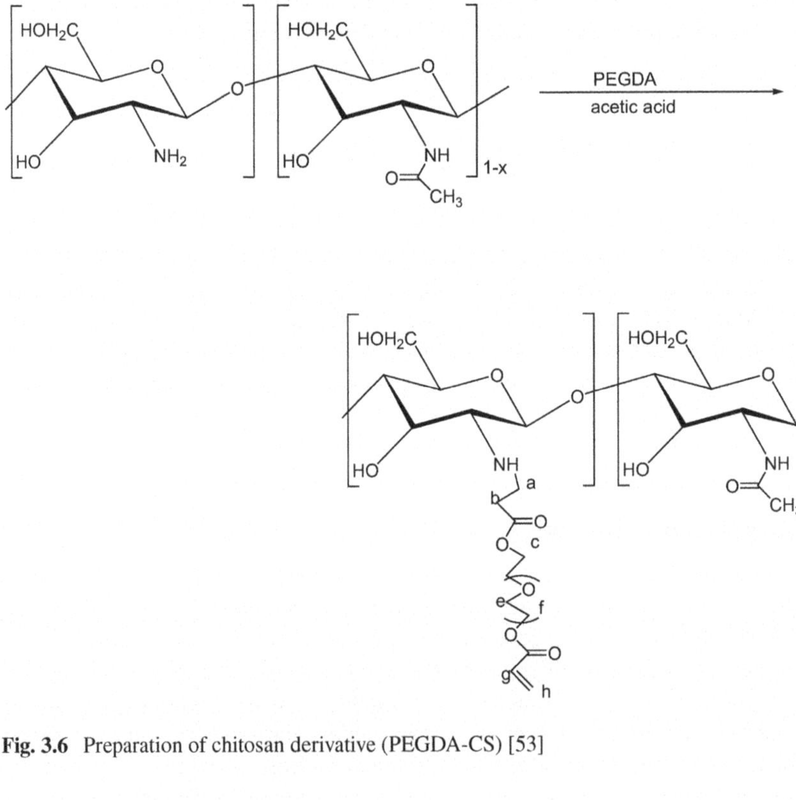

Fig. 3.6 Preparation of chitosan derivative (PEGDA-CS) [53]

3.10 Glycol Chitosan (GCS)

Glycol chitosan (GCS), a water-soluble chitosan derivative, offers dual functionality and easy chemical modification. Its versatility has led to the development of diverse biomaterials through encapsulation and conjugation, with numerous applications in biomedical research [54]. Glycol chitosan (GCS) is a water-soluble chitosan derivative, featuring hydrophilic ethylene glycol branches. Its properties include molecular weight: 20–250 kDa, degree of deacetylation: 60–82.7%, solubility: water-soluble at neutral and acidic pH. These characteristics make GCS a versatile biomaterial for various applications [55]. The reactive functional groups of GCS, such as amine and hydroxyl groups, allow for a variety of chemical changes, resulting in a variety of derivatives. These GCS derivatives self-assemble into nanostructures or hydrogels, which can be used to deliver diagnostic and therapeutic medications to fight cancer and dangerous microbes. It has been reported in the literature that GCS-based materials have advanced in cell imaging and drug delivery applications for diagnosing and treating a variety of ailments, particularly cancer and pathogen infections. Furthermore, smart drug delivery systems for passive and active tumor-targeted drug administration, stimuli-responsive drug release, and combination therapy based on GCS [54].

3.11 Chitosan Catechol

Since the introduction of mucoadhesive polymers for prolonged drug release, various attempts to produce novel mucoadhesive polymers have been attempted [56]. Poly(acrylic acid), carboxyl methyl cellulose, scleroglucan, and chitosan are examples of polymers with excellent mucoadhesive characteristics. Chitosan has been extensively studied because to its strong contact with the mucin layer through several intermolecular hydrogen bonds and electrostatic interactions with the negatively charged sialic acid and sulfate residues in mucin [57]. Furthermore, because of its biocompatibility and degradability, chitosan is thought to be ideal for therapeutic usage. Despite the fact that chitosan is mucoadhesive and its adherence has been increased after attempting series of experiments through chemical changes. Thiolation is regarded to be a particularly successful strategy for enhancing mucoadhesion because tethered thiols have the unique characteristic of forming disulfide linkages with the cysteine residues in mucin [58]. On the other hand, the disulfide bridges between the thiolated polymers and mucin are reversible. As a result, thiol group retention on mucosal surfaces in vivo is usually brief. As a result, thiolation may not be the best way to improve the mucoadhesion of current polymers. Enhancing the mucoadhesive characteristics of lectin can also be done. Lectins are proteins that are found in nature and have a function in biological recognition. By receptor-mediated adhesion, bacteria use lectins to attach themselves to the cells of the host organism. Lectins, however, bind reversibly to cell surfaces, exhibit immunogenic properties, and are highly toxic [59]. As a result, finding a new functional group that can significantly increase mucosal layer binding remains a difficult task. Because chemical conjugation of the functional group to existing mucoadhesive polymers can boost mucoadhesive characteristics synergistically, identifying the functional group is crucial. Catechol is a good choice. It occurs naturally as a side chain of L-3,4-dihydroxyphenylalanine (DOPA), an amino acid that mussels need for their strong underwater adhesion [60]. Catechol derivatives are readily available commercially for bioconjugation to amine or carboxylic acid groups, which is an advantage of employing catechol [61]. By converting catechol to quinone, it is also capable of forming irreversible covalent bonds with thiols and amines [62]. Several earlier investigations looked into catechol's mucoadhesive characteristics. In 2012, chitosan and catechol derivatives are combined to create a mucoadhesive hydrogel [63]. However, no previous research has looked at catechol as a mucoadhesion enhancer or examined its mucoadhesive qualities in vivo. The contact stage and the consolidation stage are the two phases of mucoadhesion [64]. The charge interaction between chitosan and mucin is reversible; however, it allows for fast contact with mucin, which speeds up the contact stage. In the consolidation stage, the development of catechol-mediated contacts would offer irreversible anchoring to mucin, resulting in a cooperative two-step process. Surface Plasmon Resonance (SPR) spectroscopy revealed a fourfold increase in mucoadhesion for Chi-C compared to unmodified chitosan. Additionally, Chi-C demonstrated significantly longer in vivo retention, exceeding that of chitosan and poly(acrylic acid), which are typically undetectable after 3 h post-administration. Key findings indicate Chi-C forms non-cytotoxic, catechol-mediated covalent cross-links with mucin, while catechol coupling enhances the intrinsic mucoadhesive properties of polymers. Chi-C's improved mucoadhesion and retention make it a promising biomaterial for oral drug delivery and mucosal applications.

3.11.1 Synthesis of Chi-C

Chi-C is created in the same manner as per the reports (Fig. 3.7) [65]. Chi-C, a muco-adhesive polymer with enhanced retention, offers key benefits such as a fourfold increase in muco-adhesion compared to unmodified chitosan (as shown by SPR spectroscopy), significantly prolonged in vivo retention compared to chitosan and poly(acrylic acid), and non-cytotoxic, catechol-mediated cross-links with mucin. The mechanism involves catechol coupling, which enhances intrinsic mucoadhesive properties, leading to improved oral drug delivery and enhanced mucosal applications. Chi-C's unique properties make it an attractive biomaterial for various therapeutic applications. At room temperature, the reaction mixture is rapidly stirred for 4 h, with the pH of the reaction solution maintained between 4.5 and 5.0. Dialysis is carried out for two days in acidified DDW (pH 5.0, HCl), four hours in PBS buffer, and one day in DDW, with the dialysis solvent being replaced every 36 h. For a greater degree of catechol conjugation (DOC), 6.49 mmol HCA and 6.49 mmol EDC are used, while for a lower degree of catechol conjugation, 3.25 mmol HCA and

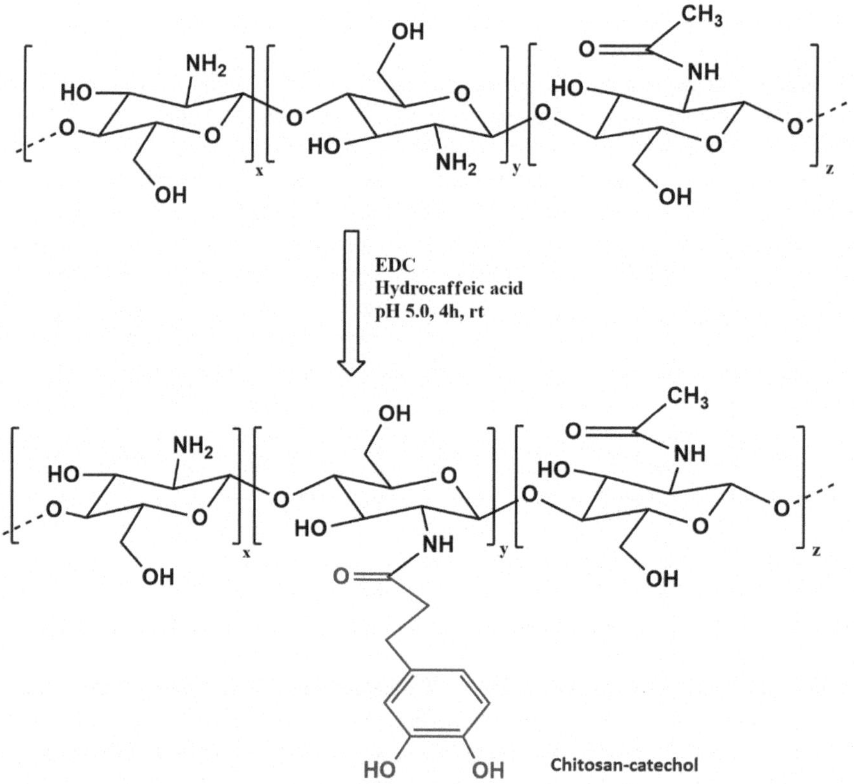

Fig. 3.7 Synthesis and structure of chitosan catechol [65]

3.25 mmol EDC are used. The final product is lyophilized and kept in a desiccator to keep it dry. Chi-DOC C's is determined using a previously reported ^{1}H-NMR approach.

3.12 Methyl Pyrrolidinone Chitosan (MPC)

Chitosan has received much interest in recent years for pharmaceutical and medical applications because it has a lot of desirable biological properties like chemical inertness, low cytotoxicity, and biodegradability, along with being widely available in nature, inexpensive, and having a lot of adaptability in use. Methyl pyrrolidinone chitosan (MPC) is a chitosan derivative in which the glucosamine units' amino groups have been partly replaced with 5-methyl-pyrrolidinone. The MPC has various features that make it an appealing molecule for usage in the pharmaceutical industry, in addition to its good biocompatibility and biodegradability. It's a very hydrophilic chemical that dissolves in water, saline, and water-alcohol mixes and produces viscous solutions or gels at physiological pH levels. The MPC solutions have been tested for mucoadhesive and penetration augmentation capabilities via buccal and vaginal mucosae [66]. The MPC has been found in several investigations to enhance organized tissue regeneration and vascularization. The MPC is also a promising solution for wound dressing fabrication since it combines great wound healing properties with antimicrobial properties [67]. Despite the fact that it has not yet been studied for the creation of drug delivery systems such as films. Buccal drug delivery, comprising bioadhesion and transmucosal drug transport, is an unique chitosan use [68]. Buccal drug delivery systems for systemic and local therapy should be tailored to prolong the medication's retention time in the oral cavity while simultaneously satisfying the criteria for adherence to the mucosa's moist surface and saliva flushing resistance. Furthermore, the buccal mucosa could be an essential route for peptide or protein medication delivery. Although chitosan solutions have been tested for protein delivery in the mouth, films or hydrogels should be preferred for longer drug release [69]. Under physiological conditions, chitosan films should breakdown slowly, which is why it must be cross-linked. The cross-linking procedure, which is used to manage the drug release rate, can also have an impact on one of the system's primary features, such as mucoadhesion [70]. For the manufacture of cross-linked chitosan films, tripolyphosphate (TPP) has been employed as a safe ionic cross-linking agent [71]. Ionic cross-linking requires the employment of multivalent counter ions as cross-linkers to form bridges between polymeric chains; since chitosan is a polycation, TPP permits ionic cross-linking even without auxiliary molecules under simple and moderate circumstances. The pH of the TPP solution has a substantial impact on the ionic reaction of the chitosan-TPP network, and adjusting the pH of the curing agent can increase the ionic cross-linking density. The MPC produced films and assessed them as protein delivery systems for use in the oral cavity [72]. Since its molecular weight (MW) is below the transparency limit of buccal mucosa, myoglobin (Mb) is selected as the standard protein and it can be detected using a spectrophotometric approach without the use of particular protein markers or reagents. It is worth mentioning that myoglobin (Mb) is a relatively small globular protein composed of a single polypeptide chain of 154 amino acids and an iron-containing heme prosthetic group identical to that of hemoglobin. The MPC films containing Mb are cast and cross-linked with varied TPP: MPC molar ratios at two distinct pH levels. The bioadhesiveness and swelling behavior of the films are assessed. The buccal bioadhesive qualities of the films are tested using a technique established by the research group and based on the use of buccal cell suspensions [70]. The swelling properties of MPC films cross-linked with different TPPs have been studied. Both placebo and MHb-loaded films, with varying MPC molar ratios, have been analyzed. A spectrophotometric approach using the model protein chromophore is used to perform in vitro release studies on MHb-loaded films [72].

3.13 Cyclodextrin Chitosan (CDCS)

With a lipophilic inner cavity and a hydrophilic outer surface cyclodextrins (CDs) have cone-shaped structure, allowing them to form non-covalent inclusion complexes with a variety of chemical and inorganic compounds [73].

Research has demonstrated that forming complexes between CDs, particularly their derivatives, and drugs can enhance aqueous solubility and oral absorption, protect unstable drugs from degradation, inhibit P-glycoprotein (P-gp) activity, overcome multidrug resistance, and modify pharmacokinetics. These benefits improve drug efficacy, stability, and bioavailability, making CD complexes attractive formulations for various therapeutic applications [74]. Chitosan (CS) has been gaining popularity as a biomaterial and pharmaceutical excipient for drug administration due to its favorable biological qualities such as low toxicity, biodegradability, strong biocompatibility, mucoadhesive property, and absorption enhancing ability across mucosal barriers [75]. The CS NPs have shown that it has a particularly promising future in pharmaceutical applications due to the amazing properties of both CS and NP dosage forms. The dextrin (DTX) oral administration is difficult due to its limited water solubility and oral absorption [76]. It's a potential method for making DTX/CD/CS NPs that combine the advantages of CD complexes and CS NPs to suit the need for DTX oral administration. Some research has been done on drug NPs made up of CS and CD [75, 76]. However, the quantity of CDs employed in these tests are restricted, and no study of DTX-loading NPs composed of CS and sulfobutylether-cyclodextrin (SBE—CD) at the same time is conducted. The DTX/CD complexes and DTX/CD/CS NPs are generated in this work utilizing DTX and SBE—CD [77]. In terms of features, oral permeability, and bioavailability, DTX/CD/CS NPs are examined and compared to pure DTX formulations and DTX/CD complexes [77].

3.13.1 Preparation of CD-g-CS

A modified version of Gonil's approach is used to make CD-g-CS [78]. The CS (0.25 g) is dissolved in acetic acid at a concentration of 1% (v/v). The above-mentioned 6-OTs—CD is added to the CS/acetic acid solution. The solution is dialyzed for 7 days using deionized water after being agitated for 48 h at 4 °C. The solution is then freeze-dried [79]. Figure 3.8 depicts the chemical interaction of cyclodextrin-g-chitosan [79].

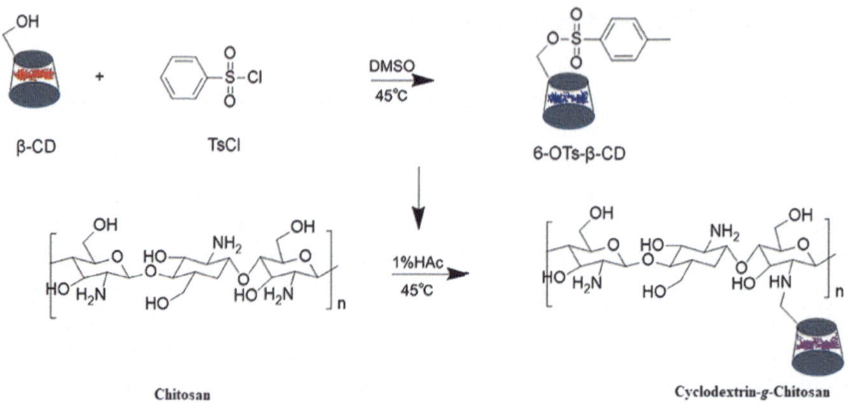

Fig. 3.8 Synthesis of cyclodextrin-g-chitosan [79]

3.14 Oleoyl-Quaternized Chitosan

Mucoadhesion is a unique strategy for enhancing the effectiveness of drug delivery systems that has been employed in the nasal, bladder, vaginal, and oral routes [80]. Mucoadhesive polymers are bioadhesive substances that may cling to mucus membranes and prolong the drug carrier's absorption time, improving therapeutic outcomes [81]. Lectin thiolated polymers, alginate-polyethylene glycol acrylate, and modified chitosan have all shown promise as mucoadhesive aids for drug delivery systems [82]. An amine-functionalized polyacrylamide nanogel is proved to be an excellent carrier for docetaxel administration in the treatment of bladder cancer: utilizing the carrier which resulted in a prolonged contact duration between the medication and the cell surface [83]. Chitosan has been extensively employed in a number of medicinal and materials applications due to its non-toxic, biodegradable, and biocompatible characteristics. In the context of drug delivery, chitosan is an important material for increasing the mucoadhesive properties of delivery systems such as solid lipid nanocarriers (SLN), nanostructure lipid carriers (NLC), liposomes, and Fe_3O_4 magnetic nanocarriers [80]. This effect is due to strong electrostatic attraction between positively charged chitosan and negatively charged mucosal surfaces. For these purposes, it is also simple to alter other chitosan-derived systems, such as glucosamine, N-acetylglucosamine, and other chitosan-derived systems. For instance, modifying the surface of solid lipid nanocarriers with chitosan oligosaccharides led to SLN systems with less pronounced burst release patterns [84]. As a way to increase particle contact duration with the mucosal surface and, thus, as a delivery method for ocular and brain diseases, the nanostructure lipid carriers (NLC) coated with modified chitosan has also been addressed [85]. Solid and liquid lipids are combined to form nanostructure lipid carriers (NLC), which have holes in the lipid crystal lattice that let medicinal molecules be loaded. The NLCs have been studied in great detail due to their ability to support prolonged drug circulation times, give high degrees of drug entrapment, and provide controlled drug release [86]. Although instability is a significant obstacle that is now being addressed in addition to expanding the range of their administration route by functionalization, the NLC has a remarkable promise as drug delivery vehicles (e.g., improved mucoadhesion). It has been shown that the xanthone alpha-mangostin (AP), which is abundant in mangosteen fruit, has anticancer and antiproliferative effects on a variety of human cancer cell types, including gastrointestinal, breast, and neck and head cancer cells [87]. Additionally, cytotoxicity tests conducted on WRL-68 normal cells in vitro and in vivo show only modest toxicity, with an IC50 of 65 g/ml. These results emphasize the anticancer potential of AP, but because of its poor water solubility, its medicinal uses are presently constrained. In this work, a mucoadhesive nanostructure lipid carrier (NLC) for the potential oral delivery of alpha-mangostin (AP) is created. A nanostructure lipid carrier coated in oleoyl-quaternized-chitosan (CS) and loaded with alpha-mangostin is created by stepwise emulsification to encapsulate AP and electrostatic coating to improve the mucoadhesive properties of the NLC particles (CS-AP-NLC).

3.14.1 Synthesis of Oleoyl-Quaternized-Chitosan

The method outlined by Sashiwa et al. [88] is used to create oleoyl chitosan. In a nutshell, 4.0 g of chitosan is dissolved in 80 mL of methane sulfonic acid, and 0.4 mL of oleoyl chloride is then forcefully stirred into the chitosan solution. The 150 g of crushed ice is added to the reaction mixture after 5 h of room-temperature stirring. The acidic mixture is dialyzed against deionized water for one day to remove the bulk of the acid and then neutralized with sodium bicarbonate before continuing the dialysis against deionized water (3 days). By freeze drying, the oleoyl chitosan product is turned into powder (Gamma 2-16 LSC, Christ, Newtown, UK). Powdered oleoyl chitosan (0.5 g) is dissolved in 1% (v/v) acetic acid (40 mL).

3.14.2 Synthesis of Chitin/Chitosan-Based Shape Memory Polyurethanes Bionanocomposites

Molecular engineering and properties of chitin-based shape memory polyurethanes and XRD studies of polyurethane elastomers based on chitin/1, 4-butane diol blends have been reported [89, 90]. XRD pattern and cytotoxicity and mechanical behavior of chitin–bentonite clay based polyurethane bionanocomposites have studied earlier [91, 92]. Molecular engineering and morphological studies of chitin-based polyurethanes using hydroxyl terminated polybutadiene have been well documented [93, 94]. Surface characteristics of chitin-based shape memory polyurethane elastomers (SMPU) and UV-irradiated chitin-based SMPU has been compared with pristine [95, 96]. Effect of diisocyanate structure on the evaluation of biocompatibility and mechanical behavior of chitin-based polyurethane elastomers and effect of chitin/1,4-butane diol blends on the evaluation of biocompatibility and mechanical behavior of polyurethane elastomers have been well studied [97, 98]. The synthesis and characterization of stable and biologically active chitin-based polyurethane elastomers and chitin–bentonite clay-based polyurethane bio-nanocomposites have been well-documented in the literature [99, 100].

3.15 Conclusion

This chapter provides a comprehensive overview of chitosan derivatization via green technology, highlighting enhanced properties and potential applications. Chitosan, a versatile biopolymer, undergoes derivatization to enhance its biological and physicochemical properties. Its intrinsic characteristics, influenced by source and isolation process, significantly impact derivative properties. Green derivatization improves chitosan's stability at basic and neutral pH, increases solubility and hydrophilic character, and improves affinity for bioactive compounds and gelling capabilities. Key reactions and techniques include: click chemistry, cycloadditions, quaternary derivatization, carboxymethyl chitosan synthesis, thiolation and enzyme inhibition, and EDTA modification. Notable derivatives include: half-acetylated chitosan, acrylated chitosan, glycol chitosan, chitosan catechol (Chi-C), and other specialized derivatives.

References

1. Zhang Y, Zhao M, Cheng Q, Wang C, Li H, Han X, Fan Z, Su G, Pan D, Li Z (2021) Research progress of adsorption and removal of heavy metals by chitosan and its derivatives: a review. https://doi.org/10.1016/j.chemosphere.2021.130927
2. Nornberg AB, de Aquino TFB, Martins CC, Luchese C, Wilhelm EA, Jacob RG, Hartwig D, Fajardo AR (2021) Organoselenium-chitosan derivative: synthesis via "click" reaction, characterization and antioxidant activity. Int J Biol Macromol 191. https://doi.org/10.1016/j.ijbiomac.2021.09.053
3. Verlee A, Mincke S, Stevens CV (2017) Recent developments in antibacterial and antifungal chitosan and its derivatives. https://doi.org/10.1016/j.carbpol.2017.02.001
4. Mourya VK, Inamdar NN (2009) Trimethyl chitosan and its applications in drug delivery. J Mater Sci Mater Med 20:1057–1079. https://doi.org/10.1007/s10856-008-3659-z
5. Kotzé AF, Lueßen HL, de Boer AG, Verhoef JC, Junginger HE (1999) Chitosan for enhanced intestinal permeability: prospects for derivatives soluble in neutral and basic environments. Eur J Pharm Sci 7. https://doi.org/10.1016/S0928-0987(98)00016-5
6. Sabaa MW (2013) Chitosan-g-copolymers: synthesis, properties, and applications. In: Polysaccharide based graft copolymers. https://doi.org/10.1007/978-3-642-36566-9_4
7. Amato A, Migneco LM, Martinelli A, Pietrelli L, Piozzi A, Francolini I (2018) Antimicrobial activity of catechol functionalized-chitosan versus *Staphylococcus epidermidis*. Carbohydr Polym 179. https://doi.org/10.1016/j.carbpol.2017.09.073

8. Argüelles-Monal W, Recillas-Mota M, Fernández-Quiroz D (2017) Chitosan-based thermosensitive materials. In: Biological activities and application of marine polysaccharides. https://doi.org/10.5772/65713
9. Yu C, Kecen X, Xiaosai Q (2018) Grafting modification of chitosan. In: Biopolymer grafting: synthesis and properties. https://doi.org/10.1016/B978-0-323-48104-5.00007-X
10. Nishimura SI, Kohgo O, Kurita K, Kuzuhara H (1991) Chemospecific manipulations of a rigid polysaccharide: syntheses of novel chitosan derivatives with excellent solubility in common organic solvents by regioselective chemical modifications. Macromolecules 24. https://doi.org/10.1021/ma00017a003
11. Kurita K, Ikeda H, Shimojoh M, Yang J (2007) N-phthaloylated chitosan as an essential precursor for controlled chemical modifications of chitosan: synthesis and evaluation. Polym J 39. https://doi.org/10.1295/polymj.PJ2007032
12. Kurita K, Ikeda H, Yoshida Y, Shimojoh M, Harata M (2002) Chemoselective protection of the amino groups of chitosan by controlled phthaloylation: facile preparation of a precursor useful for chemical modifications. Biomacromolecules 3. https://doi.org/10.1021/bm0101163
13. Montiel-Herrera M, Gandini A, Goycoolea FM, Jacobsen NE, Lizardi-Mendoza J, Recillas-Mota MT, Argüelles-Monal WM (2015) Furan–chitosan hydrogels based on click chemistry. Iran Polym J (Engl Ed) 24. https://doi.org/10.1007/s13726-015-0325-4
14. Chen Y, Ye Y, Li R, Guo Y, Tan H (2013) Synthesis of chitosan 6-OH immobilized cyclodextrin derivates via click chemistry. Fibers Polym 14. https://doi.org/10.1007/s12221-013-1058-7
15. Chen Y, Wang F, Yun D, Guo Y, Ye Y, Wang Y, Tan H (2013) Preparation of a C_6 quaternary ammonium chitosan derivative through a chitosan Schiff base with click chemistry. J Appl Polym Sci 129. https://doi.org/10.1002/app.38431
16. Guinesi LS, Cavalheiro ÉTG (2006) Influence of some reactional parameters on the substitution degree of biopolymeric Schiff bases prepared from chitosan and salicylaldehyde. Carbohydr Polym 65. https://doi.org/10.1016/j.carbpol.2006.01.030
17. Gu C, Le V, Lang M, Liu J (2014) Preparation of polysaccharide derivates chitosan-graft-poly(e-caprolactone) amphiphilic copolymer micelles for 5-fluorouracil drug delivery. Colloids Surf B Biointerfaces 116. https://doi.org/10.1016/j.colsurfb.2014.01.026
18. Sashiwa H, Kawasaki N, Nakayama A, Muraki E, Yamamoto N, Aiba SI (2002) Chemical modification of chitosan. 14:1 synthesis of water-soluble chitosan derivatives by simple acetylation. https://doi.org/10.1021/bm0200480
19. Kolb HC, Sharpless KB (2003) The growing impact of click chemistry on drug discovery. https://doi.org/10.1016/S1359-6446(03)02933-7
20. Gandini A (2013) The furan/maleimide Diels–Alder reaction: a versatile click-unclick tool in macromolecular synthesis. https://doi.org/10.1016/j.progpolymsci.2012.04.002
21. Ifuku S, Wada M, Morimoto M, Saimoto H (2012) A short synthesis of highly soluble chemoselective chitosan derivatives via "click chemistry". Carbohydr Polym 90. https://doi.org/10.1016/j.carbpol.2012.06.074
22. Gandini A (2008) Polymers from renewable resources: a challenge for the future of macromolecular materials. Macromolecules 41. https://doi.org/10.1021/ma801735u
23. Montiel-Herrera M, Gandini A, Goycoolea FM, Jacobsen NE, Lizardi-Mendoza J, Recillas-Mota M, Argüelles-Monal WM (2015) N-(furfural) chitosan hydrogels based on Diels–Alder cycloadditions and application as microspheres for controlled drug release. Carbohydr Polym 128. https://doi.org/10.1016/j.carbpol.2015.03.052
24. Abueva C, Ryu HS, Min JW, Chung PS, You HS, Yang MS, Woo SH (2021) Quaternary ammonium N,N,N-trimethyl chitosan derivative and povidone-iodine complex as a potent antiseptic with enhanced wound healing property. Int J Biol Macromol 182:1713–1723. https://doi.org/10.1016/j.ijbiomac.2021.05.153
25. Domard A, Rinaudo M, Terrassin C (1986) New method for the quaternization of chitosan. Int J Biol Macromol 8. https://doi.org/10.1016/0141-8130(86)90007-3
26. Jintapattanakit A, Mao S, Kissel T, Junyaprasert VB (2008) Physicochemical properties and biocompatibility of N-trimethyl chitosan: effect of quaternization and dimethylation. Eur J Pharm Biopharm 70. https://doi.org/10.1016/j.ejpb.2008.06.002

27. El-Sherbiny IM, Salih E, Reicha F (2016) New trimethyl chitosan-based composite nanoparticles as promising antibacterial agents. Drug Dev Ind Pharm 42. https://doi.org/10.3109/03639045.2015.1075035
28. Fazli Y, Shariatinia Z (2017) Controlled release of cefazolin sodium antibiotic drug from electrospun chitosan-polyethylene oxide nanofibrous mats. Mater Sci Eng C 71. https://doi.org/10.1016/j.msec.2016.10.048
29. Rahmani S, Mohammadi Z, Amini M, Isaei E, Taheritarigh S, Rafiee Tehrani N, Rafiee Tehrani M (2016) Methylated 4-N,N dimethyl aminobenzyl N,O carboxymethyl chitosan as a new chitosan derivative: synthesis, characterization, cytotoxicity and antibacterial activity. Carbohydr Polym 149. https://doi.org/10.1016/j.carbpol.2016.04.116
30. Patrulea V, Applegate LA, Ostafe V, Jordan O, Borchard G (2015) Optimized synthesis of O-carboxymethyl-N,N,N-trimethyl chitosan. Carbohydr Polym 122. https://doi.org/10.1016/j.carbpol.2014.12.014
31. Heuser M, Rivera C, Christian N, Cárdenas G (2009) Chitosan-copper paint types as antifouling. J Chil Chem Soc 54. https://doi.org/10.4067/S0717-97072009000400028
32. Vo DT, Sabrina S, Lee CK (2017) Silver deposited carboxymethyl chitosan-grafted magnetic nanoparticles as dual action deliverable antimicrobial materials. Mater Sci Eng C 73. https://doi.org/10.1016/j.msec.2016.12.066
33. Upadhyaya L, Singh J, Agarwal V, Pandey AC, Verma SP, Das P, Tewari RP (2015) Efficient water soluble nanostructured ZnO grafted O-carboxymethyl chitosan/curcumin-nanocomposite for cancer therapy. Process Biochem 50. https://doi.org/10.1016/j.procbio.2014.12.029
34. ten Broek RPG, Strik C, Issa Y, Bleichrodt RP, van Goor H (2013) Adhesiolysis-related morbidity in abdominal surgery. Ann Surg 258. https://doi.org/10.1097/SLA.0b013e31826f4969
35. Shariatinia Z (2018) Carboxymethyl chitosan: properties and biomedical applications. https://doi.org/10.1016/j.ijbiomac.2018.09.131
36. Bukzem AL, Signini R, dos Santos DM, Lião LM, Ascheri DPR (2016) Optimization of carboxymethyl chitosan synthesis using response surface methodology and desirability function. Int J Biol Macromol 85:615–624. https://doi.org/10.1016/j.ijbiomac.2016.01.017
37. Sadoughi F, Mansournia MA, Mirhashemi SM (2020) The potential role of chitosan-based nanoparticles as drug delivery systems in pancreatic cancer. https://doi.org/10.1002/iub.2252
38. Bernkop-Schnürch A, Hornof M, Guggi D (2004) Thiolated chitosans. https://doi.org/10.1016/S0939-6411(03)00147-4
39. Bernkop-Schnürch A, Hornof M, Zoidl T (2003) Thiolated polymers—thiomers: synthesis and in vitro evaluation of chitosan-2-iminothiolane conjugates. Int J Pharm 260. https://doi.org/10.1016/S0378-5173(03)00271-0
40. Leitner VM, Walker GF, Bernkop-Schnürch A (2003) Thiolated polymers: evidence for the formation of disulphide bonds with mucus glycoproteins. Eur J Pharm Biopharm 56. https://doi.org/10.1016/S0939-6411(03)00061-4
41. Patil S, Pandit A, Godbole A, Dandekar P, Jain R (2021) Chitosan based co-processed excipient for improved tableting. Carbohydr Polym Technol Appl 2. https://doi.org/10.1016/j.carpta.2021.100071
42. Wu L, Wang J, Gao N, Ren J, Zhao A, Qu X (2015) Electrically pulsatile responsive drug delivery platform for treatment of Alzheimer's disease. Nano Res 8. https://doi.org/10.1007/s12274-015-0750-x
43. Bernkop-Schnurch A, Kast CE (2001) Chemically modified chitosans as enzyme inhibitors
44. Zheng L, Gao Y, Du J, Zhang W, Huang Y, Wang L, Zhao Q, Pan X (2020) A novel, recyclable magnetic biochar modified by chitosan-EDTA for the effective removal of Pb(II) from aqueous solution. RSC Adv 10. https://doi.org/10.1039/d0ra07499c
45. Shahbazi A, Marnani NN, Salahshoor Z (2019) Synergistic and antagonistic effects in simultaneous adsorption of Pb(II) and Cd(II) from aqueous solutions onto chitosan functionalized EDTA-silane/mGO. Biocatal Agric Biotechnol 22. https://doi.org/10.1016/j.bcab.2019.101398

46. Sobeih MM, El-Shahat MF, Osman A, Zaid MA, Nassar MY (2020) Glauconite clay-functionalized chitosan nanocomposites for efficient adsorptive removal of fluoride ions from polluted aqueous solutions. RSC Adv 10. https://doi.org/10.1039/d0ra02340j
47. Rahmi, Iqhrammullah M, Audina U, Husin H, Fathana H (2021) Adsorptive removal of Cd(II) using oil palm empty fruit bunch-based charcoal/chitosan-EDTA film composite. Sustain Chem Pharm 21. https://doi.org/10.1016/j.scp.2021.100449
48. Juang RS, Wu FC, Tseng RL (1999) Adsorption removal of copper(II) using chitosan from simulated rinse solutions containing chelating agents. Water Res 33. https://doi.org/10.1016/S0043-1354(98)00469-2
49. Ritonga H, Nurfadillah A, Rembon FS, Ramadhan LOAN, Nurdin M (2019) Preparation of chitosan-EDTA hydrogel as soil conditioner for soybean plant (*Glycine max*). Groundw Sustain Dev 9. https://doi.org/10.1016/j.gsd.2019.100277
50. Feng F, Liu Y, Zhao B, Hu K (2012) Characterization of half N-acetylated chitosan powders and films. Procedia Eng 718–732. https://doi.org/10.1016/j.proeng.2011.12.511
51. Feng T, Du Y, Li J, Wei Y, Yao P (2007) Antioxidant activity of half N-acetylated water-soluble chitosan in vitro. Eur Food Res Technol 225:133–138. https://doi.org/10.1007/s00217-006-0391-0
52. Shitrit Y, Bianco-Peled H (2017) Acrylated chitosan for mucoadhesive drug delivery systems. Int J Pharm 517:247–255. https://doi.org/10.1016/j.ijpharm.2016.12.023
53. Ma G, Zhang X, Han J, Song G, Nie J (2009) Photo-polymeriable chitosan derivative prepared by Michael reaction of chitosan and polyethylene glycol diacrylate (PEGDA). Int J Biol Macromol 45:499–503. https://doi.org/10.1016/j.ijbiomac.2009.08.007
54. Lin F, Jia HR, Wu FG (2019) Glycol chitosan: a water-soluble polymer for cell imaging and drug delivery. https://doi.org/10.3390/molecules24234371
55. Park K, Kim JH, Nam YS, Lee S, Nam HY, Kim K, Park JH, Kim IS, Choi K, Kim SY, Kwon IC (2007) Effect of polymer molecular weight on the tumor targeting characteristics of self-assembled glycol chitosan nanoparticles. J Controlled Release 122. https://doi.org/10.1016/j.jconrel.2007.04.009
56. Khutoryanskiy VV (2011) Advances in mucoadhesion and mucoadhesive polymers. Macromol Biosci 11. https://doi.org/10.1002/mabi.201000388
57. Sogias IA, Williams AC, Khutoryanskiy VV (2008) Why is chitosan mucoadhesive? Biomacromolecules 9. https://doi.org/10.1021/bm800276d
58. Sonia TA, Sharma CP (2011) Chitosan and its derivatives for drug delivery perspective. Adv Polym Sci 243. https://doi.org/10.1007/12_2011_117
59. Miyake K, Tanaka T, McNeil PL (2007) Lectin-based food poisoning: a new mechanism of protein toxicity. PLoS ONE 2. https://doi.org/10.1371/journal.pone.0000687
60. Lee H, Dellatore SM, Miller WM, Messersmith PB (2007) Mussel-inspired surface chemistry for multifunctional coatings. Science (1979) 318. https://doi.org/10.1126/science.1147241
61. Kim K, Ryu JH, Lee DY, Lee H (2013) Bio-inspired catechol conjugation converts water-insoluble chitosan into a highly water-soluble, adhesive chitosan derivative for hydrogels and LbL assembly. Biomater Sci 1. https://doi.org/10.1039/c3bm00004d
62. Liu Y, Ai K, Lu L (2014) Polydopamine and its derivative materials: synthesis and promising applications in energy, environmental, and biomedical fields. https://doi.org/10.1021/cr400407a
63. Xu J, Soliman GM, Barralet J, Cerruti M (2012) Mollusk glue inspired mucoadhesives for biomedical applications. Langmuir 28. https://doi.org/10.1021/la3025414
64. Boddupalli BM, Mohammed ZNK, Nath RA, Banji D (2010) Mucoadhesive drug delivery system: an overview. https://doi.org/10.4103/0110-5558.76436
65. Kim K, Kim K, Ryu JH, Lee H (2015) Chitosan-catechol: a polymer with long-lasting mucoadhesive properties. Biomaterials 52. https://doi.org/10.1016/j.biomaterials.2015.02.010
66. Sandri G, Rossi S, Ferrari F, Bonferoni MC, Muzzarelli C, Caramella C (2004) Assessment of chitosan derivatives as buccal and vaginal penetration enhancers. Eur J Pharm Sci 21. https://doi.org/10.1016/j.ejps.2003.10.028

67. Berscht PC, Nies B, Liebendörfer A, Kreuter J (1995) In vitro evaluation of biocompatibility of different wound dressing materials. J Mater Sci Mater Med 6. https://doi.org/10.1007/BF00146856
68. Smith J, Wood E, Dornish M (2004) Effect of chitosan on epithelial cell tight junctions. Pharm Res 21. https://doi.org/10.1023/B:PHAM.0000012150.60180.e3
69. Şenel S, Ikinci G, Kaş S, Yousefi-Rad A, Sargon MF, Hincal AA (2000) Chitosan films and hydrogels of chlorhexidine gluconate for oral mucosal delivery. Int J Pharm 193. https://doi.org/10.1016/S0378-5173(99)00334-8
70. Genta I, Colonna C, Perugini P, Pavanetto F, Modena T, Valli M, Muzzarelli C, Conti B (2005) Evaluation of bioadhesive performance of chitosan derivatives as films for buccal application. J Drug Deliv Sci Technol 15. https://doi.org/10.1016/s1773-2247(05)50088-1
71. Shu XZ, Zhu KJ (2002) The influence of multivalent phosphate structure on the properties of ionically cross-linked chitosan films for controlled drug release. Eur J Pharm Biopharm 54. https://doi.org/10.1016/S0939-6411(02)00052-8
72. Colonna C, Genta I, Perugini P, Pavanetto F, Modena T, Valli M, Muzzarelli C, Conti B (2006) 5-methyl-pyrrolidinone chitosan films as carriers for buccal administration of proteins. AAPS PharmSciTech 7. https://doi.org/10.1208/pt70370
73. Çirpanli Y, Bilensoy E, Lale Doğan A, Çaliş S (2009) Comparative evaluation of polymeric and amphiphilic cyclodextrin nanoparticles for effective camptothecin delivery. Eur J Pharm Biopharm 73. https://doi.org/10.1016/j.ejpb.2009.04.013
74. Yin YM, de Cui F, Mu CF, Choi MK, Kim JS, Chung SJ, Shim CK, Kim DD (2009) Docetaxel microemulsion for enhanced oral bioavailability: preparation and in vitro and in vivo evaluation. J Controlled Release 140. https://doi.org/10.1016/j.jconrel.2009.08.015
75. Trapani A, Lopedota A, Franco M, Cioffi N, Ieva E, Garcia-Fuentes M, Alonso MJ (2010) A comparative study of chitosan and chitosan/cyclodextrin nanoparticles as potential carriers for the oral delivery of small peptides. Eur J Pharm Biopharm 75. https://doi.org/10.1016/j.ejpb.2010.01.010
76. Marchettini P, Stuart OA, Mohamed F, Yoo D, Sugarbaker PH (2002) Docetaxel: pharmacokinetics and tissue levels after intraperitoneal and intravenous administration in a rat model. Cancer Chemother Pharmacol 49. https://doi.org/10.1007/s00280-002-0439-1
77. Wu J, Shen Q, Fang L (2013) Sulfobutylether-β-cyclodextrin/chitosan nanoparticles enhance the oral permeability and bioavailability of docetaxel. Drug Dev Ind Pharm 39. https://doi.org/10.3109/03639045.2012.694588
78. Gonil P, Sajomsang W, Ruktanonchai UR, Pimpha N, Sramala I, Nuchuchua O, Saesoo S, Chaleawlert-Umpon S, Puttipipatkhachorn S (2011) Novel quaternized chitosan containing β-cyclodextrin moiety: synthesis, characterization and antimicrobial activity. Carbohydr Polym 83. https://doi.org/10.1016/j.carbpol.2010.08.080
79. Ding WY, di Zheng S, Qin Y, Yu F, Bai JW, Cui WQ, Yu T, Chen XR, Bello-Onaghise G, Li YH (2019) Chitosan grafted with β-cyclodextrin: synthesis, characterization, antimicrobial activity, and role as absorbefacient and solubilizer. Front Chem 7. https://doi.org/10.3389/fchem.2018.00657
80. Yostawonkul J, Surassmo S, Iempridee T, Pimtong W, Suktham K, Sajomsang W, Gonil P, Ruktanonchai UR (2017) Surface modification of nanostructure lipid carrier (NLC) by oleoyl-quaternized-chitosan as a mucoadhesive nanocarrier. Colloids Surf B Biointerfaces 149. https://doi.org/10.1016/j.colsurfb.2016.09.049
81. Mythri G, Kavitha K, Kumar MR, Jagadeesh Singh SD (2011) Novel mucoadhesive polymers—a review. https://japsonline.com/admin/php/uploads/209_pdf.pdf
82. Grabovac V, Guggi D, Bernkop-Schnürch A (2005) Comparison of the mucoadhesive properties of various polymers. https://doi.org/10.1016/j.addr.2005.07.006
83. Lu S, Neoh KG, Kang ET, Mahendran R, Chiong E (2015) Mucoadhesive polyacrylamide nanogel as a potential hydrophobic drug carrier for intravesical bladder cancer therapy. Eur J Pharm Sci 72. https://doi.org/10.1016/j.ejps.2015.03.006
84. Ying XY, Cui D, Yu L, Du YZ (2011) Solid lipid nanoparticles modified with chitosan oligosaccharides for the controlled release of doxorubicin. Carbohydr Polym 84. https://doi.org/10.1016/j.carbpol.2011.01.037

85. Gartziandia O, Herran E, Pedraz JL, Carro E, Igartua M, Hernandez RM (2015) Chitosan coated nanostructured lipid carriers for brain delivery of proteins by intranasal administration. Colloids Surf B Biointerfaces 134. https://doi.org/10.1016/j.colsurfb.2015.06.054
86. Li S, Su Z, Sun M, Xiao Y, Cao F, Huang A, Li H, Ping Q, Zhang C (2012) An arginine derivative contained nanostructure lipid carriers with pH-sensitive membranolytic capability for lysosomolytic anti-cancer drug delivery. Int J Pharm 436. https://doi.org/10.1016/j.ijpharm.2012.06.040
87. Pedraza-Chaverri J, Cárdenas-Rodríguez N, Orozco-Ibarra M, Pérez-Rojas JM (2008) Medicinal properties of mangosteen (*Garcinia mangostana*). https://doi.org/10.1016/j.fct.2008.07.024
88. Sashiwa H, Shigemasa Y, Roy R (2002) Chemical modification of chitosan 8: preparation of chitosan-dendrimer hybrids via short spacer. Carbohydr Polym 47. https://doi.org/10.1016/S0144-8617(01)00166-7
89. Barikani M, Zia KM, Bhatti IA, Zuber M, Bhatti HN (2008) Molecular engineering and properties of chitin based shape memory polyurethanes. Carbohydr Polym 74(3):621–626. https://doi.org/10.1016/j.carbpol.2008.04.023
90. Zia KM, Bhatti IA, Barikani M, Zuber M, Bhatti HN (2009) XRD studies of polyurethane elastomers based on chitin/1,4-butane diol blends. Carbohydr Polym 76(2):183–187. https://doi.org/10.1016/j.carbpol.2008.10.005
91. Zia KM, Zuber M, Barikani M, Jabbar A, Khosa MK (2010) XRD pattern of chitin based polyurethane bio-nanocomposites. Carbohydr Polym 80(2):539–543. https://doi.org/10.1016/j.carbpol.2009.12.017
92. Zia KM, Zuber M, Barikani M, Hussain R, Jamil T, Anjum S (2011) Cytotoxicity and mechanical behavior of chitin–bentonite clay based polyurethane bio-nanocomposites. Int J Biol Macromol 49(5):1131–1136. https://doi.org/10.1016/j.ijbiomac.2011.09.010
93. Zia KM, Mahmood K, Zuber M, Jamil T, Shafiq M (2013) Chitin based polyurethanes using hydroxyl terminated polybutadiene. Part I: molecular engineering. Int J Biol Macromol 59:320–327. https://doi.org/10.1016/j.ijbiomac.2013.04.069
94. Zia KM, Qureshi NA, Mujahid M, Mahmood K, Zuber M (2013) Chitin based polyurethanes using hydroxyl terminated polybutadiene. Part II: morphological studies. Int J Biol Macromol 59:313–319. https://doi.org/10.1016/j.ijbiomac.2013.04.042
95. Zia KM, Zuber M, Barikani M, Bhatti IA, Khan MB (2009) Surface characteristics of chitin-based shape memory polyurethane elastomers. Colloids Surf B Biointerfaces 72(2):248–252. https://doi.org/10.1016/j.colsurfb.2009.04.011
96. Zia KM, Zuber M, Mahboob S, Sultana T, Sultana S (2010) Surface characteristics of UV-irradiated chitin-based shape memory polyurethanes. Carbohydr Polym 80(1):229–234. https://doi.org/10.1016/j.carbpol.2009.11.015
97. Zia KM, Zuber M, Bhatti IA, Barikani M, Sheikh MA (2009) Evaluation of biocompatibility and mechanical behavior of chitin-based polyurethane elastomers. Part-II: effect of diisocyanate structure. Int J Biol Macromol 44(1):23–28. https://doi.org/10.1016/j.ijbiomac.2008.11.002
98. Zia KM, Zuber M, Bhatti IA, Barikani M, Sheikh MA (2009) Evaluation of biocompatibility and mechanical behavior of polyurethane elastomers based on chitin/1,4-butane diol blends. Int J Biol Macromol 44(1):18–22. https://doi.org/10.1016/j.ijbiomac.2008.09.011
99. Zuber M, Zia KM, Mahboob S, Hassan M, Bhatti IA (2010) Synthesis of chitin–bentonite clay based polyurethane bio-nanocomposites. Int J Biol Macromol 47(2):196–200. https://doi.org/10.1016/j.ijbiomac.2010.04.022
100. Amjed N, Bhatti IA, Zia KM, Iqbal J, Jamil Y (2020) Synthesis and characterization of stable and biological active chitin-based polyurethane elastomers. Int J Biol Macromol 154:1149–1157. https://doi.org/10.1016/j.ijbiomac.2019.11.097

Chapter 4
Modification of Chitosan Nanoparticles (CSNPs)

4.1 Modification of CSNPs

The CSNPs' backbone can be modified to amend its qualities such as solubility, mucoadhesion, and constancy. The $-NH_2$ and $-OH$ groups of chitosan (CS) act as active sites for transformation [1] as shown in Fig. 4.1.

4.2 Physical Modification

The physical amendment of CSNPs is attained by blending with a physical mixture of two or more than two polymers. This is among one of the most straightforward techniques to modify the polymer. The eminence and performance of the blend can be varied, relying on the proportion of polymer to be blended. Blending is the most convenient way to alter the properties of a polymer for a particular purpose [2]. There are several general hydrophilic polymers that can be mixed with CS to attain oral drug deliverance i.e., polyvinyl alcohol (PVA), polyvinyl pyrrolidine (PVP), and polyethyl ether (PEE). A mixture of CS and PVA enhances the mechanical and barrier (liquid vapor-permeability) characteristics of CS films [3]. The intermolecular interactions amid CS and PVA results in a PVA-CS blend, which exhibits better mechanical qualities (tensile strength) of the CS film for controlled drug deliverance. Hence, blended CSNPs have been characterized by utilizing both chemical and spectroscopic routes.

K. M. Zia, *Chitosan*, Smart Nanomaterials Technology,
https://doi.org/10.1007/978-981-96-2629-8_4

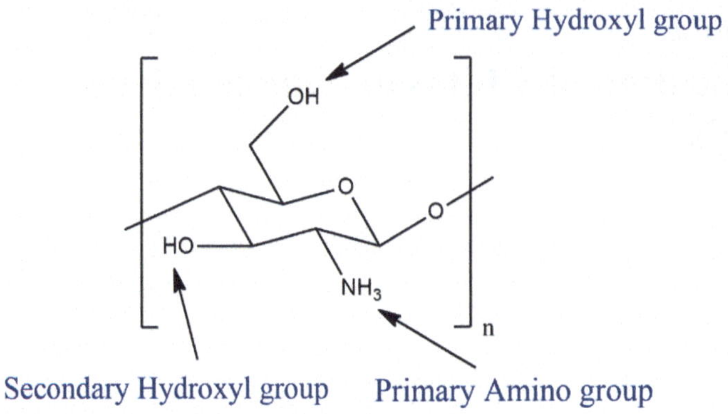

Fig. 4.1 Functional groups in chitosan nanoparticles (CSNPs)

4.3 Chemical Modification and Applications of CSNPs

The CSNP can be chemically modified from the base polymer. The reactive groups of CS (amino and hydroxyl) are utilized for this rationale, and variety of basic conditions and temperatures have been used [4]. The chemical modification includes replacement reactions, series elongation (cross-linking with glutaraldehyde GA), and copolymerization of grafts and polymer networks. Figure 4.2 shows a demonstration of the cross-linking response of CS with glutaraldehyde (GA) [5]. Thus, the modified CS derivatives exhibit improved reprocess ability upon regeneration.

The CSNP modification with fatty acylated chloride leads to the creation of CS derivatives as indicated in Fig. 4.3. It has superior properties and increases the adsorption capacity and adsorption sites [6].

The CSNPs could also be modified by grafting process with aniline and acetaldehyde. In this process, aniline produces a bright yellow and viscous CS solution; at the same time, acetone produces a colorless solution of low viscosity. The reaction mixture is stirred at 69.0°C for 19 hours, after which a 3.0 mol/L NaOH solution is gradually added while stirring until the pH reaches 7, resulting in the formation of a pink precipitate. Separate the precipitate from the supernatant, rinse it with uncontaminated water and ethanol, and then dry in an oven at 55.0 °C for at least 7 h. The transformation increased the particular surface zone of CS and its competence for fluorescent dyes to 62.90 mg [7]. Table 4.1 indicates the most prominent cross-linking agents and their relevancies.

Schematic (Fig. 4.4) shows the transformation of CSNP by grafting process with aniline-NH_2 and acetaldehyde.

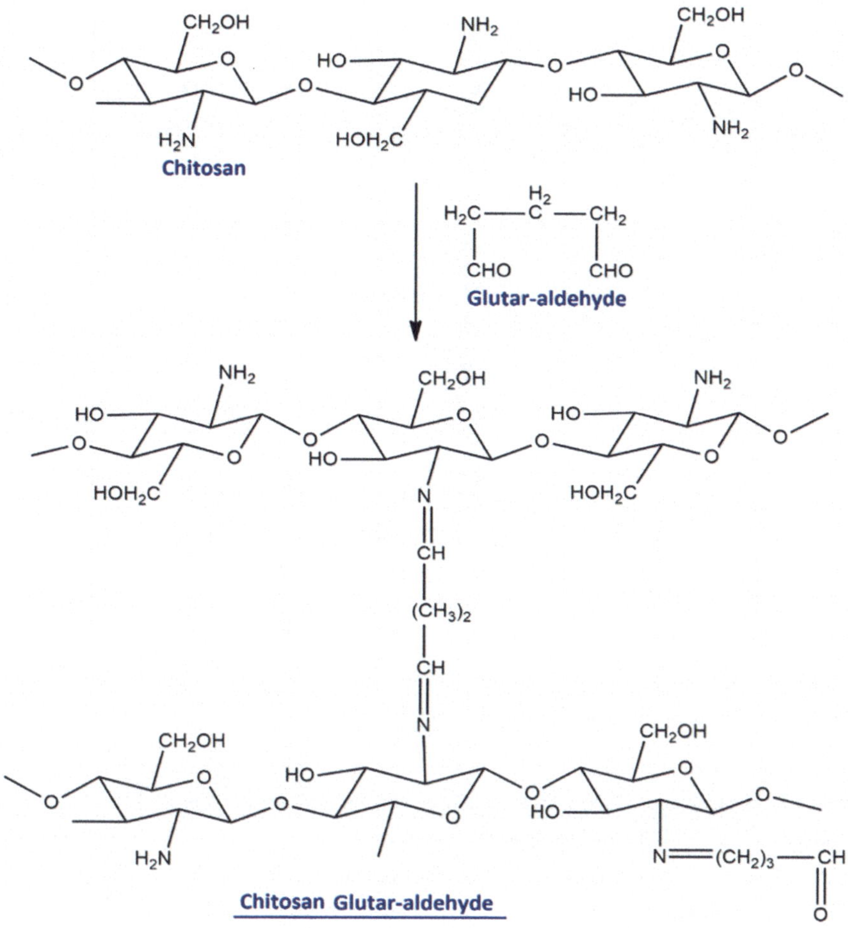

Fig. 4.2 Schematic depiction of the cross-linking response of CSNPs with GA [5]

4.3.1 Applications

4.3.1.1 Wastewater Treatment

Chitosan contains two functional groups (amino and hydroxyl), making these NPs attractive for removing a wide sort of pollutants such as dyes, heavier metals, and insect killers in wastewater. In addition, NPs have large surface area, which allows it to have a higher capacity than the commonly used microabsorbents. The applications of CSNPs in wastewater treatment are presented in Table 4.2. In contrast to applications in other fields, ultrafine NPs below 100.0 nm are selected for wastewater treatment, resulting in bigger surface vicinity that can absorb further impurities. The CSNP-modified derivatives which have improved electrostatic and magnetic properties are

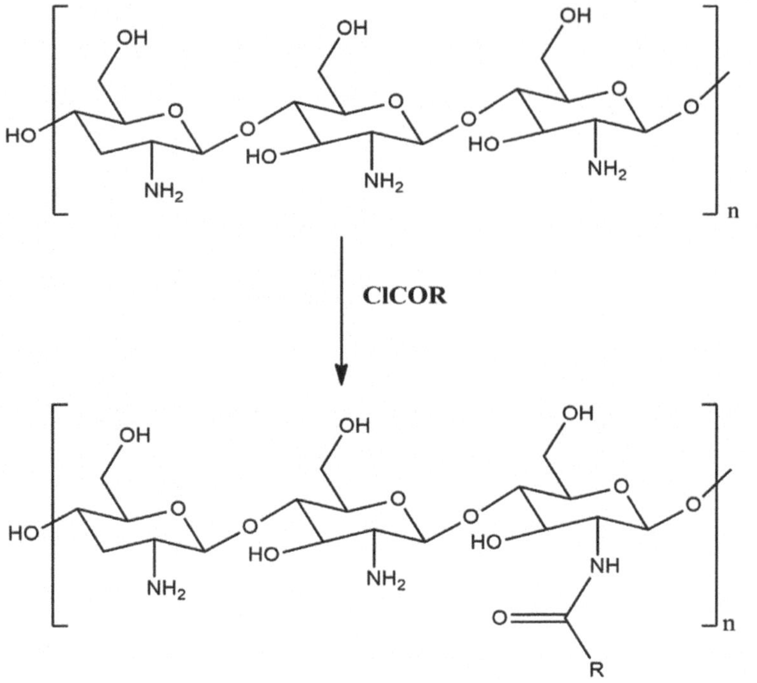

Fig. 4.3 CSNPs' derivatization with fatty acyl chlorides [6]

Table 4.1 Most common CSNPs, cross-linking agents, and their applications [8–11]

Materials	Cross-linking agent	Application	References
CSNPs-PVA	GA-H_2SO_4	Food packaging	[8]
CSNPs-microspheres	Genipin	Indomethacin is discharged at high rate of dissolution	[9, 10]
CSNPs-gelatin films	Proanthocyanidin	Tissue engineering	[11]

commonly utilized for treatment of wastewater. Amination is performed by grafting process which is ethylene-diamine, hexa-diamine, or diethylene-triamine. These –NH_2 groups are involved in the electrostatic interaction with pollutants. When utilized in combination with magnetic properties, CSNP has been applied to enhance the removal of heavy metals and dangerous dyes [12, 13].

4.3.1.2 Agriculture

Utilization of CSNPs for agricultural applications has been enhanced so rapidly by the need for eco benign agricultural and ecological chemicals such as fertilizers and insect killers. As similar to in medication, the CSNP is mainly utilized as nanocarriers

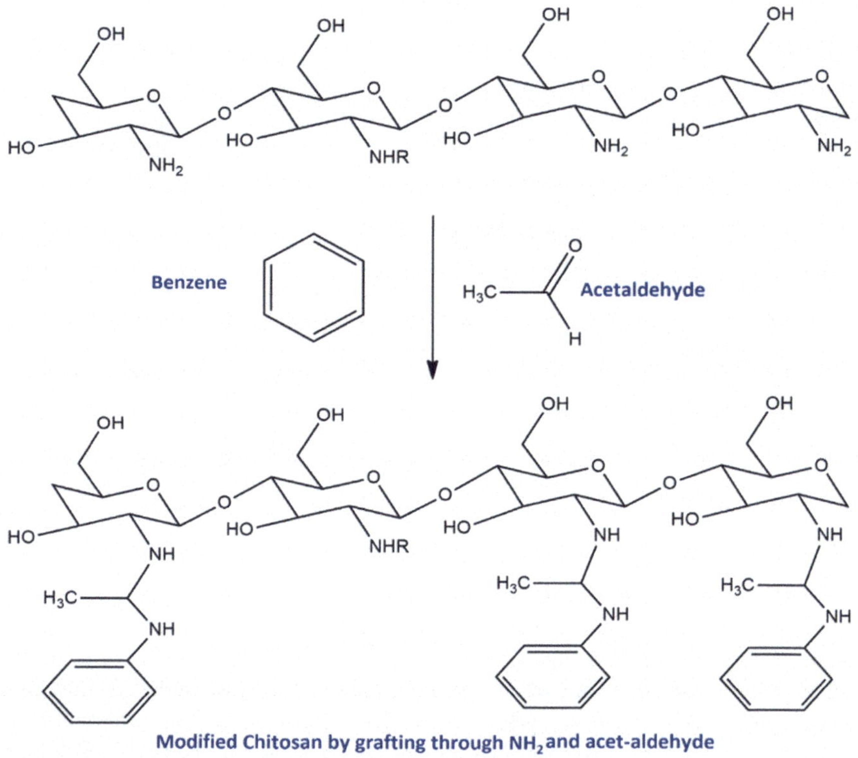

Fig. 4.4 Single-step modification of CS by grafting process through NH_2 and acetaldehyde [5]

Table 4.2 Application of CSNPs in treatment of wastewater [12–15]

Aim of utilize	Compounds	NPs' formation technique	Particle size (nm)	Modifications	References
Heavier-metal elimination	CSNPs; Fe_3O_4	Co-precipitation and cross-linking	– 30	Conjunction with magnetic particles (MPs)	[12, 13]
Heavy-metal elimination	CSNPs; acrylic acid, lead nitrate	Cross-linking	200	Direct grafting	[12, 13]
Dye amputation	CSNPs, β-cyclo-dextrin; Fe_3O_4	Co-precipitation emulsification and cross-linking	– 100	Conjunction with MP	[14, 15]
Dye amputation	CSNPs, Fe_3O_4	Co-precipitation and cross-linking	78.82	Conjunction with MP	[14, 15]

Table 4.3 Application of CSNPs in agricultural field [12–15]

Aim of utilization	Compounds	NPs' formation process	Particle size (nm)	Charge on the surface (mV)	Modifications
Herbicide delivery	CSNPs, paraquat	Ionic-gelation	– 301	+ 46.0	Spontaneous nanoencapsulation
Insecticide delivery	CSNPs, PAA	Polymerization and cross-linking	52.78	–	–
Fertilizer delivery	CSNPs, fertilizers K, P, and N	Ionic gelation	– 501.0	50.0	Nanoencapsulation after NPs preparation
Fungicides delivery	CSNPs, *Cymbopogon martinii* essential oil (CMEO)	Self-assemblage	454.9–479.9	+ 037.20 to + 039.30	Spontaneous nanoencapsulation
Plant growth agent deliverance	CSNPs, gibberellic acid (GA3)	Ionic (pre)gelation	~ 450 ~ 195	29.00 ± 0.3 + 27.00 ± 3.0	Spontaneous nanoencapsulation

to improve the stability of active ingredients. With this effect, agricultural chemicals can be given at minor doses and need less treatment, as well as the risk of environmental pollution and toxic impacts on other non-targeted creatures reduced. Table 4.3 indicates a broad overview of agricultural CSNPs relevance's. In general, intermediate and extremely positive NPs provides better constancy in the preferred liquid medium to guarantee slow discharge and permanently energetic ingredients in the soil. Naturally occurring nanoencapsulation during ion gelation has been utilized to encapsulate agrochemicals or energetic ingredients, e.g., fundamental oils in CSNPs [15].

4.3.1.3 Gene Therapy

In vivo application, cancer cure is by far one of the most significant applications of gene therapy and many medical trials utilizing gene therapy are under-study. About 70.0% of all cancer gene therapy experiments are carried out with local administration. The majority of clinical trials of gene therapy are in initial Phase-I, while only a few trials are in Phase-II. More than two-thirds of gene therapy medical trials utilize viral vectors; only 13.0% employed lipid carriers and about 09.1% utilized unprotected DNA. Cationic polymers including CS derivatives have only been utilized in animal sculpts and have not progressed to medical trials owing to their low conversion affectivity. The CS itself has some anticancerous activity, and its positive charge can deactivate the negative charge of cancerous cells, leading to selective absorption. Therefore, CSNP can augment the drug meditation at the lump site and recover the therapeutic effect. The more cytotoxic doxorubicin/methoxy PEG-engrafted CSNPs are able to enter cells and effectively inhibit tumor cell proliferation [16, 17].

4.3.1.4 Nervous Tissue Engineering

In fully mature neurons, nerve damage complicates successful healing and can cause dysfunctionality in the nervous system and some other parts of the body. The rationale for this dysfunctionality is basically due to fully grown neurons do not endure cell separation. In such sort of cases, the modified CSNPs can be utilized as a promising scaffold. New biomaterials have been developed for neural regeneration by immobilization of laminin peptides utilizing molecularly aligned CS derivatives [18]. Aligned CS-polycaprolactone fiber scaffolds are compared to arbitrarily oriented fiber scaffolds and cast films to examine the impact of fiber orientation on neuronal functionality. Biodegradable (poly-D-lactic acid) chondroitin-sulfate CS nerve leads have probably excellent biocompatibility and involuntary qualities which are appropriate for surgical procedures.

4.3.1.5 Bioimaging

Bio-based imaging is one of the most vital diagnostic tool to inspect and envisage the biochemical process persuaded by drugs in cells. Modified CSNPs are found to have significant functions as biomarkers. It has been found that imaging agents' incorporation, such as Fe_3O_4 for MRI into the self-accumulated NPs, improve imaging directed at hepatocyte, with the particle's acting as MRI agents. Scientists found numerous inorganic materials having metals into modified CSNPs. Such NPs are discovered to have a surplus of biomedical relevances [6].

4.3.1.6 Wound Healing

Chronic wounds have a significant socioeconomic impact due to the high risk of resultant infection and in the nastiest circumstances; it can still be so life threatening. The attendance of bacteria, which are resistant to antibiotic, limits the utilization of antibiotics in the cure of chronic lesions. The finest way to solve this predicament is to utilize NP-based carriers containing anti-infective, which target the microbial population to the spot of infection. Modified NPs containing CS have been successfully used as carriers of anti-infective agents for chronic wound healing [19].

Alternatively, in circumstances where auto-logos skin grafts are not accessible, biofilm-forming agents such as CS can be utilized to support and commence normal lesion healing in the skin and epidermis. It is an ideal dressing that protects wounds from bacterial infections and supports healing process. The CS-based materials can self-heal and result in little or no scarring. Bandages utilized in the initial stages of lesion healing promote wound healing, provide dissolving antibacterial agents, and support fibroblast growth. The CS-based hydrogel fulfills all these prerequisites. It has been studied that CS-based dressings caused minimal side effects, where fibrous encapsulation is slightly negligible or nearly absent. In addition, CS accelerates wound healing when applied as powder, nano and microparticle, granule, sponge, or a mixture with further materials. The CS-based hydrogel scaffolds permeated with fibroblast augmentation factor β have been manufactured with better swelling and drug discharge qualities. The pH-responsive hydrogels of carboxymethyl CS have been synthesized using fusion-emulsion combined with ultrasonic technique and are used as transporters and drug delivery devices [20].

4.3.1.7 Gold-CSNPs for Heavy-Metal Sensing

An innovative strategy utilizing CS-coated gold NPs to detect heavy metal ions was proposed in 2005. The CS is polycationic and can therefore be attached to the surface of −ve charged gold NPs by electrostatic interaction. The utilization of CS provides sufficient steel resistance to ensure colloidal stability and functionalize the NPs that can be utilized as sensors. The chelating properties of CS and the optical properties of gold NPs have also been employed to detect fewer amounts of heavy metal ions in water [21].

4.3.1.8 Chitosan in Liver Tissue Engineering (LTE)

Tissue engineering (TE) has made significant strides in developing an artificial liver, offering new hope for treating liver diseases. This innovative approach holds immense potential for revolutionizing liver therapy and improving patient outcomes. Since very few people donate liver, the need is urgent therapy for acute and persistent liver conditions. Owing to the structural resemblance of CS to glycosaminoglycans, it has been utilized as a substrate for hepatocyte culture [22]. Cross-linked CS collagen matrix utilizing 1ethyl3 (3dimethylaminopropyl) carbodiimide (EDC) and N-hydroxysuccinimide (NHS) in one 2–4,6 morpho-lino-ethane-sulfonic-acid buffer system (MES) is utilized as a scaffold for liver tissue (LT) technology. This matrix shows excellent blood and cell affinity in the process of platelet-deposition and hepatocyte culture trials. In the case of short-term crops, alginate-galactosylated CS substrates have been utilized to augment hepatocyte attachment [23]. A perfect extracellular matrix is required in LTE for hepatocytes to sustain specific high-grade liver function. Scaffolds for LTE are manufactured by utilizing oxidized form of alginate cross-links covalently with galactosylated CS by a Schiff base reaction without the utilization of external chemical cross-linking agents. Synthetic scaffold pops out extremely porous surfaces with typical pore sizes from 52 to 150.0 μm and vents internal structure. The inner structure has a pore size of 100.0–250.0 μm. The reduction and degradation rates increase as the amount of oxidized alginate in the scaffold increases. Moreover, in vitro biocompatibility tests indicated that the hepatocytes cultured on the scaffolds has a distinctive spheroidal morphological structure and made aggregates (multicellular) and integrated immaculately with scaffolds. Researchers manufactured alginate and galactosylated CS-based extremely porous hydrogel scaffolds for LTE. Synthetic scaffolds (size 16–200 mm) are utilized as extracellular matrix for liver cells. Another high porosity, water-soluble silk/fibroin/CS-based substrate has been developed by using freeze-drying and controlled synthetic scaffolds mechanical properties as well as pore sizes in the range of 100.0–150.0 μm. Manufactured scaffolding is found to indicate significant cellular activities in hepatocellular carcinoma of the liver [24]. In a similar way, self-assembled homogenous scaffolds of fibroin and CSNPs silk are also manufactured, which support hepatocyte combination [25].

4.3.1.9 Carrier of Protein Drugs

Protein-based drugs are simply degraded by utilizing enzymes in vivo and have meager permeability and permanence. However, CSNPs may well guard proteins and promote drug-biofilm contact, thereby improving bioavailability, but it is hard to control the sudden discharge of high weight proteins [26]. Scientists utilized insulin and the β-cyclodextrin cation to form a composite enclosed in alginate-CSNPs [27]. The binding rates and drug intake are reported 87.01 and 09.45%, correspondingly, and the discharge of insulin into the simulated intestinal fluid is 41.10%. Insulin is well protected in the NP-core, avoiding degradation in the simulated gastric juice, as well as the insulin composition upon discharge. The CS glycol NPs modified with 5β-cholanic acid and polypeptide (ArgGlyAsp) are straightforwardly encapsulated into NPs with a drug load more than 85.0% [28]. These NPs can reduce lump cell growth and significantly reduce microvascular density.

4.3.1.10 Anticancer Drug Delivery

The CS-based NPs have been successfully utilized for controlled discharge of antineoplastic agents. Researchers tested CS-loaded hydrogel in a murine breast cancer xenograft model. Tumor succession is found to be lessened, and thus, drug release is able to reduce tumor re-occurrence and metastatic spread by as much as 69.1% [29]. Nanoparticles of salicylic acid and CS oligosaccharides have been shown to be so effective in delivering paclitaxel. Similarly, CS arachidic acid oligosaccharide conjugates have been successfully manufactured and utilized to grow self-assembled NPs for doxorubicin delivery. Doxorubicin is conjugated to CS by an acid cleavable hydrazone bond to form a pH-activated internal precursor. Doxorubicin-CS can self-accumulate into NPs, which initially accumulate in lump cells and then quickly free doxorubicin as shown in Fig. 4.5 [30]. The CSNPs can also load some other drugs including antivirus drugs, antiallergic drugs, and hormone drug.

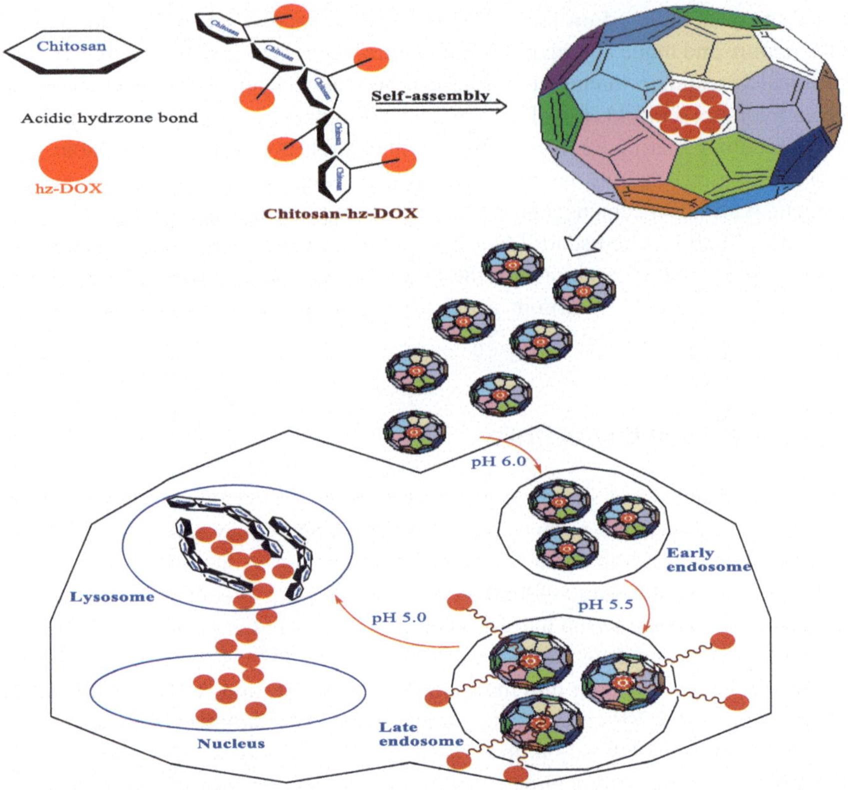

Fig. 4.5 Illustration of the formation of CS-doxorubicin (CS-hz-DOX) pro-drug NPs and their intracellular uptake

4.3.1.11 Ocular Discharge of Drugs

The CS and its various derivatives based NPs are too utilized for the ocular release of drugs. The CSNPs utilizing ionic gelation with tripolyphosphate (TPP) is employed for the ocular discharge of dorzolamine hydrochloride (Dorzo) and pramipexole hydrochloride (Prami). Furthermore, it has also been scrutinized that one of the drugs, namely Dorzo, is dissolved in the CSNPs. Moreover, mucoadhesive qualities exhibited by CSNPs are lessened by more rise in the drug content. In a further study, NPs of CS and their derivatives with succinic anhydride and 1-carboxybenzaldehyde are also utilized for the ocular release of timolol drug [31]. In this situation, drug is encapsulated throughout the ion gelation of drug carrier utilized and TPP. The mass of the nanoparticles (NPs) is determined by the ratio of chitosan (CS) to tripolyphosphate (TPP). The particle size of NPs is enhanced with rising amount of TPP.

4.3.1.12 Delivery of Antibiotics

A pH-responsive CS-PVP-controlled drug discharge system has been utilized for antibiotic deliverance. The matrix is manufactured by cross-linking a CS and PVP intermingle with GA to yield a semi-IPN. Interpenetrating polymer networks (IPNs) are a combination of two or more polymers in networks where a partial interlacing on the molecular scale is present in the matrix. Scientists investigated drug delivery into the inner ear through the round window membrane using chitosan (CS). A lot of researchers have investigated the controlled discharge of antibiotics utilizing diverse CS-based matrices and estimated their antimicrobial activity.

4.3.1.13 Other Applications of CSNPs

The CSNPs and its derivatives can carry novel solutions and new techniques to a variety of scientific disciplines. Advances and uses are not as widespread and highly developed as in medication and agriculture, but there is also research in cosmetology, food technology, and dentistry [32]. The CSNPs have been proposed as a hauler for drugs utilized in skin and hair care [14, 33]. Nanoencapsulation of retinol provided protection against degradation, while its complexation with succinic acid in chitosan nanoparticles (CSNPs) enhanced antioxidant activity compared to pure retinol [14]. Pioneering study performed by del Carpio utilized CSNP instead of EDTA in root canal cure. The researchers have demonstrated that the antibacterial and chelating properties of CSNP make it a suitable nanomaterial for dental applications. In addition, CSNP can be utilized to protect steel from corrosion. The CSNP is also considered as a filler material [34]. In addition, there are several studies on the antibacterial activity of CSNPs and their potential utilize in (edible) food packaging [35, 36]. Table 4.4 shows some other applications of CSNPs in pharmaceutics and medicine.

4.4 Modification of CSNPs with Polymeric Compounds

Eco-friendly and renewable thermoplastic polyester derived from polylactic acid (PLA), lactic acid is a traditional petroleum-based plastic due to its renewable raw material production, high tensile strength and UV blocking properties, and excellent aroma. The PLA seems to be one of the most promising alternatives to barriers in standard plastic production lines. Surface amendment is an imperative tool for increasing the hydrophobicity of CSNPs and promoting uniform distribution within the polymer matrix. Chloride-type fatty acid (dodecanoyl chloric acid) bound to CSNP by acylation. These amended NPs have improved hydrophobicity and have shown higher antifungal activity when extruded with PLA. This is considered an interesting clue to the further improvement of NP reinforced packaging materials. Figure 4.6 shows the overall manufacturing, functionalization, and decomposition

Table 4.4 Applications of CSNPs in pharmaceutics and medicine [14]

Aim of utilization	Compounds	NPs' formation process	Particle size (nm)	Charge on surface (mV)	References
Insulin (INS) delivery	CSNPs, hydroxypropyl methylcellulose phthalate (HPMCP), INS	Ionic gelation	257	+ 30.3 ± 0.71	[14, 27]
		Self-assembly	210.1–526.5	+ 2.82 + 24.62	
	CSNPs, INS	Ionic gelation	46.2 ± 2.7	+ 9.3 ± 1.4	
Cyclosporin A (Cs: A)	CSNPs, Cs: A	Ionic gelation	293 ± 9	+ 37.6 ± 0.8	[14]
	CSNPs, Cs: A	Spray-drying	317.20 ± 78.69	+ 22.0 ± 0.30 to + 30.5 ± 0.51	
Malignant cells therapy	Thiolated CSNPs, curcumin Thiolated CSNPs, 5-fluorouracil	Ionic gelation	149.9 ± 20	+ 48.2 ± 5	[27]
Circulatory system regulator	Thiolated CSNPs, heparin	Ionic gelation technique	324.1	+ 36.1	[14, 27]

step of CSNP-enhanced PLA nanocomposites. Considerable evolution has been made in industry and academic world, particularly in the manufacture of PLA. The PLA nanocomposites enriched with CSNP play an important part in biodegradable food packaging. These plastics have qualities analogous to polymers which are petroleum-based and have additional efficiency, e.g., antibacterial and antioxidant [14].

4.5 Modification of CSNPs with Organic Compounds

4.5.1 Modification with Aldehydes

The CSNPs (0.49 g) are suspended in distilled water for at least 1 h. Some drops of acetic acid are added and sonicated for an additional 02 h until complete dissolution is achieved. Aldehyde is added drop by drop (0.49 g), while stirring the mixture, and it is stirred for another 3 h. A pale yellowish material is formed which is separated from the mixture by filtration technique. The attained product is dried under vacuum at 39.9 °C for 24.0 h/1 day [37].

The aldehyde-amended CS derivatives have been characterized by utilizing both chemical and spectroscopic processes. Such sort of chemical-tests demonstrated

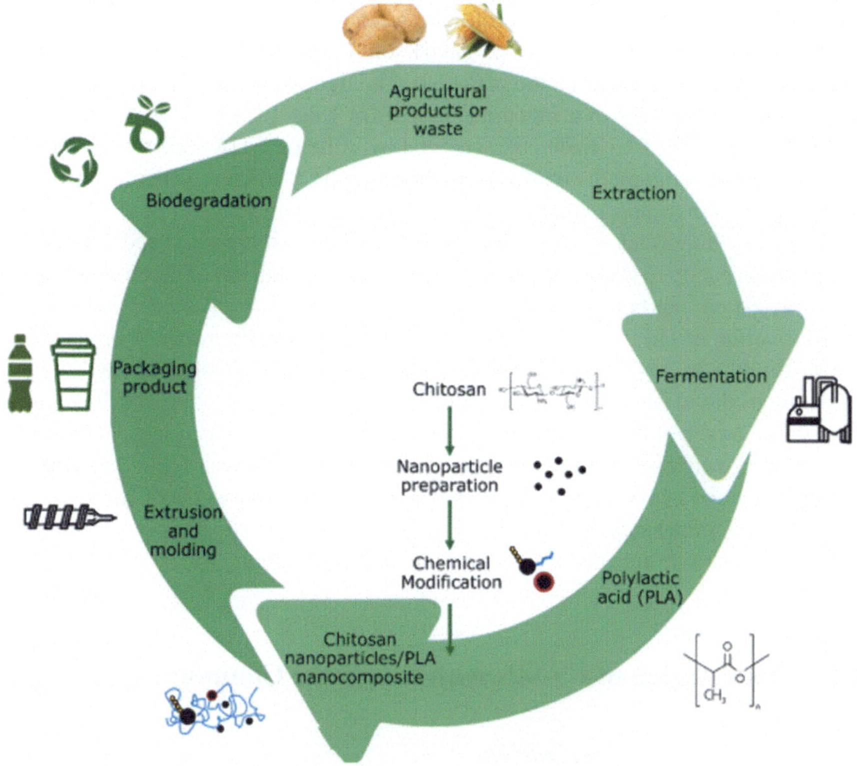

Fig. 4.6 Life cycle of CSNPs reinforced PLA nanocomposites [14]

the amended conversion of CS to the relevant aldehydes. Salicylaldehyde-modified chitosan nanoparticles (CSNPs) tested positive for phenolic hydroxyl groups when reacted with a ferric chloride solution. Meanwhile, cinnamaldehyde-modified CSNPs exhibited a positive test for unsaturation when reacted with ferric chloride solution, bromine water, and potassium permanganate, resulting in color discharge in all cases. In addition, all CSNPs converted to aromatic aldehydes have a smoky flame when burned owing to the residual aromatics in the aldehyde molecule. The FTIR spectroscopy is also utilized for characterization to explicate the transformations that arise in the chemical structure.

4.5.2 Modification with Acids

The CSNPs (0.49 g) are suspended in slight quantity of purified water for approximately 60 min. Some drops of acetic acid are added and sonicated for 120 min until homogeneity is achieved. About 02 ml of acetic anhydride is added dropwise, while stirring the reaction mixture for half an hour. The response mixture is heated approximately for 60 min, and then, NaOH solution is carefully added subsequent to cooling in order to maintain the pH at 8.6. The light yellowish material formed is filtered off and dried out in a vacuum at 4.0 °C for 1 day [37].

Modified CSNPs have been characterized by utilizing FTIR spectroscopic techniques. It shows no major distinction from the original CS utilized in chemical amendment owing to their approximately analogous functional groups in both. However, there are few important variations between CSNPs and its reacetylated derivative, such as its capacity to absorb water and absorb heavy metal ions. Therefore, product characterization can help us to demonstrate the modification has been made. The transformation of chitin to CS improves solubility in slightly acidic aqueous solutions owing to the regeneration of free amine groups throughout chitin hydrolysis. In addition, the elasticity of all the chains will obviously be improved. It mainly contributes to the gelling nature of CS unlike chitin, which has a lesser attraction to water for diffusion into the polymer matrix and thus have a lesser capacity to absorb water and heavy metal ions. However, CSNPs reactivity generally affects the hydrophilicity of the substrate and the availability of free amine groups to participate in chelation with diverse metal ions owing to site segregation.

4.6 Modification of CSNPs with Inorganic Compounds

4.6.1 Sulfated CS Derivatives

Structural diversity of products is obtained during the sulfation of CS. This may be narrated to the different reactivity of the three functional groups of the starting polymer, with dissimilar degrees of freedom of individual assemblies. Sulfated CS is a water-soluble anionic CS derivative with antiviral [18], antibacterial, and bone-forming activity [38]. This derivative also exhibits antiobesity activity by blocking the adhesion of human malignant melanoma cells and promoting the inhibition of antifat production. Therefore, sulfated CS is attained by adding a sulfated complex. The sulfated complex is obtained by dropping $HCl–SO_3$ into pre-cooled (4.0 °C) DMF with stirring [39, 40]. The reaction assortment is stirred till the solution attained room temperature. The CS is added to DMF and stirred at room temperature for about 12 h. Excess solvent is removed by filtration technique to give solvated CS. The solvated polysaccharide is added to the sulfated complex and the reaction mixture is carried out at room temperature for approximately 5 h with constant stirring. The final assortment is neutralized with 20.0% (w/v) NaOH and precipitated with methanol in an ice bath.

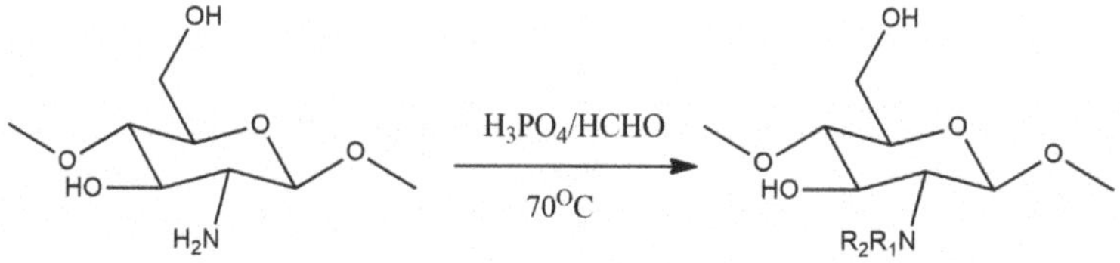

Fig. 4.7 Phosphorylation of CS utilizing phosphoric and formaldehyde

4.6.2 Phosphorylated CS Derivatives

The phosphorylation modification of CS is of great curiosity for orthopedic relevance, owing to the cation exchange qualities of the phosphate functions. Phosphate groups bind with the calcium ions, which could then induce the development of a layer of calcium phosphate, which promote osteoclastogenesis of polymer-based implantable devices. There are many techniques for the manufacture of phosphorylated derivatives of CS; some of them include phosphorylation by phosphorus pentoxide and methane-sulfonic acid. For the mixture of CS in methane-sulfonic acid, phosphorus pentoxide is added, and the mixture is stirred at 05 °C for 23 h. The attained product is precipitated with the ether and then it is centrifuged. The filtrate is then washed with ether and acetone and then dried. The CS N-methylene phosphoric acid is another derivative with various relevance. It is manufactured by blending phosphoric acid with water; CS is added drop by drop while stirring for one hour. Then, the reactor temperature is increased to 71 °C and formaldehyde (34.9%) is added dropwise for one hour with reflux reaction [41]; heating lasts for six hours (Fig. 4.7).

4.6.3 N-Substitutions of CS Without Protecting Agent

Quaternary ammonium (QM) CS is one of the most significant hydrophilic modifications of CS because of its antimicrobial activity and anticoagulant activity. In such sort of chemical modifications, N-substitution is carried out by turning the amino group (–NH_2) into a salt of QM, introducing QM compounds by a chemical response with CS. This process is commonly utilized owing to the uncomplicated modification involved in obtaining the derivatives which have developed better biological properties and water solubility owing to a permanent positive charge on the polymer backbone [42, 43].

N,N,N-trimethyl chitosan (TMCs) chloride is one of the most significant quaternary CS derivatives owing to its exceptional solubility in aqueous solution and being used as an absorption enhancer, antibacterial agent, and gene vector. TMC can be gained by two main routes. These are (1) one-step process and (2) two-step process.

The one-step process is utilized for obtaining TMCs which is carried out by the reaction of CS with methyl iodide (CH_3I) under highly alkaline conditions at room temperature utilizing N-methyl-2-pyrrolidone (NMP) as a solvent. Meanwhile, in

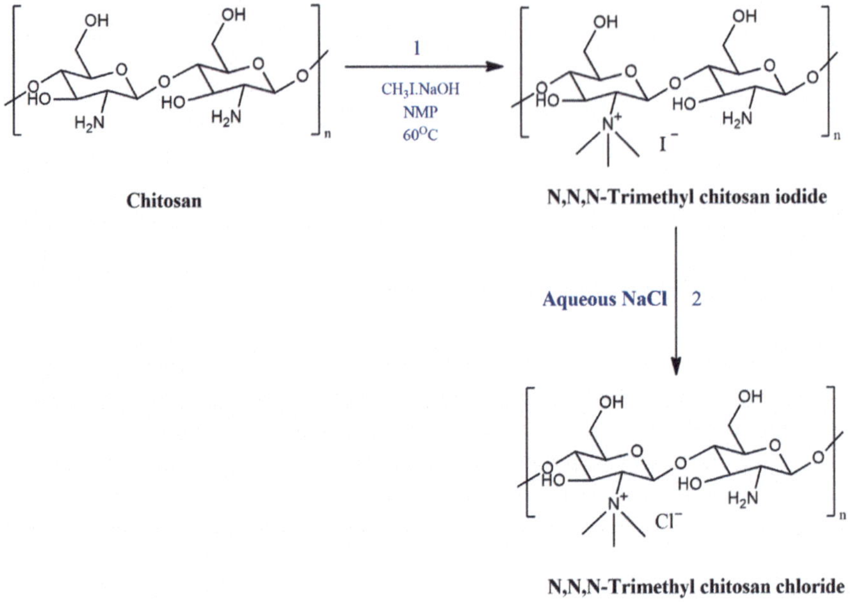

Fig. 4.8 Preparation of TMCs by the two-step reductive process

the two-step process, N-methyl-CS or N, N-dimethyl CS is first gained by reacting CS with formaldehyde under strongly acidic situations in order to get a Schiff base, followed by a reaction with CH_3I for the methylation method. Also, the two-step process has the benefit of evading O-methylation of the attained TMCs unlike the one-step process (Fig. 4.8) [44]. The obtained TMCs have better properties of bioadhesion, biocompatibility, good solubility, and viscosity and enhanced the swelling index [45].

4.6.4 N-Substitutions Utilizing Protective Agents

The protection of the –OH groups in the C_3 and C_6 of CS utilizing chemo-selective technique is one of the best strategies for the preparation of homogeneous N-quaternized CS derivatives without O-methyl substitution, and O-silylation has also been proposed to protect the –OH groups of CS in the attendance of amines, by transformation into silyl-ethers [46]. Researchers also studied the protection of the –OH groups of CS utilizing di-tert-butyldimethylsilyl (di-T-BDMS) owing to its constancy under acidic situations and easiness of its elimination under highly basic or mild acid situations without impacting the functional group.

Three derivatives of CS (N,N,N-trimethyl CS, N-PEGylated CS, and N-2-aceto-PEGylated CS) have been developed utilizing TBDMS CS as an intermediate,

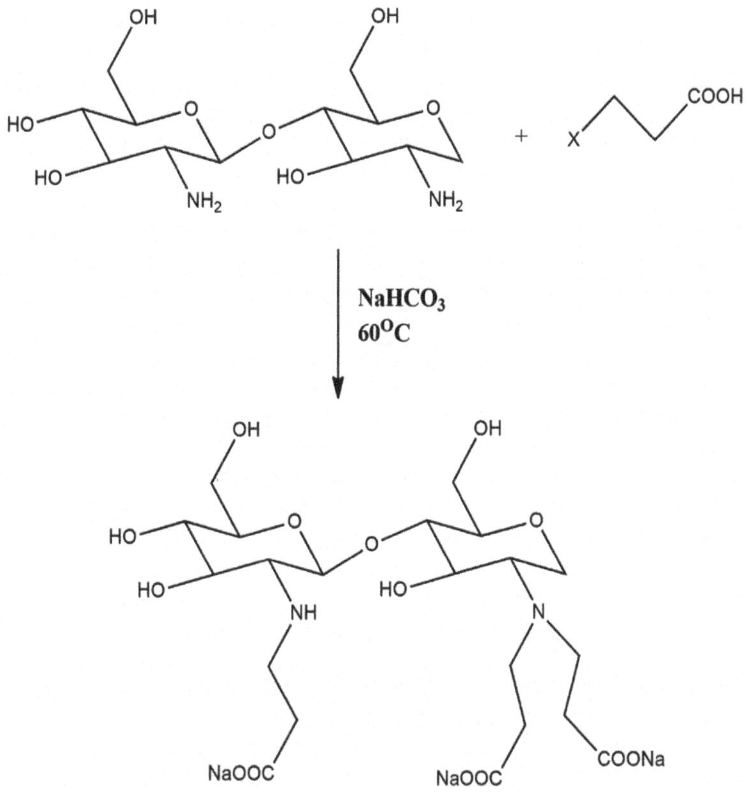

Fig. 4.9 Carboxyethylation of CS with 3-halo-propionic acid

protecting 99.99% of –OH groups [47]. These products displayed better water solubility and bioactivity. In addition, N-(2-carboxy-ethyl) CS under region-selective condition protects the –OH and also manages the alkaline media [48]. Under neutral or moderate alkaline situations (pH 8.0–9.0), it is possible to attain 98.9% N-carboxyethylation (Fig. 4.9).

4.6.5 O-Substitutions Utilizing Protective Agents

Usually, the chemical transformation of CS is performed on the 2NH_2 group because the reactivity of this group is higher than that of the 3–OH and 6–OH groups, resulting in a new structure and biological activity [49]. However, this response interferes with the amine group (–NH_2) of CS, possibly affecting its bioactive properties (i.e., its antibacterial activity), easily protected to redevelop free amine groups, enhancing the solubility of CS [50, 51]. After the response, the phthaloyl group can be eliminated by introducing an electron withdrawing agent (e.g., –NO_2, –Cl) to the phthaloyl aromatic ring, removing the amine group to yield three 6N-substituted products by a multistep reaction (Fig. 4.10). The results indicated that increased antifungal activity against Rhizoctonia grain is, *Fusarium oxysporum* and *Botrytis cinerea*. In addition, antibacterial activity against *Staphylococcus*, *Escherichia coli*, and *Salmonella typhi* is also enhanced.

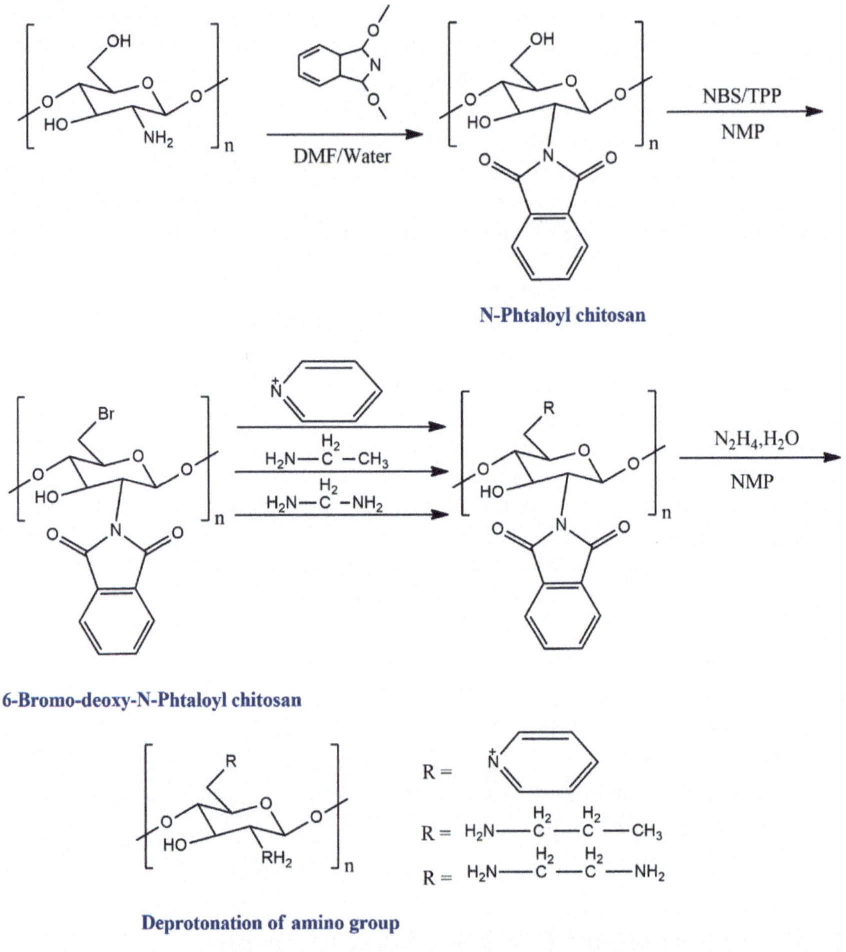

Fig. 4.10 Preparation of 6-N-substituted CS dimethylformamide (DMF); N-bromosuccinimide (NBS); and tripolyphosphate (TPP)

4.6.6 N,O-Substitutions

The substitution of both of the functional groups of CSNPS (–NH_2 and OH) is commonly utilized to provide an amphipathic character to CS, but it can be further utilized to refine the hydrophobic and hydrophilic character of CS. Carboxylation is one of the key chemical modifications used to enhance the properties of chitosan (CS). This process is carried out through two main routes: C_6-hydroxyl oxidation and C_2-amino substitution, both of which convert insoluble chitosan into a soluble form. The N,O-carbonylated CS has been manufactured by oxidation and substitution reactions. Specifically, the oxidation process is done in an aqueous solution utilizing the Tempo-mediated technique (Tempo; 2,2,6,6-tetra-methyl-piperidine-1-oxyl), and for the substitution response, the product is mixed in water having 3-chloro-propionic acid (Fig. 4.11) [52]. The N,O-carboxymethyl CS is manufactured by carboxymethylation of CS in a hydroalcoholic media at 51.0 °C utilizing chloroacetic acid. The N,O-carboxymethyl CS destabilizes marine diesel, indicating fine adsorption at sea-water alkalinity.

The N-benzoyl-O-acetyl-CS is prepared by utilizing selective partial-acylation of CS with benzoyl-chloride and acetic acid under highly-intense ultrasound (Fig. 4.12) [53]. The product gained indicated good foaming stability interrelated to its amphipathic character and revealed an expanded antimicrobial spectrum against *E. coli.*

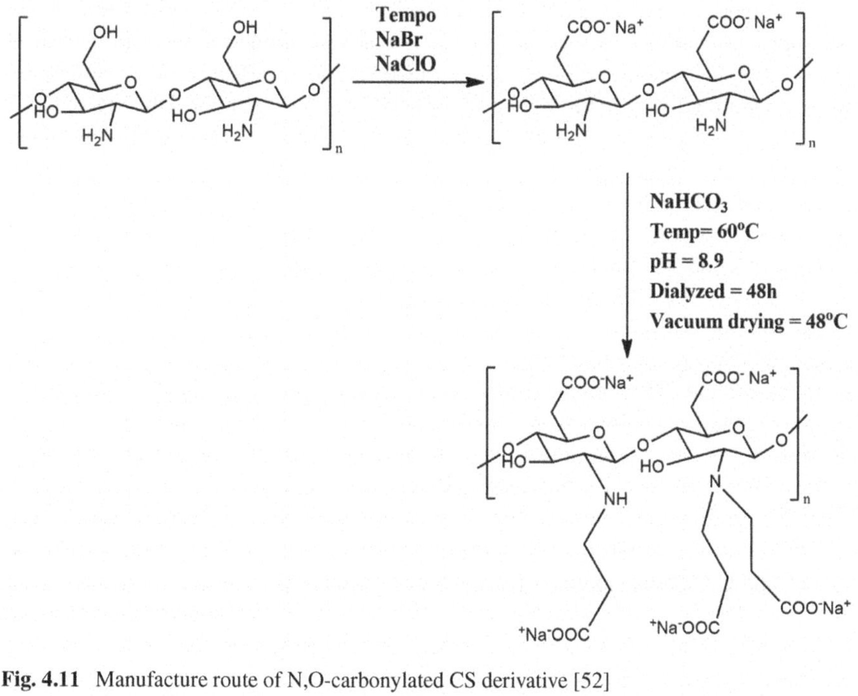

Fig. 4.11 Manufacture route of N,O-carbonylated CS derivative [52]

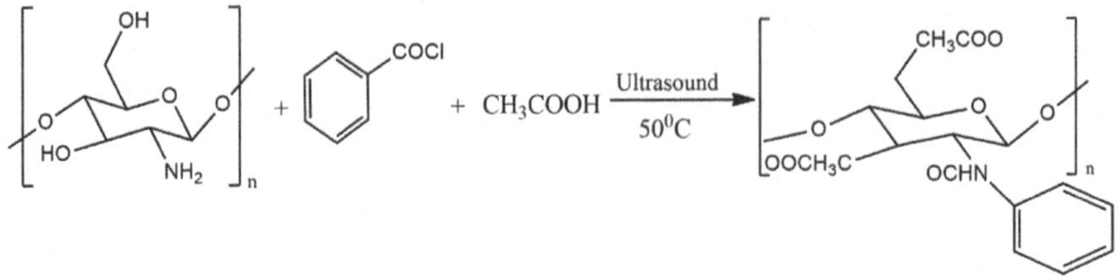

Fig. 4.12 Preparation by ultrasound of N-benzoyl-O-acetyl-CS [53]

4.7 Chitosan Diacetate (CDA) and Chitosan Triacetate (CTA) Nanoparticles (NPs)

Utilization of shellfish industrial waste for isolation, purification, and characterizations of chitin from crustacean's sources has been reported in the literature [54]. Molecular engineering of novel, biodegradable, thermally stable chitin-based polyurethane elastomers (PUEs) [55, 56] and synthesis and characterization of chitin/curcumin blended PUEs have also been well documented [57]. The conversions of chitin into chitosan and subsequent derivatives have been well documented [1–6].

The chitosan diacetate (CDA) and chitosan triacetate (CTA) particles are mostly in nanosize and are approximately 100.0 nm in diameter and spherical in form. The nanosized drug deliverance systems are of particular significance. It enhances particle dispersion in biological fluids, regardless of the administration route, while also improving cellular membrane penetration and accumulation through increased permeability and retention effects. The surface charge of NPs has a very significant impact on their delivery in biological liquids and the zeta-potential. The NPs of CDA and CTA showed a net negative charge on the surface. The mean zeta-potentials of the CDA and CTA-NPs are 28.0 ± 03 mV and 33.1 ± 05 mV, respectively. Though the divergence in zeta-potential amid CDA and CTA is small, the high negative charge in CTA can be demonstrated by the fact that the CTA-based polymer has three functional groups of acetate in comparison with the only two acetate functional groups in the CDA-based polymer. It is anticipated that only a few of the acetate groups participate in cross-linking and numerous others are free, resulting in a negative charge on its surface, which is represented by the zeta-potential. The CTA-NPs also show a higher charge than CDA-NPs, which can be elaborated by the actuality that CTA has 3 groups of acetate (in comparison to only 2 groups in the CDA), which may possibly lead to stronger cross-linking which resulting in an articulate NP matrix.

The novel CDA and CTA-NPs drug deliverance system can effectively encapsulate anticancer drugs, chiefly those with encouraging physicochemical properties, which are soluble in water and ionized into positive charge species. Drug encapsulation in CDA and CTA-NPs enhances their cellular growth by a number of mechanisms. The CTA-NPs improve the oral bioavailability of doxorubicin in rats. Based on these outcomes, the CTA and CDA-NP systems have immense potential as anticancer drug deliverance systems. Table 4.5 shows that encapsulation efficiency process goes hand

Table 4.5 Encapsulation efficiency of CDA-NPs and CTA-NPs [58]

NPs	Form of drug	Drug load (mg/100 mg)	Encapsulation efficacy
CDA-doxorubicin	Water soluble and weak base	28.0	59.0
CTA-doxorubicin	Water soluble and weak base	37.0	81.0
CDA-pemetrexed	Low soluble and weak acid	6.0	19.0
CTA-pemetrexed	Low soluble and weak acid	14.0	31.0
CDA-vinblastine	Less soluble and weak base	28.0	64.0
CTA-vinblastine	Less soluble and weak base	29.0	66.0
CDA-clopidogrel-l	Low soluble and weak base	20.0	60.0
CTA-clopidogrel-l	Low soluble and weak base	25.0	65.0

in hand with the drug load outcomes, where elevated water solubility and the form of the drug (which is being slightly weak base) augmented the effectiveness of NPs. The studies inferred that the value of encapsulation effectiveness for doxorubicin is 81.0% in CTA-NPs and 59.0% in CDA-NPs; it has only 31.0 and 19.0% for pemetrexed in CTA and CDA-NPs, correspondingly. The distinctions amid doxorubicin and pemetrexed, CTA, and CDA-NPs can also be elaborated by the form of the drug (water solubility; weak base against weak acid) and amount of acetate groups, which are available for cross-linking correspondingly [58].

4.8 Conclusion

Owing to the increase in human population, there is a serious need of utilizing bio-based materials. Chitosan is basically a unique cationic polysaccharide, it is obtained from chitin, and it has been extensively studied and is widely utilized in many areas, owing to its biodegradability, biocompatibility, non-toxic, and non-adhesive nature. This chapter explores the modification of chitosan nanoparticles (CSNPs) through both physical and chemical approaches. Physical modification is achieved using the blending technique, while chemical modification involves various methods incorporating inorganic, organic, and polymeric compounds. These modifications enhance the properties and overall efficiency of CSNPs. Hence, CS-based NPs and their derivatives are biodegradable, bio-based, and biocompatible. The CSNPs have property profiles that can be exploited effectively in numerous areas. The presence of $-NH_2$ and –OH groups on its backbone allows CS to be functionalized, so as to be incorporated with the required qualities. The foremost aim of this chapter is to talk about the physical and chemical modifications of CSNPs with various organic, inorganic, and polymeric compounds. The CSNPs' modification has been done in order to enhance its properties. Hence, this modified form of CSNPs has been utilized in various applications owing to their numerous advantages, such as in pharmaceutical field, wastewater treatment, agricultural field, dentistry, cosmetic industries, and also food packaging. The CS triacetate and CS diacetate NPs have also been studied. Both CDA and CTA-NPs have been extensively utilized for anticancer drug deliverance system.

References

1. Mohammed MA, Syeda JTM, Wasan KM, Wasan EK (2017) An overview of chitosan nanoparticles and its application in non-parenteral drug delivery. https://doi.org/10.3390/pharmaceutics9040053
2. Luzi F, Torre L, Kenny JM, Puglia D (2019) Bio- and fossil-based polymeric blends and nanocomposites for packaging: structure-property relationship. https://doi.org/10.3390/ma12030471
3. Park SY, Jun ST, Marsh KS (2001) Physical properties of PVOH/chitosan-blended films cast from different solvents. Food Hydrocolloids 15. https://doi.org/10.1016/S0268-005X(01)00055-8
4. Shukla SK, Mishra AK, Arotiba OA, Mamba BB (2013) Chitosan-based nanomaterials: a state-of-the-art review. https://doi.org/10.1016/j.ijbiomac.2013.04.043
5. Mahaninia MH, Wilson LD (2017) Phosphate uptake studies of cross-linked chitosan bead materials. J Colloid Interface Sci 485. https://doi.org/10.1016/j.jcis.2016.09.031
6. Mittal H, Ray SS, Kaith BS, Bhatia JK, Sukriti, Sharma J, Alhassan SM (2018) Recent progress in the structural modification of chitosan for applications in diversified biomedical fields. https://doi.org/10.1016/j.eurpolymj.2018.10.013
7. Vafakish B, Wilson LD (2019) Surface-modified chitosan: an adsorption study of a "tweezer-like" biopolymer with fluorescein. Surfaces 2. https://doi.org/10.3390/surfaces2030035
8. Tripathi S, Mehrotra GK, Dutta PK (2009) Physicochemical and bioactivity of cross-linked chitosan-PVA film for food packaging applications. Int J Biol Macromol 45. https://doi.org/10.1016/j.ijbiomac.2009.07.006
9. Mi FL, Sung HW, Shyu SS (2001) Release of indomethacin from a novel chitosan microsphere prepared by a naturally occurring crosslinker: examination of crosslinking and polycation-anionic drug interaction. J Appl Polym Sci 81. https://doi.org/10.1002/app.1602
10. Mi FL, Tan YC, Liang HC (2001) In vitro evaluation of a chitosan membrane cross-linked with genipin. J Biomater Sci Polym Ed 12. https://doi.org/10.1163/156856201753113051
11. Kim S, Nimni ME, Yang Z, Han B (2005) Chitosan/gelatin-based films crosslinked by proanthocyanidin. J Biomed Mater Res B Appl Biomater 75. https://doi.org/10.1002/jbm.b.30324
12. Bhatnagar A, Sillanpää M (2009) Applications of chitin- and chitosan-derivatives for the detoxification of water and wastewater—a short review. https://doi.org/10.1016/j.cis.2009.09.003
13. Moradi Dehaghi S, Rahmanifar B, Moradi AM, Azar PA (2014) Removal of permethrin pesticide from water by chitosan-zinc oxide nanoparticles composite as an adsorbent. J Saudi Chem Soc 18. https://doi.org/10.1016/j.jscs.2014.01.004
14. Yanat M, Schroën K (2021) Preparation methods and applications of chitosan nanoparticles; with an outlook toward reinforcement of biodegradable packaging. React Funct Polym 161. https://doi.org/10.1016/j.reactfunctpolym.2021.104849

15. Perez JJ, Francois NJ (2016) Chitosan-starch beads prepared by ionotropic gelation as potential matrices for controlled release of fertilizers. Carbohydr Polym 148. https://doi.org/10.1016/j.carbpol.2016.04.054
16. Wang JJ, Zeng ZW, Xiao RZ, Xie T, Zhou GL, Zhan XR, Wang SL (2011) Recent advances of chitosan nanoparticles as drug carriers. https://doi.org/10.2147/ijn.s17296
17. Bartoş A, Bartoş D, Szabo B, Breazu C, Opincariu I, Mironiuc A, Iancu C (2016) Recent achievements in colorectal cancer diagnostic and therapy by the use of nanoparticles. https://doi.org/10.3109/03602532.2015.1130052
18. Matsuda A, Kobayashi H, Itoh S, Kataoka K, Tanaka J (2005) Immobilization of laminin peptide in molecularly aligned chitosan by covalent bonding. Biomaterials 26. https://doi.org/10.1016/j.biomaterials.2004.07.032
19. Ramasamy P, Shanmugam A (2015) Characterization and wound healing property of collagen-chitosan film from *Sepia kobiensis* (Hoyle, 1885). Int J Biol Macromol 74. https://doi.org/10.1016/j.ijbiomac.2014.11.034
20. Sweeney IR, Miraftab M, Collyer G (2014) Absorbent alginate fibres modified with hydrolysed chitosan for wound care dressings—II. Pilot scale development. Carbohydr Polym 102:920–927
21. Mathew M, Sureshkumar S, Sandhyarani N (2012) Synthesis and characterization of gold–chitosan nanocomposite and application of resultant nanocomposite in sensors. Colloids Surf B Biointerfaces 93:143–147
22. Lindahl U, Höök M (1978) Glycosaminoglycans and their binding to biological macromolecules. https://doi.org/10.1146/annurev.bi.47.070178.002125
23. Chung TW, Lu YF, Wang SS, Lin YS, Chu SH (2002) Growth of human endothelial cells on photochemically grafted Gly–Arg–Gly–Asp (GRGD) chitosans. Biomaterials 23. https://doi.org/10.1016/S0142-9612(02)00231-4
24. She Z, Jin C, Huang Z, Zhang B, Feng Q, Xu Y (2008) Silk fibroin/chitosan scaffold: preparation, characterization, and culture with HepG2 cell. J Mater Sci Mater Med 19. https://doi.org/10.1007/s10856-008-3526-y
25. She Z, Liu W, Feng Q (2009) Self-assembly model, hepatocytes attachment and inflammatory response for silk fibroin/chitosan scaffolds. Biomed Mater 4. https://doi.org/10.1088/1748-6041/4/4/045014
26. Amidi M, Mastrobattista E, Jiskoot W, Hennink WE (2010) Chitosan-based delivery systems for protein therapeutics and antigens. https://doi.org/10.1016/j.addr.2009.11.009
27. Zhang N, Li J, Jiang W, Ren C, Li J, Xin J, Li K (2010) Effective protection and controlled release of insulin by cationic β-cyclodextrin polymers from alginate/chitosan nanoparticles. Int J Pharm 393. https://doi.org/10.1016/j.ijpharm.2010.04.006
28. Kim JH, Kim YS, Park K, Kang E, Lee S, Nam HY, Kim K, Park JH, Chi DY, Park RW, Kim IS, Choi K, Chan Kwon I (2008) Self-assembled glycol chitosan nanoparticles for the sustained and prolonged delivery of antiangiogenic small peptide drugs in cancer therapy. Biomaterials 29. https://doi.org/10.1016/j.biomaterials.2007.12.038
29. Azab AK, Kleinstern J, Doviner V, Orkin B, Srebnik M, Nissan A, Rubinstein A (2007) Prevention of tumor recurrence and distant metastasis formation in a breast cancer mouse model by biodegradable implant of 131I-norcholesterol. J Controlled Release 123. https://doi.org/10.1016/j.jconrel.2007.07.014
30. Chen C, Zhou JL, Han X, Song F, Wang XL, Wang YZ (2014) A prodrug strategy based on chitosan for efficient intracellular anticancer drug delivery. Nanotechnology 25. https://doi.org/10.1088/0957-4484/25/25/255101
31. Siafaka PI, Titopoulou A, Koukaras EN, Kostoglou M, Koutris E, Karavas E, Bikiaris DN (2015) Chitosan derivatives as effective nanocarriers for ocular release of timolol drug. Int J Pharm 495. https://doi.org/10.1016/j.ijpharm.2015.08.100
32. Aranaz I, Acosta N, Civera C, Elorza B, Mingo J, Castro C, Gandía MDLL, Caballero AH (2018) Cosmetics and cosmeceutical applications of chitin, chitosan and their derivatives. https://doi.org/10.3390/polym10020213

33. Wu X, Price GJ, Guy RH (2009) Disposition of nanoparticles and an associated lipophilic permeant following topical application to the skin. Mol Pharm. https://doi.org/10.1021/mp9001188
34. Mondal K, Ghosh T, Katiyar V (2021) Nanotechnology in antimicrobial edible packaging: a candidate for prolong shelf life of food commodities. In: Materials horizons: from nature to nanomaterials. https://doi.org/10.1007/978-981-33-6169-0_12
35. Coma V, Bartkowiak A (2019) Potential of chitosans in the development of edible food packaging. In: Chitin and chitosan: properties and applications, pp 349–369
36. Ju S, Zhang F, Duan J, Jiang J (2020) Characterization of bacterial cellulose composite films incorporated with bulk chitosan and chitosan nanoparticles: a comparative study. Carbohydr Polym 237. https://doi.org/10.1016/j.carbpol.2020.116167
37. Sobahi TRA, Abdelaal MY, Makki MSI (2014) Chemical modification of chitosan for metal ion removal. Arab J Chem 7. https://doi.org/10.1016/j.arabjc.2010.12.011
38. Pokhrel S, Yadav PN (2019) Functionalization of chitosan polymer and their applications. https://doi.org/10.1080/10601325.2019.1581576
39. Yang Y, Xing R, Liu S, Qin Y, Li K, Yu H, Li P (2020) Chitosan, hydroxypropyltrimethyl ammonium chloride chitosan and sulfated chitosan nanoparticles as adjuvants for inactivated Newcastle disease vaccine. Carbohydr Polym 229. https://doi.org/10.1016/j.carbpol.2019.115423
40. Pires NR, Cunha PLR, Maciel JS, Angelim AL, Melo VMM, De Paula RCM, Feitosa JPA (2013) Sulfated chitosan as tear substitute with no antimicrobial activity. Carbohydr Polym 91. https://doi.org/10.1016/j.carbpol.2012.08.011
41. Zhao D, Xu J, Wang L, Du J, Dong K, Wang C, Liu X (2012) Study of two chitosan derivatives phosphorylated at hydroxyl or amino groups for application as flocculants. J Appl Polym Sci 125:E299–E305
42. Chen Y, Li J, Li Q, Shen Y, Ge Z, Zhang W, Chen S (2016) Enhanced water-solubility, antibacterial activity and biocompatibility upon introducing sulfobetaine and quaternary ammonium to chitosan. Carbohydr Polym 143. https://doi.org/10.1016/j.carbpol.2016.01.073
43. Sajomsang W (2010) Synthetic methods and applications of chitosan containing pyridylmethyl moiety and its quaternized derivatives: a review. https://doi.org/10.1016/j.carbpol.2009.12.037
44. Verheul RJ, Amidi M, van Steenbergen MJ, van Riet E, Jiskoot W, Hennink WE (2009) Influence of the degree of acetylation on the enzymatic degradation and in vitro biological properties of trimethylated chitosans. Biomaterials 30. https://doi.org/10.1016/j.biomaterials.2009.03.013
45. Madera-Santana TJ, Herrera-Méndez CH, Rodríguez-Núñez JR (2018) An overview of the chemical modifications of chitosan and their advantages. https://doi.org/10.1680/jgrma.18.00053
46. Kulkarni AD, Patel HM, Surana SJ, Vanjari YH, Belgamwar VS, Pardeshi CV (2017) N,N,N-trimethyl chitosan: an advanced polymer with myriad of opportunities in nanomedicine. https://doi.org/10.1016/j.carbpol.2016.10.041
47. Sahariah P, Árnadóttir B, Másson M (2016) Synthetic strategy for selective N-modified and O-modified PEGylated chitosan derivatives. Eur Polym J 81. https://doi.org/10.1016/j.eurpolymj.2016.05.020
48. Skorik YA, Gomes CAR, Vasconcelos MTSD, Yatluk YG (2003) N-(2-carboxyethyl)chitosans: regioselective synthesis, characterisation and protolytic equilibria. Carbohydr Res 338. https://doi.org/10.1016/S0008-6215(02)00432-9
49. Xie M, Hu B, Wang Y, Zeng X (2014) Grafting of gallic acid onto chitosan enhances antioxidant activities and alters rheological properties of the copolymer. J Agric Food Chem 62. https://doi.org/10.1021/jf503207s
50. Chen C, Tao S, Qiu X, Ren X, Hu S (2013) Long-alkane-chain modified N-phthaloyl chitosan membranes with controlled permeability. Carbohydr Polym 91. https://doi.org/10.1016/j.carbpol.2012.08.042
51. Kurita K, Yoshida Y, Umemura T (2010) Finely selective protections and deprotections of multifunctional chitin and chitosan to synthesize key intermediates for regioselective chemical modifications. Carbohydr Polym 81. https://doi.org/10.1016/j.carbpol.2010.02.042

52. Liu H, Liu X, Yue L, Jiang Q, Xia W (2016) Synthesis, characterization and bioactivities of N,O-carbonylated chitosan. Int J Biol Macromol 91. https://doi.org/10.1016/j.ijbiomac.2016.05.051
53. Cai J, Dang Q, Liu C, Fan B, Yan J, Xu Y, Li J (2015) Preparation and characterization of N-benzoyl-O-acetyl-chitosan. Int J Biol Macromol 77. https://doi.org/10.1016/j.ijbiomac.2015.03.007
54. Ilyas HN, Zia KM, Rehman S, Ilyas R, Sultana S (2021) Utilization of shellfish industrial waste for isolation, purification, and characterizations of chitin from crustacean's sources in Pakistan. J Polym Environ 29:2337–2348
55. Mahmood K, Zia KM, Aftab W, Zuber M, Tabasum S, Noreen A, Zia F (2018) Synthesis and characterization of chitin/curcumin blended polyurethane elastomers. Int J Biol Macromol 113:150–158
56. Zia KM, Barikani M, Bhatti IA, Zuber M, Bhatti HN (2008) Synthesis and characterization of novel, biodegradable, thermally stable chitin-based polyurethane elastomers. J Appl Polym Sci 110(2):769–776
57. Zia KM, Barikani M, Zuber M, Bhatti IA, Sheikh MA (2008) Molecular engineering of chitin based polyurethane elastomers. Carbohydr Polym 74(2):149–158
58. Khdair A, Hamad I, Alkhatib H, Bustanji Y, Mohammad M, Tayem R, Aiedeh K (2016) Modified-chitosan nanoparticles: novel drug delivery systems improve oral bioavailability of doxorubicin. Eur J Pharm Sci 93. https://doi.org/10.1016/j.ejps.2016.07.012

Chapter 5
Regulatory Status and Clinical Trials of Chitosan and Derivatives

5.1 Introduction

Chitosan-based medicinal products must undergo a thorough safety assessment before being approved for commercial use in human medicine. The creation of novel biomedical goods entails a number of factors, including biopolymer characteristics, manufacturing both the quantity and quality, and therapeutic impacts. National and international legislative organizations, including the Food and Drug Administration (FDA) in the USA, must approve the use of chitosan in human trial. Characterization and functionality, product specifications, and analysis using validated methods have all been important regulatory issues in the commercialization of chitosan for biomedical uses. Chitosan items that are being considered for premarketing approval or other uses should be thoroughly examined for biocompatibility. Potential medications and biologics, such as gene products, can be delivered using chitosan as an excipient or a delivery system. The Chitosan has been studied as a potential drug for neuronal disorders, blood pressure control, and body weight loss. In addition, a number of animal studies and clinical trials of ^{166}Ho-chitosan for various cancers are examined.

5.2 Regulatory Status of Chitosan and Derivatives

Chitosan possesses impressive biological characteristics, such as being non-toxic, having mucoadhesive qualities, compatibility with blood, biodegradability, and exhibiting properties like antitumor, antioxidant, and antimicrobial effects. These characteristics give chitosan an extremely appealing natural materials for various biological applications [1] (Fig. 5.1).

K. M. Zia, *Chitosan*, Smart Nanomaterials Technology,
https://doi.org/10.1007/978-981-96-2629-8_5

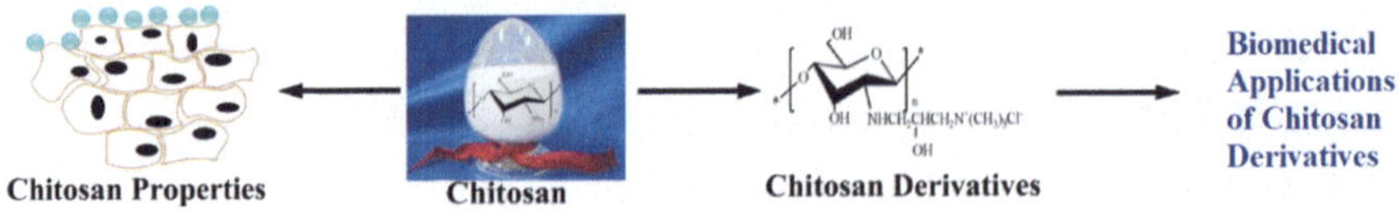

Fig. 5.1 Chitosan and its derivative, properties, and applications [1]

5.2.1 *Regulatory Aspects*

Chitosan has previously been used in biomedical applications, but those usage demanded regulatory permission or approval of the combined or finished item. For these legislative applications, comprehensive and detailed information on chitosan, the production process, and the chitosan material's finished product characteristics is required [2].

The Food and Drug Administration (FDA's) Centre for Devices and Radiological Health (CDRH) reviews and regulates medical equipment's. The CDRH has used the premarket notification (510(k) route) to clear medical devices based on substantial equivalence [2], as well as the premarket approval (PMA) method to clear medical devices grounded on well-being and efficiency studies. The 510(k) procedure may lead to the creation of a fresh premarket marketing plan, like the de novo approach. Class III low- and moderate-risk devices are subjected to the de novo technique after the 510(k) process determined that are not substantially identical with previously approved devices. For the remaining medical devices that don't have a comparable product, producers have sought out products clearance utilizing the PMA procedure. Some devices have been identified as having a significant risk, necessitating examination under the IDE methodology if clinical performance data is needed in the USA. The manufacturer is free from some restrictions in this situation per Section 520(g) of the Drug, Food, and Cosmetics Act and 21 CFR 812.2, permitting them to conduct tests on their goods without an approved or certified commercial application, meaning that the research is not subject to the law's efficacy standards. The regulatory bodies apply Code of Federal Regulations (CFR) where applicable. Humanitarian Device Exemptions (HDEs), which are defined as illnesses or conditions that impact fewer than 4000 persons in the USA each year, have been granted for devices created for orphan or atypical population reasons. An HDE product merely needs to demonstrate that it is secure and offers a foreseeably beneficial benefit, in contrast to PMA goods which must demonstrate relative efficacy. The CDRH is required to acknowledge national and worldwide consensus standards in medical device evaluations for 510(k)s, IDEs, HDEs, and PMAs by the FDA Modernization Act (FDAMA) of 1997 [2].

A standard is referred to as a "recognized consensus standard" if FDA has approved it for use entirely or partially in fulfilling a regulatory requirement. An accessible and broad consensus approach is used to produce a "consensus standard" by a private enterprise standards organization. Even if the FDA–CDRH does not adopt a consensus norm, if it is appropriate, it may be used in the review process. Medical device makers have the choice of conforming to widely accepted consensus standards or handling relevant challenges in a different way [3]. The FDA must perform regulatory examination and monitoring for goods that are subject to premarket assessment.

As a pharmaceutical excipient, chitosan has no regulatory status and may not be sold for use in food or approved drugs unless it can be qualified through FDA approval mechanisms for food components or finished new drug dosage forms. The laws governing the manufacture of healthcare gadgets or combined items containing CDRH lead, as well as regulatory pathways for combination goods, are briefly discussed. Regulatory Status and Relevant Jurisdictional Issues for Chitosan and Derivatives Links to resources on FDA regulatory rules that apply to chitosan item manufacturers have also been provided. The review, clearance, and approval processes for human medical devices differ from those for medications. The FDA–CDRH is in charge of reviewing and regulating medical equipment. Depending on the level of risk, pose, medical devices are categorized as classes (I, II, and III). Medical equipment classified as Class I is the safest, while medical equipment classified as Class III is the most dangerous. General restrictions have been placed on Class I articles (21 CFR 860(c)) (1). In addition to conventional controls, there are additional controls for Class II items under 21 CFR 860(c) (2). The majority of Class I medical devices are free from 510(k) review requirements as long as these don't fulfill the exemptions' requirements. Premarket Notification, often known as the "510(k) Process," is used to approve both non-exempt Class I and Class II medical devices (21 CFR 807 Subpart E). The 510(k) procedure compared a medical device to a predicate device that has either been approved by CDRH for interstate commercial distribution or is a pre-amendment device, meaning that it had been in use prior to May 28, 1976 (the pre-amendment date, 21 CFR 807 and Section 360(k) of Title 21 of the United States Code, or 21 USC 360(k)). A study on the equivalent of product quality, safety, and effectiveness has been undertaken [2] to determine significant equivalent.

The European Union's evaluation and approving procedures for pharmaceutical and bioactive substances are different from those in the USA. Due to variations in medical device categorization, the evaluation and regulating procedures in the USA and the European Union are different. When used in biological uses such matrices for tissue engineering, products for treating wounds, and drug delivery, chitosan and chitosan derivatives are being used increasingly often. Building a substance and product security rating, material characterization, and manufacturing consistency all are necessary for the legal approval of chitosan and its derivatives for use in various applications [2].

5.2.2 Chitosan Food and Drug Administration (FDA) Approval

Pharmaceuticals, biological products, medical equipment, and combination items (such as drug biologic, drug device, and biologic device) are all subject to FDA evaluation and regulation in the USA [2]. The Office of Regulatory Affairs (ORA) and the Office of the Commissioner are just two examples; the National Centre for Toxicological Research (NCTR) (OC), the Centre for Tobacco Products (CTP), and the Centre for Veterinary Medicine (CVM) are all housed inside the FDA. The Code of Federal Regulations (CFR) contains FDA rules. The use of all medical items, including medications, biologicals, technology, hybrid goods, and tissues, is governed by Title 21 of the CFR (21 CFR). The collection of significant regulatory rules that follows is useful for comprehending legislative obligations even if it is not exhaustive: Specifically, 21 CFR sections 200–299, 300–369, and 3.2 apply to combination items (e). Devices are governed by 21 CFR Parts 800–898; biologics by 21 CFR Parts 600–680; human tissues, cells, and products derived from them are governed by 21 CFR Parts 1270–1271; and recalls are governed by 21 CFR Part 7. Good laboratory practices (21 CFR Part 58) and good guiding procedures (21 CFR Part 10) are also necessary for nonclinical laboratory studies [2].

Employees of the FDA who work in regulated industries gain from guideline publications. Businesses may utilize FDA guidance documents to identify what information has to be sent to the organization. The FDA's interpretation of a regulatory issue is presented in guidance papers or combination of issues pertaining to the creation, manufacture, testing, labeling, and marketing of regulated goods, as well as the handling, content, review, or mechanisms for inspection, enforcement, and acceptance of submissions. Guidance papers, which adhere to 21 CFR 10.115's Good Guidance Practices, offer the FDA's present perspective on regulatory issues and practices. Although guidance documents, unlike rules and regulations, cannot be enforced, these are meant to be transparent about the requirements for managing or supplying challenges. Before the document is finalized, industry has the chance to offer feedback on the suggested drafting guiding principles. Sponsors can use alternative strategies that adhere to current rules and regulations while still delivering the same information that the agency and industry value. Federal government organizations, such as the Centre for Drug Evaluation and Research of the FDA, have been advised wherever possible, conduct out regulatory tasks using voluntary consensus standards, whether local or global. Performance characteristics, testing techniques, production methods, item standards, research procedures, compliance standards, ingredient demands, marking, and additional technical or policy requirements are all examples of standard-setting operations.

Regulation of CDRH-lead-containing medical devices and combination products, as well as the use of chitosan and its by-products, are just a few examples of the subjects covered and briefly discussed. Links to materials on FDA regulatory requirements that apply to chitosan product makers have also been discussed. The evaluation, clearance, and approval procedures for human medical devices vary from those for pharmaceuticals. Medical device regulation and evaluation are handled by FDA–CDRH which are divided into three categories (Classes I, II, and III) based on the level of risk it pose. Medical equipment classified as Class I is the safest, while medical equipment classified as Class III is the most harmful. General limits have been placed on Class I items (21 CFR 860(c)) (1). Special controls are in place for Class II items in addition to traditional demands—21 CFR 860(c) (2). Unless these meet the exemptions' conditions, for the vast majority of Class I medical equipment, the 510(k) review procedure is often not necessary. However, as stated in 21 CFR 807 Subpart E, non-exempt Class I and II medical equipment can be approved using the "510(k) procedure," often known as premarket notification. A medical device's 510(k) process compared it to a predicate device that is either a pre-amendment gadget or one that CDRH has approved for interstate commercial distribution, i.e., before May 28, 1976, the date of the pre-amendment, as defined by 21 CFR 807 and Article 360(k) of the 21st section of the U.S. Code (21 USC 360(k)). To assess significant equivalents in product quality, safety, and efficacy, a study of equivalence of product quality, safety, and efficacy is conducted [2].

5.2.3 Regulatory Status: As a Pharmaceutical Excipient

The additives are crucial elements of pharmaceutical formulations, and a variety of substances have a wealth of research trial experience. On the other hand, new excipient development has been slowed, partly due to costs, but also due to regulatory hurdles in proving that these are secure for usage by people [4]. Safety evaluation recommendations for these compounds have been published by the International Pharmaceutical Excipients Council (IPEC), a business group that promotes excipients [5, 6], and a report on excipient testing standards has been published in Japan [7]. The Food and Drug Administration (FDA) of the USA issued nonclinical research guidelines for excipient production [5], the key point being that excipients should be thoroughly studied as potential toxins. Despite the lack of specific recommendations in Europe, the toxicity and pharmacokinetics of a new excipient used in the pharmaceutical industry should be investigated. Furthermore, before being approved for use in Europe, a novel excipient must have undergone nonclinical studies that is appropriate and an evaluation of risk and benefit [5].

Despite widespread interest in chitosan as a pharmaceutical excipient, no commercially available drugs appear to contain it. It is, however, being studied for the use in a variety of formulations for drugs, including drug administration, gene delivery, rapid release forms, colonic drug delivery systems, mucoadhesive dosage forms, and improved peptide delivery [8]. Investigations have been conducted on gels, films, emulsions, wet granulations wet agents, coatings, tiny particles, and microcapsules. Chitosan has been utilized as a drug carrier in oral nanoparticles and as a key component in an intranasal vaccine formulation [9–11].

Although chitosan is not used as a pharmaceutical excipient, it has been shown to cause human exposure when utilized as a food supplement in obesity treatment plans and hypercholesterolemia [12, 13]. Supplements containing chitosan are marketed as fat trappers and magnets [14]. Chitosan-containing medicinal goods, such as the hemostatic Celox for hemorrhage therapy and bandages like HemCon or QuikClot for bleed management, are readily accessible in Europe and the USA [15]. It's also being studied as a coating or substance for contact lenses [12]. Other studies have looked into chitosan's role in healing wounds [16], and Tegasorb, a cut healing product with chitosan has been briefly marketed [12], but it does not appear to be available anymore. It' has also been shown to help with periodontal disease [17]. In the food industry in the USA, chitosan has been regarded as generally recognized as secure (GRAS) [5]. Chitosan is a food ingredient in Japan [18], Italy [12], and Finland [12]. Chitosan is utilized as a basic material in the cosmetics industry [19]. Chitosan is a chelating and flocculants compound that is used in manufacturing, beverage clarification, and crop protection [12].

5.2.4 *Evidence for Possible Carcinogenic Activity*

International agency for research on cancer (IARC) considers there to be strong evidence that both tobacco smoking and eating processed meat can cause cancer, so both are listed as "carcinogenic to humans." But smoking is much more likely to cause cancer than eating processed meat, even though both are in the same category. In the public literature, there is no evidence of chitosan's carcinogenic potential. When rats were given up to 5% of their chow (equivalent to 1935–2244 mg/kg/day) for two years, the chitosan component N-acetyl-glucosamine exhibited no signs of carcinogenicity [20].

5.2.5 *The Twenty-First Century Revival of Chitosan in Service to Bio-Organic Chemistry*

Natural goods served as the source and inspiration for the majority of today's pharmacopoeia, offering a wide and unique supply of bioactive chemicals for medication development [21]. Chitosan is a naturally derived cationic polymer that was discovered in 1811 and revived biologically in the first decade of the twenty-first century. Antibacterial, antifungal, analgesic, and anticancer properties that are both biocompatible and biodegradable properties have all been discovered in recent scientific studies. These traits define its present and potential uses in pharmaceutics, agriculture, and the food sector as well as in human and veterinary medicine [22].

Chitosan is a powerful additive that has great promise for the development of innovative pharmacological dosage forms due to its polyelectrolyte properties, gel-forming abilities, and high adsorption capacity [23]. As controlled release carriers for synthetic pharmaceuticals, proteins, enzymes, and naturally occurring bioactive substances, new chitosan gels, suspensions, microparticles, globes, pills, and drips have recently been studied. The discovery of various natural polyphenols, flavonoids, saponins, tannins, and other bioactive compounds with diverse bioactivities, such as antioxidant, anti-inflammatory, anticancer, antimicrobial, and antidiabetic properties, as well as their applicability and efficacy for therapeutic and pharmacological uses, have been the focus of contemporary scientific research [24]. The main obstacles to their practical application, according to scientific literature, are difficulties in preserving their stability and important biological qualities, as well as their high susceptibility to a range of circumstances and factors (temperature, light, oxygen, pH, enzymes that oxidize, humidity). As a consequence, it has been shown that nano and microencapsulation supports are useful methods for stopping the degradation of naturally occurring bioactive components [25]. Another area of science that has made extensive use of chitosan formulations is innovative cell therapies and technologies. In cell-based regenerative treatment and tissue engineering, chitosan scaffolds in two and three dimensions have been used to support a variety of cell types [26]. Expanding scientific understanding and desires for the distinctive industrial potential of chitosan-based systems encounter a number of challenges that modern science must conquer and address, including a lack of long-lasting strength, constrained solubility, and occasionally insufficient mechanical strength [27].

Biopolymer nano and microformulations are increasingly used in the sectors of biomedicine, bioorganic chemistry, pharmaceutics, agrochemistry, and the food sector. This covers a variety of recently created and enhanced analytical and physicochemical tools for effective and efficient synthesis that is kind to the environment. Yaneva et al. showed major improvements in the processes for creating groundbreaking chitosan micro and nanocarriers and transporters [27]. Particular attention is placed on the capacity of encapsulating naturally occurring bioactive substances and releasing them in a controlled in vitro/in vivo manner under a variety of biological/physiological conditions. Our expectation of their intended applications develops along with our knowledge of the physical, chemical, and biological properties of chitosan formulations and chitosan-based hybrid systems. It provides recommendations for objectively assessing the potential, applicability, and adaptability of newly built hybrid systems based on chitosan.

A thorough integrative approach is used to investigate the bioprecursors and created chitosan-carrier micro/nanohybrid systems, which incorporates cutting-edge scientific discoveries based on intricate, novel, precise, and trustworthy morphological, structural, spectral, chemical, and biochemical analyses [27].

5.3 Clinical Studies on Chitin, Chitosan, and Their Derivatives: Case Studies

5.3.1 *A Chart of the Clinical Studies on Chitosan and Its Derivatives*

Recent developments in clinical research on chitosan and its by-products as therapeutic supplies, as well as the potential of chitosan and its compounds as novel inhibitors of brain degeneration have been comprehensively reviewed (Fig. 5.2). On the clinicaltrials.gov website as of November 8, 2016, 53 clinical studies on chitosan or chitin have been conducted [28]. Most of them (16 out of 53) are concerned with figuring out how chitosan-based dressings may treat wounds and prevent bacterial recolonization, especially in cases of diabetic neuropathic foot ulcers (four cases). In addition, eleven clinical trials focused on hemostasis, including postpartum hemorrhage, dental surgery, and other surgical procedures, because positively charged chitosan attracts platelets and blood cells with negative charges, increasing clot formation. These two groups account for roughly half of all chitosan and derivatives clinical trials (Fig. 5.2). Other clinical trials looked into chitosan's use in the treatment of dry eye syndrome (DES), chronic renal disease, bone disorders, metabolic diseases, and immunological diseases (Fig. 5.2). In addition, only two clinical trials were conducted to see if using a nerve tube in addition to small-incision repair of acute sensory nerve damage affects recuperation and functionality outcomes. Chitosan has been shown to be biocompatible, with positive effects on Schwann cell survival and orientation, neuronal cell survival and differentiation, and the prevention of painful neuromas. Despite extensive studies on the neuroprotective effects of chitosan and its by-products, clinical studies on neuropathy are rare (Fig. 5.2). Chitosan and its metabolites should therefore be studied more in human trials or animal studies as potential treatments for neurodegenerative diseases [28].

5.3.2 *An Overview of the Protective Effects of Chitosan and Acetylated Chitosan Oligosaccharides Against Neuronal Disorders*

5.3.2.1 Potential Applications of Chitosan and Its Derivatives in Alzheimer's Disease

Chitosan oligosaccharides (COSs) are a type of chitosan breakdown product that occurs when chitin, the main component of crustacean exoskeletons, is deacetylated. Chitosan oligosaccharides (COSs) have shown neuroprotective effects due to their demonstrated abilities to counteract amyloid accumulation, inhibit acetylcholinesterase activity, reduce neuroinflammation, and prevent apoptosis in the

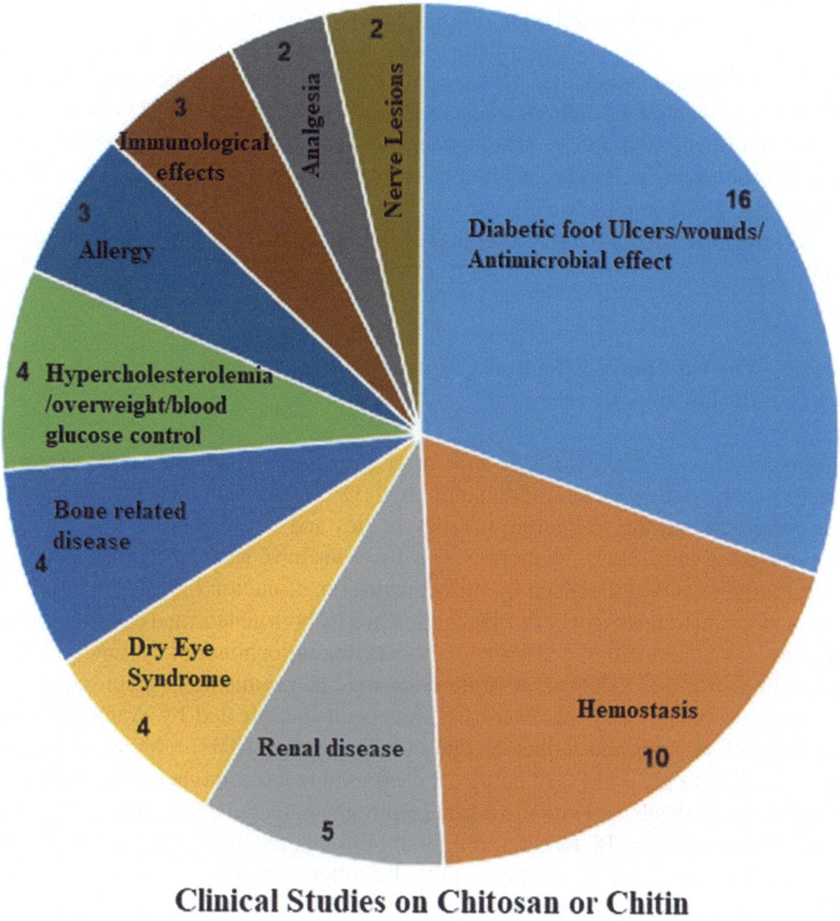

Fig. 5.2 Chart of the clinical studies on chitosan and its derivatives [28]

nervous system [29–31]. Hao and colleagues found that the amount of cell death brought on by glutamate is greatly reduced when PC12 cells were first exposed to peracetylated chitosan oligosaccharides (PACOs) [32]. The amount of PACOs used is directly correlated with this decrease in cell death [32]. Release of lactate dehydrogenase and production of reactive O_2 species and mitochondrial membrane potential loss are all significantly reduced after pre-treatment with PACOs. PACOs also inhibited apoptosis. These substances seem to effectively prevent brain cell death triggered by glutamate by reducing the Bax/Bcl-2 ratio and inhibiting caspase-3 activation [32]. Chitosan oligosaccharides (COS) are reported to improve cognitive functions and lessen learning and memory impairments in rats induced with A1-42 when given orally at dosages of 200, 400, or 800 mg/kg [33]. The COS' neuroprotective properties have been linked to its ability to reduce oxidative stress. The COS has also been shown to reduce inflammation and modulate the inflammatory response by inhibiting the production of pro-inflammatory cytokines [34]. In an A1 42-induced Alzheimer's disease model, the COSs therefore lessen cognitive deficits via reducing oxidative stress and neuroinflammatory reactions. Additionally, the COSs may have fibril-destabilizing and anti-A fibrillogenesis capabilities. Dai et al. [33] revealed that chitosan oligosaccharides (COSs) effectively mitigated the neurotoxic effects induced by A1-42 in rat cortical neurons. Based on these findings, it suggests that COSs could have a potential role in both preventing and treating Alzheimer's disease (AD) [33].

5.3.2.2 The Inhibitory Effects of Chitosan and Its Derivatives Against Parkinson's Disease

Parkinson's disease (PD) is the second most common neurodegenerative disease, which is characterized by loss of dopaminergic (DA) neurons in the substantia nigra pars compacta and the formation of Lewy bodies and Lewy neurites in surviving DA neurons in most cases. Numerous investigations have shown that DA neurons' death is significantly influenced by mitochondrial malfunction [35]. Mitochondrial dysfunction and the emergence of PD are impacted by both genetic and environmental variables associated with the disease [36]. Targeting mitochondrial dysfunction as a treatment goal in neurodegenerative illnesses may be promising as a consequence. In a dose-dependent way, chitosan (CS) has been investigated by Wang et al. to increase cell survival and reduce lactate dehydrogenase (LDH) release caused by dibutyltin (DBT) [37]. In PC12 cells, CS decreased reactive oxygen species (ROS) production, cell death, and mitochondrial membrane potential disturbance [37]. As a result, CS may prevent PC12 cells from going through the mitochondria-dependent pathway that DBT causes apoptosis [37]. In cultured hippocampus neurons, the COSs have also been shown to be effective in reducing glutamate-induced neurotoxicity [29]. A concentration-dependent mechanism for COS pre-treatment may reduce glutamate-induced neuronal cell death. Ca^{2+} intracellular calcium concentration increases and caspase-3 activation caused by glutamate are both reduced by COSs [29]. By preventing a rise in Ca^{2+}, hippocampal neurons in vitro may be protected by COSs against glutamate-induced cell death in neurons. The effectiveness of chito-oligosaccharides and chitosan in the treatment of Parkinson's disease has also been shown [28].

5.3.2.3 The Inhibition Effects of Chitosan and Its Derivatives Against Huntington's Disease

Huntington's disease (HD), which is thought to be a dopamine-predominant disorder, has no effective treatment [38]. The majority of current HD medications are designed to alleviate the disease's primary symptoms, and the patient's standard of life can be enhanced [39]. Both glutamate neurotoxicity and reactive oxygen species (ROS)-induced neuronal damage, on the other hand, may perform significant pathophysiological functions in the development of the Huntington's disease. As a result, drugs that block glutamate and reactive oxygen species (ROS) neurotoxicity may be used to treat HD. Chito-oligosaccharides protect rats' cortical neurons from Cu(II)-induced neurotoxicity [40]. The preliminary treatment with COSs may greatly lessen the harmful effects of Cu(II) on mouse cortex neurons in a dose-dependent manner. Cu(II)-induced increases in intracellular reactive oxygen species are found to be reduced by COSs (ROS). By lowering inside the cell's ROS production, COSs may defend against Cu(II) induced neurodegeneration [40]. The COSs could thus be used to treat HD by reducing glutamate neurotoxicity and generation of reactive O_2 compounds (ROS) in neurons [28].

5.3.2.4 The Inhibitory Effects of Chitosan and Its Derivatives Against Other Neuronal Disorders

Chitosan is an effective neuroprotector after damage of spinal cord since it has been shown to seal injured nerve cell membranes [41]. In ex vivo testing, surface administration of chitosan following full dissection or contraction of the guinea-pig's spinal column accelerated the encapsulation of neuronal membranes, and nerve impulse conduction across their length in living creatures are used to repair the spinal cords' somatosensory induced current records [41]. Additionally, when used on spinal cord samples taken from outside the body, chitosan showed a special affinity for damaged regions and acted as an inhibitor of the production of reactive oxygen species (ROS) and the ensuing peroxidation-induced damage to lipid membranes. Therefore, chitosan therapy may be employed as a cutting-edge therapeutic method to prevent catastrophic behavior loss following severe cerebral and spinal cord injuries. Gong and colleagues [42] examined how chito-oligosaccharides affected after peripheral nerve damage, nerve regeneration. It has been reported that the number of myelinated nerve fibers that have been regrown has been enhanced after COS therapy, muscle contraction potentials, the size of a muscle fiber's cross section, and the thickness of the myelin sheaths that has grown back [42]. After the common peroneal nerves of the rabbit are crushed, the COSs aided in peripheral nerve regrowth. As a consequence, the COSs should be further investigated as possible neuroprotective substances to boost injured peripheral nerve regeneration [42]. In a rat sciatic nerve crush damage model, Jiang et al. discovered that COS treatment promoted regeneration of peripheral nerves with the predicted recovery of function, indicating that COS may be developed as a practical protective medication for uses in peripheral nerve healing [43]. Since it has been demonstrated to seal ruptured membranes of nerve cells, chitosan is an effective neuroprotector after spinal cord injuries [41]. Using somatosensory induced potential recordings, Cho et al. [41] found that adding chitosan to the guinea pig spinal cord following full transection or compression helped with the restoration of nerve impulse conduction in vivo and the sealing of the membranes of neurons in external examination [41].

Additionally, chitosan specifically targeted damaged regions in vitro spinal cord specimens, functioning as an inhibitor of peroxidation of lipids in the membrane and the production of reactive O_2 species (ROS). Chitosan therapy might thus be employed as a cutting-edge therapeutic approach to stop catastrophic behavior loss following severe brain and spinal cord injuries. Muscle action potentials, the quantity of the myelinated neurons that have grown again, and the cross-sectional area of muscle fibers, and the nerves' renewed myelin sheath thickness are all improved by COS treatment, according to research by Gong and colleagues into the impact of chito-oligosaccharides on nerve healing after regional nerve damaging [42]. The COSs encouraged peripheral nerve regrowth when the rabbit's common peroneal nerves are cut out. As a result, further study into COSs as potential neuroprotective drugs to encourage damaged peripheral nerve regeneration is needed [42]. In a rat sciatic nerve crush injury model, based on the findings from Jiang et al.'s study, treatment with chitosan oligosaccharides (COS) not only facilitated the regeneration of peripheral nerves but also led to the anticipated improvement in functionality. These results suggest that COS has the potential to be developed into a viable neuroprotective medication for repairing peripheral nerves [43].

5.3.3 Clinical Trial Evaluating the Safety and Efficacy of the Use of Chitosan Derivatives

In order to learn how chitosan-N-acetyl-cysteine (C-NAC) eye drops alter tear film thickness in people with dry eye syndrome (DES), Schmidl et al. [44] undertook a randomized, double-blind clinical research. In Cohort I, a group of 21 people, one eye received C-NAC eye drops and the other received a placebo regular saline solution. The average tear film thickness (TFT) increased significantly after C-NAC treatment in Cohort I, going from 3.9 ± 0.5 μm before administration to 4.8 ± 1.1 μm just 10 min later. The increase is significantly different from placebo over time ($P < 0.0001$) and remained stable until 24 h post-dose. In Cohort II, TFT increased with eye drops once (QD) or twice (BID) instillation, with no significant difference between regimens. In both groups, Ocular Surface Disease Index scores improved, fewer patients presented with corneal damage, and symptoms of ocular discomfort/conjunctival redness are reduced. It has been reported that a single C-NAC instillation significantly increased mean TFT in dry eye syndrome (DES) patients as soon as 10 min after instillation and that the effect lasted for 24 h. Comparable to what was observed after five days of twice-daily instillation, the increase in TFT following a single instillation. Over 60% of patients has a reduction or elimination of their corneal injury. ClinicalTrials.gov has conducted clinical experiment on "CHITOWOUND" to evaluate the efficacy and safety of utilizing chitosan gel in patients with chronic wounds. Updated information can be viewed there on.

5.3.4 *Quantitative Assessment of the Effects of Chitosan Intervention on Blood Pressure Control*

In order to investigate the effects of chitosan, a dietary fiber recognized for lowering dietary fat absorption and assisting in weight and blood lipid management, Huang et al. combined analyses of the existing randomized controlled trials (RCTs). The effect of chitosan on blood pressure is still being studied; thus, the study sought to find out how it would impact systolic blood pressure (SBP) and diastolic blood pressure (DBP) [45]. The researchers used computerized searches to find relevant placebo-controlled randomized controlled trials (RCTs) that involved human subjects in databases like Medline, Cochrane Library, Scopus, and EMBASE. The weighted mean difference (WMD) and related 95% confidence intervals (CIs) are computed using fixed-effect and random-effect models, respectively. As part of the research, the researchers assessed statistical heterogeneity, looked into predetermined subgroups, looked into publication bias, performed sensitivity analyses, and ran meta-regression investigations. In eight trials with ten research groups, a total of 617 people were chosen. The findings showed that neither systolic blood pressure (SBP) nor diastolic blood pressure (DBP) were significantly affected by chitosan, with a weighted mean difference (WMD) of − 1.41 mmHg, 95% confidence interval (CI): − 3.29 to 0.47 and $P = 0.14$, respectively. The consumption of chitosan did, however, result in a significant decrease in diastolic blood pressure (DBP) in both the short-term (less than 12 weeks) and higher-dose (greater than 2.4 g per day) groups, according to subgroup analysis. Additionally, Egger's tests and funnel plot analyses, with *P*-values of 0.36 for SBP and 0.43 for DBP, respectively, showed no evidence of publication bias in the study. This meta-analysis shows that chitosan consumption significantly lowers diastolic blood pressure (DBP), especially when given in greater dosages and over shorter periods of time. It is important to keep in mind, nevertheless, that the investigation did not discover any effects on systolic blood pressure (SBP). Due to the small number of eligible randomized controlled trials (RCTs) included in this meta-analysis, caution should be used when interpreting the results. Therefore, there is a critical need for larger, better-designed RCTs on this topic [45].

5.3.5 *Study of Chitosan for Pharmacologic Manipulation*

Strong relationships exist between enduring lifestyle decisions and cancer risk and prevention. Diet, ongoing stress, smoking, social circumstances, level of physical exercise, and economic status have all been linked to cancer development or prevention. As factors influencing cancer survival rates and how people respond to therapy, these behaviors are coming under more and more investigation. Diet, in particular, can influence the development of cancer through a variety of methods, such as by increasing or decreasing the synthesis of advanced glycation end products

(AGEs). As a result of sugar metabolism and oxidative stress, the reactive metabolites advanced glycation end products (AGEs) are produced. Furthermore, African Americans accumulated more AGE than European Americans (EAs) among Patient-Controlled Analgesia (PCA). Medicines that limit AGE accumulation could be used to treat cancer because AGEs increase cancer survival and proliferation. Chitosan, a poorly absorbed polysaccharide derived from shellfish or mushrooms, binds AGE-modified proteins in aqueous solutions directly. There is no evidence that chitosan can lower AGE levels in the blood or tissue of people who have aging-related diseases. Researchers are now running a Phase 1 chitosan trial with tests in prostate cancer patients receiving androgen deprivation therapy (ADT). Researchers believe that by acting as an absorbent and selectively targeting pre-existing advanced glycation end products (AGEs), chitosan can lower the levels of these compounds in the body by preventing their absorption from the stomach. This intervention resulted in a stunning 80% decrease in tissue AGE levels as measured by skin auto-fluorescence, as well as a considerable reduction in plasma carboxymethyl lysine (CML) AGE adduct levels. These findings offer strong evidence that chitosan may be a secure and effective method of AGE reduction. The four clinical study participants' plasma and stool biofluid samples have been collected, and analysis of these samples is now being done. On the basis of the results, it can be predicted that whether the chitosan intervention has any observable effects on systemic inflammation, intestinal permeability, or the variety of the microbiome because this mechanistic inquiry is still under progress.

One can learn more about the clinical trial conducted by Michael Lilly and his research group, titled Phase 1b/2 Study of Chitosan for the Pharmacologic Control of Advanced Glycation End (AGE) product Levels in Prostate Cancer Patients, on ClinicalTrials.gov.[1]

5.3.6 A Meta-Analysis on Randomized Controlled Clinical Trials Evaluating the Effect of the Dietary Supplement Chitosan on Weight Loss, Lipid Parameters, and Blood Pressure

Chitosan is a dietary supplement that has been studied for its potential to help overweight and obese people lose weight in adjunctive ways, but the results have been mixed. Blood pressure and serum lipid profiles are indicated as potential additional benefits of chitosan supplementation in the published literatures. Research on body

[1] A website and database devoted to clinical research trials and the distribution of its results is called ClinicalTrials.gov. Its main goal is to provide the general public, researchers, and healthcare professionals with thorough information about clinical research endeavours. Remember that not all research listed on this page have had their safety and scientific validity reviewed or endorsed by the U.S. government.

weight, BMI, triglyceride levels, systolic and diastolic blood pressure, total cholesterol, high-density lipoprotein cholesterol, and low-density lipoprotein cholesterol was done by Moraru et al. in overweight and obese people [46]. Up until December 2017, researchers searched MEDLINE and Cochrane for clinical trials that looked at how using chitosan as a dietary supplement affected overweight and obese people. The literature has also reported pertinent publication in a meta-analysis's references list. The 14 randomized control studies (RCTs) were completed to investigate the effects on blood pressure, serum lipids, and body weight. The results indicate that after 52 weeks of administration, chitosan shows a moderate reduction in body weight, with a loss of 1.01 kg (95% confidence interval: 1.67–0.34). Systolic and diastolic blood pressure showed the greatest improvements compared to the placebo group, with improvements of 2.68 mmHg (95% confidence interval: 4.19–1.18) and 2.14 mmHg, respectively (95% confidence interval: 4.14–0.14). Based on a meta-analysis of 14 randomized controlled trials (RCTs), it has been discovered that using chitosan as a dietary supplement can improve blood lipid profiles and cardiovascular markers as well as cause modest weight loss in the short and medium term [46].

5.3.7 *Single-Blind, Placebo-Controlled Randomized Clinical Study of Chitosan for Body Weight Reduction*

Chitosan is a dietary fiber that has been studied in clinical trials for weight loss because of its potential to lower fat absorption; however, the results have been mixed. The study's main goal is to evaluate the efficacy and safety of using chitosan generated from fungi to treat obesity without putting dietary restrictions in place. In a Phase IV, multicenter, single-blind, placebo-controlled clinical research, daily supplements were given to 96 overweight and obese participants for a total of 90 days. These supplements consisted of five identical placebo capsules and five chitosan capsules, each containing 500 mg of chitosan. In a 2:1 ratio, the trial subjects were assigned to receive either chitosan ($n = 64$) or a placebo ($n = 32$). At the 45-day and 90-day points, measures of body weight, body composition, anthropometric parameters, HbA1C levels, and lipid profiles were taken in order to evaluate the success of the intervention. A short form-36 quality-of-life (QoL) questionnaire is used to look into calorie intake patterns and lifestyle changes. Analysis of well-being-related factors is used to assess safety, and the study closely tracked any negative outcomes. After 45 days, the mean weight reduction in the chitosan group was about − 1.78 kg, and after 90 days, it was − 3.10 kg, which was substantially different from the placebo group (p 0.0001). After 90 days, compared to the placebo group, BMI was 10.91 times lower. Along with an improvement in quality of life, there is also a drop in body composition and anthropometric indices. Chitosan has also shown promise in lowering HbA1C levels in people with initial elevated levels (below 6%). The participants' average daily caloric intake in both groups shows that their eating patterns stayed the same. Lipid levels did not change, and any side effects that were noted include mild and unrelated to the medicine used during the trial. Fungal-derived chitosan helped to reduce mean body weight by as much as 3 kg throughout the course of the 90-day research. Significant changes in body composition, anthropometric measures, and HbA1C levels are found, indicating overall advantages for overweight people. Additionally, the quality-of-life rating increased. It's significant that every study participant agreed that the intervention is secure and well-tolerated [47].

5.3.8 *Animal Experiments and Clinical Trials of ^{166}Ho-Chitosan for Various Cancers*

The ^{166}Ho-chitosan is a 400–500 kD MW ^{166}Ho and N-glucosamine combination that chelates metal ions and degrades slowly in vivo. In mice, intraperitoneally administered ^{166}Ho-chitosan is equally bound to the peritoneal wall (94%), with a surface dose of 81 Gy/uCi/cm^2 calculated using Monte Carlo simulation (EGS 4 code). The ^{166}Ho-chitosan is utilized intraperitoneally as an auxiliary in the treatment of ovarian cancer with a diffuse dispersion of malignant cells across the peritoneal surfaces, including the diaphragm. More than 90% of the ^{166}Ho-chitosan is stuck to the peritoneal wall, with 97–99% of its remaining within the peritoneal cavity. A partial response is seen in four out of five ovarian cancer patients, with no major adverse effects. After intracavitary radiation therapy with ^{166}Ho-chitosan, five of the eight cysts reduced in size due to wall weakness, and two of the eight showed growth retardation. This approach is intended for treating large, multiple, inoperable, or chemotherapy resistant primary liver tumors. This solution is used for intra-arterial injection since it gelled with a neutral pH. In primary liver cancer, radioactivity is disseminated across the territory of chosen hepatic artery branches, and partial responses are seen in two cases. In a large solitary liver tumor that is not operable due to its location, direct intratumoral injection of the radioisotope has reported a moderate response. The ^{18}FDG PET is useful tool for tracking radionuclide therapy progress and directing treatment planning. In one case of a large solitary metastatic stomach cancer in the liver, two weeks after intratumoral injection of1 ^{166}Ho-chitosan, an ^{18}FDG PET is performed, which revealed cold defects that matched the distribution of ^{166}Ho-chitosan, and the second injection, is guided by the PET imaging. For the treatment of malignancies, several techniques of administering ^{166}Ho-chitosan might be utilized. Clinical investigations have proven the efficiency of the ^{166}Ho-chitosan complex as a radiopharmaceutical for the trans-arterial radiation therapy of liver cancer. A workable method for practical formulation of the ^{166}Ho-chitosan complex in a hospital radio-pharmacy employing a two-vial kit system has been established. It has been established how to easily prepare the ^{166}Ho-Chitosan combination for use in treating liver cancer [48]. Schematic for the formulation of ^{166}Ho-chitosan complex using a two-vial kit system is presented in Fig. 5.3.

Recent research has highlighted the diverse applications of chitosan-based materials in drug delivery, agriculture, and biomedical science [49]. Chitosan

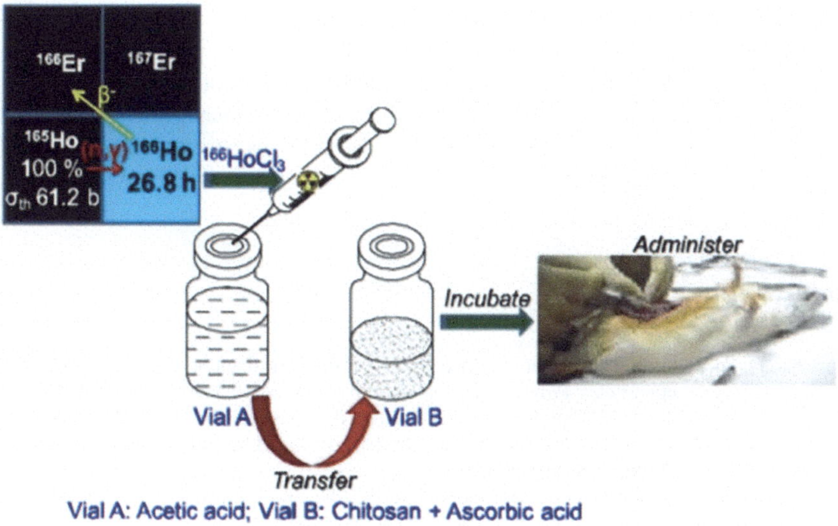

Fig. 5.3 Formulation of ^{166}Ho-chitosan complex using a two-vial kit system [48]

microparticles have been explored for sustainable agriculture by delivering bioactive compounds to plants while also serving as protective carriers for antioxidant compounds, enhancing their stability and bioavailability [50, 51]. In pharmaceuticals, chitosan-based microspheres have been developed for controlled drug release, improving therapeutic efficacy [52]. Biomedical applications include curcumin-conjugated chitosan microspheres for potential disease treatment and chitosan hydrogel capsules with enhanced adsorption capabilities [53]. Additionally, advancements in green extraction techniques have optimized the recovery of valuable plant-based compounds [54]. In cancer research, chitosan has shown promise in regulating advanced glycation endproducts (AGEs), offering potential therapeutic benefits [55].

5.4 Conclusion

A thorough safety analysis of the chitosan substance is essential before human-use medical items using it could be released on the market. The creation of new biomedical goods involves many factors, including the composition of the biopolymer, production methods, quantity, and therapeutic outcomes. Chitosan research involving humans requires authorization from global level and national level regulatory organizations, such as the FDA in the USA. Chitosan's characterization, functionality, product specifications, and application of authorized analytical techniques have all presented considerable chitosan entry into the biomedical industry has significant regulatory obstacles. Chitosan by-products must go through a thorough evaluation of their biocompatibility before being evaluated for pre-launching approval or other applications. As already indicated, chitosan can be used to deliver prospective medications and biologics, such as gene products, either as an excipient or as a carrier. Researchers have looked into chitosan and its derivatives as potential treatments for neurological conditions, blood pressure control, and weight management. The usage of ^{166}Ho-chitosan for various tumor types has also been investigated in a number of animal studies and human trials.

References

1. Zhao D, Yu S, Sun B, Gao S, Guo S, Zhao K (2018) Biomedical applications of chitosan and its derivative nanoparticles. Polymers 10(4):462. https://doi.org/10.3390/polym10040462
2. Dornish M, Kaplan DS, Arepalli SR (2012) Regulatory status of chitosan and derivatives. In: Chitosan-based systems for biopharmaceuticals: delivery, targeting and polymer therapeutics. https://doi.org/10.1002/9781119962977.ch24
3. Lee MH, Whittlesey KJ, Dang JM, Colehour M, Arcidiacono J, Lazarus E, Kaplan DS, Fink D, Durfor CN, Batra A, Hilbert SL, Grayeski DL, McFarland R, Witten C (2013) Overview of the FDA regulatory process. In: Essentials of stem cell biology, 3rd edn. https://doi.org/10.1016/B978-0-12-409503-8.00041-X
4. Katdare A, Chaubal M (2006) Excipient development for pharmaceutical, biotechnology, and drug delivery systems. https://doi.org/10.1201/9781420004137
5. Baldrick P (2010) The safety of chitosan as a pharmaceutical excipient. Regul Toxicol Pharmacol 56. https://doi.org/10.1016/j.yrtph.2009.09.015
6. Steinberg M, Borzelleca JF, Enters EK, Kinoshita FK, Loper A, Mitchell DB, Tamulinas CB, Weiner ML (1996) A new approach to the safety assessment of pharmaceutical excipients. Regul Toxicol Pharmacol 24. https://doi.org/10.1006/rtph.1996.0120
7. Uchiyama M (1999) Regulatory status of excipients in Japan. Ther Innov Regul Sci 33. https://doi.org/10.1177/009286159903300104
8. Rowe et al (2009) Handbook of pharmaceutical excipients, 6th edn. Revue des Nouvelles Technologies de l'Information, E.28
9. Wilson B, Samanta MK, Santhi K, Kumar KPS, Ramasamy M, Suresh B (2010) Chitosan nanoparticles as a new delivery system for the anti-Alzheimer drug tacrine. Nanomed Nanotechnol Biol Med 6. https://doi.org/10.1016/j.nano.2009.04.001
10. Read RC, Naylor SC, Potter CW, Bond J, Jabbal-Gill I, Fisher A, Illum L, Jennings R (2005) Effective nasal influenza vaccine delivery using chitosan. Vaccine 23. https://doi.org/10.1016/j.vaccine.2005.04.021
11. Illum L, Jabbal-Gill I, Hinchcliffe M, Fisher AN, Davis SS (2001) Chitosan as a novel nasal delivery system for vaccines. https://doi.org/10.1016/S0169-409X(01)00171-5
12. Illum L (1998) Chitosan and its use as a pharmaceutical excipient. https://doi.org/10.1023/A:1011929016601
13. Berry L (2007) Martindale: the complete drug reference, 35th edn. Nurs Stand 21. https://doi.org/10.7748/ns2007.01.21.20.31.b571
14. Gades MD, Stern JS (2003) Chitosan supplementation and fecal fat excretion in men. Obes Res 11. https://doi.org/10.1038/oby.2003.97
15. Millner RWJ, Lockhart AS, Bird H, Alexiou C (2009) A new hemostatic agent: initial life-saving experience with celox (chitosan) in cardiothoracic surgery. Ann Thorac Surg 87. https://doi.org/10.1016/j.athoracsur.2008.09.046
16. Kojima K, Okamoto Y, Kojima K, Miyatake K, Fujise H, Shigemasa Y, Minami S (2004) Effects of chitin and chitosan on collagen synthesis in wound healing. J Vet Med Sci 66. https://doi.org/10.1292/jvms.66.1595
17. Boynueğri D, Özcan G, Şenel S, Uç D, Uraz A, Öğüş E, Çakilci B, Karaduman B (2009) Clinical and radiographic evaluations of chitosan gel in periodontal intraosseous defects: a pilot study. J Biomed Mater Res Part B Appl Biomater 90B. https://doi.org/10.1002/jbm.b.31307

18. Tago K, Naito Y, Nagata T, Morimura T, Furuya M, Seki T, Kato H, Ohara N (2007) A ninety-day feeding, subchronic toxicity study of oligo-N-acetylglucosamine in Fischer 344 rats. Food Chem Toxicol 45. https://doi.org/10.1016/j.fct.2006.12.027
19. Minami S, Oh-Oka M, Okamoto Y, Miyatake K, Matsuhashi A, Shigemasa Y, Kukumoto Y (1996) Chitosan-inducing hemorrhagic pneumonia in dogs. Carbohydr Polym 29. https://doi.org/10.1016/0144-8617(95)00157-3
20. Takahashi M, Inoue K, Yoshida M, Morikawa T, Shibutani M, Nishikawa A (2009) Lack of chronic toxicity or carcinogenicity of dietary N-acetylglucosamine in F344 rats. Food Chem Toxicol 47. https://doi.org/10.1016/j.fct.2008.12.002
21. Coşge Şenkal B, Uskutoğlu T, Cesur C, Özavci V, Doğan H (2019) Determination of essential oil components, mineral matter, and heavy metal content of *Salvia virgata* Jacq. Grown in culture conditions. Turk J Agric For 43. https://doi.org/10.3906/tar-1812-84
22. Chien RC, Yen MT, Mau JL (2016) Antimicrobial and antitumor activities of chitosan from shiitake stipes, compared to commercial chitosan from crab shells. Carbohydr Polym 138. https://doi.org/10.1016/j.carbpol.2015.11.061
23. Domínguez-Delgado CL, Fuentes-Prado E, Escobar-Chávez JJ, Vidal-Romero G, Rodríguez-Cruz IM, Díaz-Torres R (2016) Chitosan and pluronic® F-127: pharmaceutical applications. In: Encyclopedia of biomedical polymers and polymeric biomaterials. https://doi.org/10.1081/e-ebpp-120050057
24. Simeonov E, Yaneva Z, Chilev C (2017) Investigation of the mechanism and kinetics of extraction from plant materials
25. Hussain SA, Hameed A, Nazir Y, Naz T, Wu Y, Suleria HAR, Song Y (2018) Microencapsulation and the characterization of polyherbal formulation (PHF) rich in natural polyphenolic compounds. Nutrients 10. https://doi.org/10.3390/nu10070843
26. Intini C, Elviri L, Cabral J, Mros S, Bergonzi C, Bianchera A, Flammini L, Govoni P, Barocelli E, Bettini R, McConnell M (2018) 3D-printed chitosan-based scaffolds: an in vitro study of human skin cell growth and an in-vivo wound healing evaluation in experimental diabetes in rats. Carbohydr Polym 199. https://doi.org/10.1016/j.carbpol.2018.07.057
27. Yaneva Z, Ivanova D, Nikolova N, Tzanova M (2020) The 21st century revival of chitosan in service to bio-organic chemistry. https://doi.org/10.1080/13102818.2020.1731333
28. Hao C, Wang W, Wang S, Zhang L, Guo Y (2017) An overview of the protective effects of chitosan and acetylated chitosan oligosaccharides against neuronal disorders. https://doi.org/10.3390/md15040089
29. Zhou S, Yang Y, Gu X, Ding F (2008) Chitooligosaccharides protect cultured hippocampal neurons against glutamate-induced neurotoxicity. Neurosci Lett 444. https://doi.org/10.1016/j.neulet.2008.08.040
30. Nidheesh T, Salim C, Rajini PS, Suresh PV (2016) Antioxidant and neuroprotective potential of chitooligomers in *Caenorhabditis elegans* exposed to Monocrotophos. Carbohydr Polym 135. https://doi.org/10.1016/j.carbpol.2015.08.055
31. Lee SH, Park JS, Kim SK, Ahn CB, Je JY (2009) Chitooligosaccharides suppress the level of protein expression and acetylcholinesterase activity induced by Aβ25–35 in PC12 cells. Bioorg Med Chem Lett 19. https://doi.org/10.1016/j.bmcl.2008.12.019
32. Hao C, Gao L, Zhang Y, Wang W, Yu G, Guan H, Zhang L, Li C (2015) Acetylated chitosan oligosaccharides act as antagonists against glutamate-induced PC12 cell death via Bcl-2/Bax signal pathway. Mar Drugs 13. https://doi.org/10.3390/md13031267
33. Dai X, Chang P, Zhu Q, Liu W, Sun Y, Zhu S, Jiang Z (2013) Chitosan oligosaccharides protect rat primary hippocampal neurons from oligomeric β-amyloid 1–42-induced neurotoxicity. Neurosci Lett 554. https://doi.org/10.1016/j.neulet.2013.08.046
34. Jia S, Lu Z, Gao Z, An J, Wu X, Li X, Dai X, Zheng Q, Sun Y (2016) Chitosan oligosaccharides alleviate cognitive deficits in an amyloid-β1–42-induced rat model of Alzheimer's disease. Int J Biol Macromol 83. https://doi.org/10.1016/j.ijbiomac.2015.11.011
35. Hu Q, Wang G (2016) Mitochondrial dysfunction in Parkinson's disease. Transl Neurodegener 5. https://doi.org/10.1186/s40035-016-0060-6
36. Migliore L, Coppedè F (2009) Genetics, environmental factors and the emerging role of epigenetics in neurodegenerative diseases. https://doi.org/10.1016/j.mrfmmm.2008.10.011
37. Wang X, Miao J, Yan C, Ge R, Liang T, Liu E, Li Q (2016) Chitosan attenuates dibutyltin-induced apoptosis in PC12 cells through inhibition of the mitochondria-dependent pathway. Carbohydr Polym 151. https://doi.org/10.1016/j.carbpol.2016.06.053

38. Spokes EGS (1980) Neurochemical alterations in Huntington's chorea. Brain 103. https://doi.org/10.1093/brain/103.1.179
39. Handley OJ, Naji JJ, Dunnett SB, Rosser AE (2006) Pharmaceutical, cellular and genetic therapies for Huntington's disease
40. Xu W, Huang HC, Lin CJ, Jiang ZF (2010) Chitooligosaccharides protect rat cortical neurons against copper induced damage by attenuating intracellular level of reactive oxygen species. Bioorg Med Chem Lett 20. https://doi.org/10.1016/j.bmcl.2010.03.105
41. Cho Y, Shi R, Borgens RB (2010) Chitosan produces potent neuroprotection and physiological recovery following traumatic spinal cord injury. J Exp Biol 213. https://doi.org/10.1242/jeb.035162
42. Gong Y, Gong L, Gu X, Ding F (2009) Chitooligosaccharides promote peripheral nerve regeneration in a rabbit common peroneal nerve crush injury model. Microsurgery 29. https://doi.org/10.1002/micr.20686
43. Jiang M, Zhuge X, Yang Y, Gu X, Ding F (2009) The promotion of peripheral nerve regeneration by chitooligosaccharides in the rat nerve crush injury model. Neurosci Lett 454. https://doi.org/10.1016/j.neulet.2009.03.042
44. Schmidl D, Werkmeister R, Kaya S, Unterhuber A, Witkowska KJ, Baumgartner R, Höller S, O'Rourke M, Peterson W, Wolter A, Prinz M, Schmetterer L, Garhöfer G (2017) A controlled, randomized double-blind study to evaluate the safety and efficacy of chitosan-N-acetylcysteine for the treatment of dry eye syndrome. J Ocul Pharmacol Ther 33. https://doi.org/10.1089/jop.2016.0123
45. Huang H, Zou Y, Chi H (2018) Quantitative assessment of the effects of chitosan intervention on blood pressure control. Drug Des Dev Ther 12. https://doi.org/10.2147/DDDT.S148064
46. Moraru C, Mincea MM, Frandes M, Timar B, Ostafe V (2018) A meta-analysis on randomised controlled clinical trials evaluating the effect of the dietary supplement chitosan on weight loss, lipid parameters and blood pressure. https://doi.org/10.3390/MEDICINA54060109
47. Trivedi V, Satia M, Deschamps A, Maquet V, Shah R, Zinzuwadia P, Trivedi J (2016) Single-blind, placebo controlled randomised clinical study of chitosan for body weight reduction. Nutr J 15. https://doi.org/10.1186/s12937-016-0122-8
48. Lohar S, Jadhav S, Chakravarty R, Chakraborty S, Sarma HD, Dash A (2020) A kit based methodology for convenient formulation of 166-Ho-chitosan complex for treatment of liver cancer. Appl Radiat Isot 161:109161. https://doi.org/10.1016/j.apradiso.2020.109161
49. Martin-Saldaña S, Chevalier MT, Iglesias MJ, Colman SL, Casalongué CA, Álvarez VA, Chevalier AA (2018) Salicylic acid loaded chitosan microparticles applied to lettuce seedlings: recycling shrimp fishing industry waste. Carbohydr Polym 200. https://doi.org/10.1016/j.carbpol.2018.08.019
50. Cabral BRP, de Oliveira PM, Gelfuso GM, Quintão TDSC, Chaker JA, Karnikowski MGDO, Gris EF (2018) Improving stability of antioxidant compounds from *Plinia cauliflora* (jabuticaba) fruit peel extract by encapsulation in chitosan microparticles. J Food Eng 238. https://doi.org/10.1016/j.jfoodeng.2018.06.004
51. Sharma M, Choudhury PK, Dev SK (2017) Formulation and in-vitro-in-vivo evaluation of alginate-chitosan microspheres of glipizide by ionic gelation method. Asian J Pharm Clin Res 10. https://doi.org/10.22159/ajpcr.2017.v10i7.18725
52. Saranya TS, Rajan VK, Biswas R, Jayakumar R, Sathianarayanan S (2018) Synthesis, characterisation and biomedical applications of curcumin conjugated chitosan microspheres. Int J Biol Macromol 110. https://doi.org/10.1016/j.ijbiomac.2017.12.044
53. Chatterjee S, Tran HN, Godfred OB, Woo SH (2018) Supersorption capacity of anionic dye by newer chitosan hydrogel capsules via green surfactant exchange method. ACS Sustain Chem Eng 6. https://doi.org/10.1021/acssuschemeng.7b03929
54. Simeonov E, Yaneva Z, Chilev C (2018) Kinetics of green solid-liquid extraction of useful compounds from plant materials: kinetics coefficients and modeling. Green Process Synth 7. https://doi.org/10.1515/gps-2016-0179
55. Brown SG, Sutton B, Biotidara T, Turner D, Wilson R, Lilly M (2020) Abstract D062: mechanistic insights into phase 1b/2 study of chitosan for pharmacologic manipulation of AGE (advanced glycation endproducts) levels in prostate cancer patients. https://doi.org/10.1158/1538-7755.disp19-d062

Chapter 6
Characterization Techniques for the Chitosan Nanoparticles and Derivatives

6.1 Chitosan Nanoparticles and Derivatives

Over the last few years, researchers have dedicated much attention to health and medical technology aiming to get development in this field which has improved living standards, life duration, and economic productivity by providing parameters against various diseases [1]. Nowadays, pharmaceutical, healthcare, agriculture, beauty, food, and nutrition industries widely depend on biopolymers [1–3] produced from starch, cellulose, insulin, alginates, and chitin not only because of their natural origin but also of their non-toxicity, biodegradability and biocompatibility nature [4]. Chitosan (CS) which is a product of the deacetylation of chitin composed of (β-1,4)-N-glucosamine and (β-1,4)-N-acetyl glucosamine units also belongs to a class of biopolymer [5]. Chitosan is heavily suited for chemical modification as compared to its parent chemical compound chitin which lacks free amino and hydroxyl groups located in the backbone of the chitosan [6]. The highly functional backbone makes chitosan (CS) an ideal genus for the encapsulation of particular entities in addition to the protection of the conjugate quantum dots composite materials [7].

The chemical structure of chitosan consists of (1,4) linked 2-amino-2-deoxy β-D-glucan. Chitosan can be procured by the simple alkaline or enzymatic deacetylation of the waste material of the ocean food industry called chitin (Fig. 6.1). Chitosan has come up with many advantages based on which it has been widely used to formulate beads, films, microspheres, and nanoscale particles for application in biomedical and pharmaceutical fields [8].

The Physical and chemical properties of chitosan can be improved by many chemical reactions which will also expand the applicability of chitosan and its derivatives in various practical fields (Fig. 6.2). Diglycidyl ethers of glycols [9], epichlorohydrin [10], and dialdehydes [11] are some examples of cross-linker reagents use for the modification of chitosan. Chitosan has also found it uses on a large scale in biomedical devices such as antimicrobial [12], antioxidant [13], antitumor activity [14], anti-inflammatory [15], protein immobilization [16], and wound dressing [17]. It

K. M. Zia, *Chitosan*, Smart Nanomaterials Technology,
https://doi.org/10.1007/978-981-96-2629-8_6

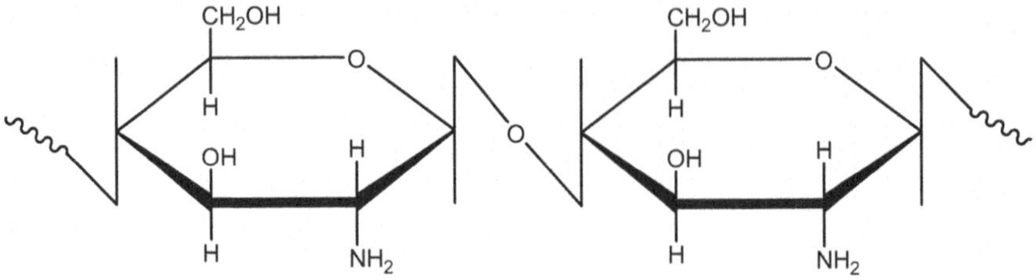

Fig. 6.1 Chemical structure of chitosan showing the repeating subunits

has been further investigated that chemically modified chitosan provides an efficient and viable delivery system that will be able to conduct control delivery of the necessary component to specific cells [18]. There are many polymer-based drug carriers such as polyethylene glycol but these carriers require additional organic components to function systematically [19].

The efficiency of a drug delivery system is linked to its charge and size, chitosan (CS) has cationic nature which facilitates its interaction with the negatively charged cell membrane which supports the drug delivery by a very simple endocytosis mechanism [20, 21]. In addition, the positive charge on chitosan helps to form different complexes with biological molecules such as peptides, proteins, and DNA which have negative charges, this electrostatic interaction results in the formulation of a perfect therapeutic delivery system. Pathological features and target organs are some factors that decide the size of the required nanoparticle, the particles having a size below 200 nm are in demand because it get reach very easily to the target sites [22]. After making conjugation with the small molecules, chitosan-based nanoparticles fall within the required particle size range, so are good candidates for drug delivery systems [23].

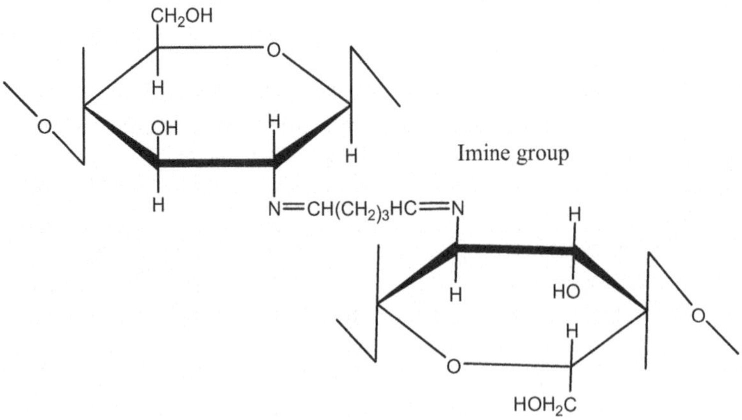

Fig. 6.2 Cross-linked chitosan

Chitosan has become one of the potent candidates for the production of biomaterials [24, 25], and this is due to its remarkable properties including non-allergenicity, non-toxicity, and compatibility that it shows with other materials [26]. Owing to these characteristics, the United States Food and Drug Administration pointed out that chitosan is a safe material for usage in foods, pharmaceuticals, and drug applications [1].

The medical field has shown a great interest in chitosan nanoparticles and their derivatives as it possess unique physiochemical properties due to their small size, high permeability toward biological membranes, biodegradability, cost-effectiveness, wide range of antimicrobial activities, and non-toxicity to humane are some elements that make them preferable to many applications [27].

Moreover, the tight junction present between the epithelial cells can be transiently opened by the adhering property that CS shows toward the mucosal surface. The movement of the macromolecules across the nasal and intestinal barriers is difficult due to their large size and structural properties, CS has been reported to enhance the degree of movement of macromolecules across these blockades [28]. The intrinsic capacity of the positively charged CS that permits it to act as a systematic system for drug delivery can be described by the easiness with which it forms complexes with negative charge carriers like DNA and other proteins [29, 30].

However, there are some limitations such as pH-dependent solubility and precipitation. These limitations cause a hurdle in CS use in the biomedical and pharmaceutical fields. For example, protonation of the primary amino groups of CS plays a key role in its solubilization, it does not get solubilize in a medium that has high pH or is alkaline, and CS tends to solubilize in an aqueous solution having pH 6.5. To be a suitable carrier of DNA and proteins drugs, CS is supposed to not lose its charge when it comes to neutral and physiochemical environments. The CS has been modified in a great number of research works by quaternization of the amino groups to improve its water solubility.

Several research groups have studied and described a quaternized derivative of CS named as N-trimethyl chitosan chloride (TMC) and found it to be improved in characteristics such as absorption enhancing effect [30, 31]. TMC with its improved properties has been found to have a pronounced potential to work in a neutral and physiological environment as an absorption enhancer, where CS is unable to stand and thus can deliver such protein and gene drugs that belong to a hydrophilic class of compounds.

The outstanding properties of the quaternized derivative of CS make them proficient to be applied in the pharmaceutical field for many applications in blood-contact problems like thrombosis, hemolysis, and embolization [32]. Many cationic polymers have adverse effects on the plasma membrane along with their undesirable interaction with the negatively charged cell components, blood corpuscles, and proteinaceous parts and stir hemolysis [29, 33]. There is a need to reduce the interaction between the cationic polymers and anionic components of the cell to a substantial level, and this can be achieved by modifying the CS derivatives with a biocompatible poly(ethylene glycol) (PEG) [34].

The PEG is non-ionic, flexible, biodegradable, and hydrophilic. The PEG-coated nanoparticles are recognized as applicable in therapeutic applications for their great potential in site-specific drug delivery in addition to the controlled release of drugs [32, 35, 36]. The nanoparticle having PEG chains attached to its surface can move rapidly in an aqueous medium as compared to those nanoparticles which have not carried corona formed by PEG chains. Moreover, nanoparticles containing PEG chains show large excluded volumes and upon losing the configurational entropy the bounded PEG chains create steric repulsion [37].

The hydrophilic nature of PEG attracts the water molecules toward itself and the presence of bulk water molecules around it results in the formation of a hydrated outer shell, due to this shell reticuloendothelial system (RES) becomes unable to quickly uptake the nanoparticles inside which assists in extending the half-lives of drugs, improve the drug biodistribution in the target tissues. Some researchers' synthesized PEGylated trimethyl chitosan copolymer by grafting the primary amino groups of trimethyl chitosan by PEG, the resulting copolymer showed low toxicity and high solubility.

6.2 Chitosan-Based Nanomaterials and Their Physicochemical Properties

At neutral and alkaline pH, chitosan shows poor solubility in organic and inorganic solvents. Chitosan shows high solubility due to the fast protonation of the NH_2 of the glucosamine repeating unit at the C2 position in the acidic environment as compared to the neutral vicinity [38]. Adsorption capacities, weak solubility, and stability of the chitosan can be improved in an acidic medium by some effective chemical and physical modifications. To enhance the absorption capacity, certain cross-linkers such as glutaraldehyde, dextran sulfate, and TPP are used [39]. The negatively charged cross-linker will be electrostatically attracted by the positively charged chitosan molecule, and these intermolecular interactions are important for the synthesis of nanoparticles by the ionic gelation process [40]. The stability of nanomaterial is also controlled by the viscosity, which can be adjusted by idealizing the chain length and molecular weight. Therefore, nanomaterial constitutes of longer chain length with high molecular weight chitosan are found to be more stable, while the stability reduces appreciably when chitosan is used to form nanomaterial is of low molecular weight with shorter chain length. However, with an increase in molecular weight, not only stability increases but also the swelling property gets improved [39]. Chitosan-derived nanomaterials exhibit pH-responsive behavior, enabling the development of modulated formulations with enhanced properties. This versatility allows for tailored applications, leveraging the unique characteristics of these materials. Maintenance of temperature is also important for the synthesis of the nanomaterial because the particle size and formation pathways are greatly influenced by the monitoring of the temperature of the reaction mixture. On a general note, a smaller

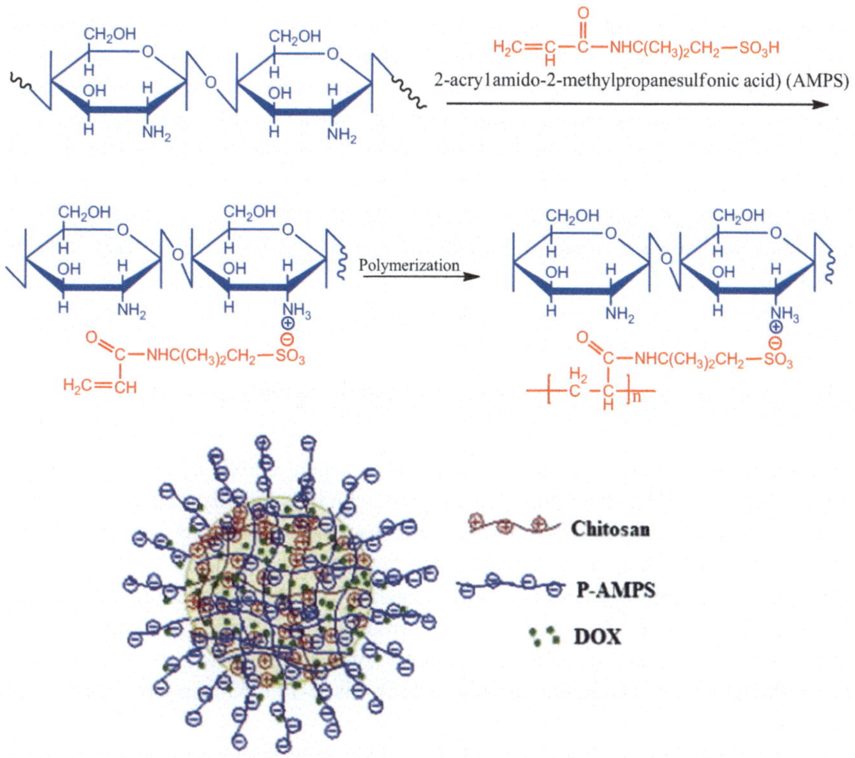

Fig. 6.3 Preparation of chitosan nanoparticles by polyelectrolyte complex method

size and compact-shaped particle can be formed by increasing the flexibility and specific volume of chitosan which in turn increase by increasing the temperature of the medium. On the other hand, when the reaction mixture is kept at a low temperature, stability and particle size increase as more and more chitosan molecules become stiff in no time [41]. Chitosan nanomaterial is quite different from bulk chitosan as it has a smaller size, charged, and high surface area and thus shows improved chemical properties and enhanced biological activities. The reactivity of C-2 amine groups and C-6 hydroxyl groups of chitosan does play an important role in the derivatization process to improve the properties of the end product [39]. Preparation of chitosan nanoparticles by polyelectrolyte complex method has been given (Fig. 6.3).

6.3 Characterization of Chitosan Nanoparticles

The characterization of the nanoparticles is very important not only to get a better understanding of the mechanism by which the chitosan nanoparticles are formed but also to know about the biological properties. In addition, a plethora of reports is available in which chitosan nanoparticles are well characterized to comprehend their physicochemical properties Chitosan nanoparticles are an efficient and active system for the target site delivery of interesting drugs/genes. To get a quint-essential delivery system, there is a need to functionalize chitosan in different ways so that a perfect targeted delivery is achieved. Such challenges require a set of proper characterization techniques to see if up-to-the-mark nanoparticles have been made or not. More essentially, physicochemical characterization including surface morphology, size, shape, and surface charge of nanoparticles is suggested to be analyzed in this regard [40].

6.4 Spectroscopic Analysis (UV–Visible Spectroscopy)

The ionic gelation method is an easy method to prepare CSNPs [41]. It is regarded as the simplest and most compatible method which is heavily dependent on the preparation of CSNPs. In this approach electrostatic interactions are involved in the formation of the nanoparticles; these interactions are originated by the presence of the positive charges on CS chains and negative charges on the cross-linkers like tripolyphosphate (TPP). Nanoparticles are formed by the interaction of chitosan with polyphosphate ions having different diameters which depend on the mutual ratio among them. Structural representation of the chitosan-TPP nanosphere is presented (Fig. 6.4).

Among the various methods available for the characterization of the nanoparticles prepared by the ionic gelation method, ultraviolet–visible spectroscopy is also an important technique. The nanoparticles prepared by the ionic gelation method can be identified by the characteristic peak at 226 nm which is a differential proof of the presence of the amido group in chitosan. Further, in another studies a peak at 201 nm for the chitosan and a peak at 310 nm has been obtained for the CSNPs. Furthermore, the absorption spectrum can also be used to make ensure the stability of the CSNPs. For this purpose, the absorption of the solution measures after 8 weeks, if it results no significant changes in the absorption pattern it supports the fact that there is no agglomeration occurs and the solution is stable.

6.5 Fourier Transform Infrared Spectroscopy (FTIR)

The chemical properties of chitosan-based nanomaterials can be investigated by performing different analyses including identification of groups and interactions which will specify the presence of different functional groups and how it impart unique and desirable properties by making out new interactions with the biomaterial. With the measurements of the types of bonds, the researcher could get an idea to modify the material to bring off the best functionalizing material. Fourier transform infrared (FTIR) spectroscopy is helpful in this regard as it is a two-way chemical

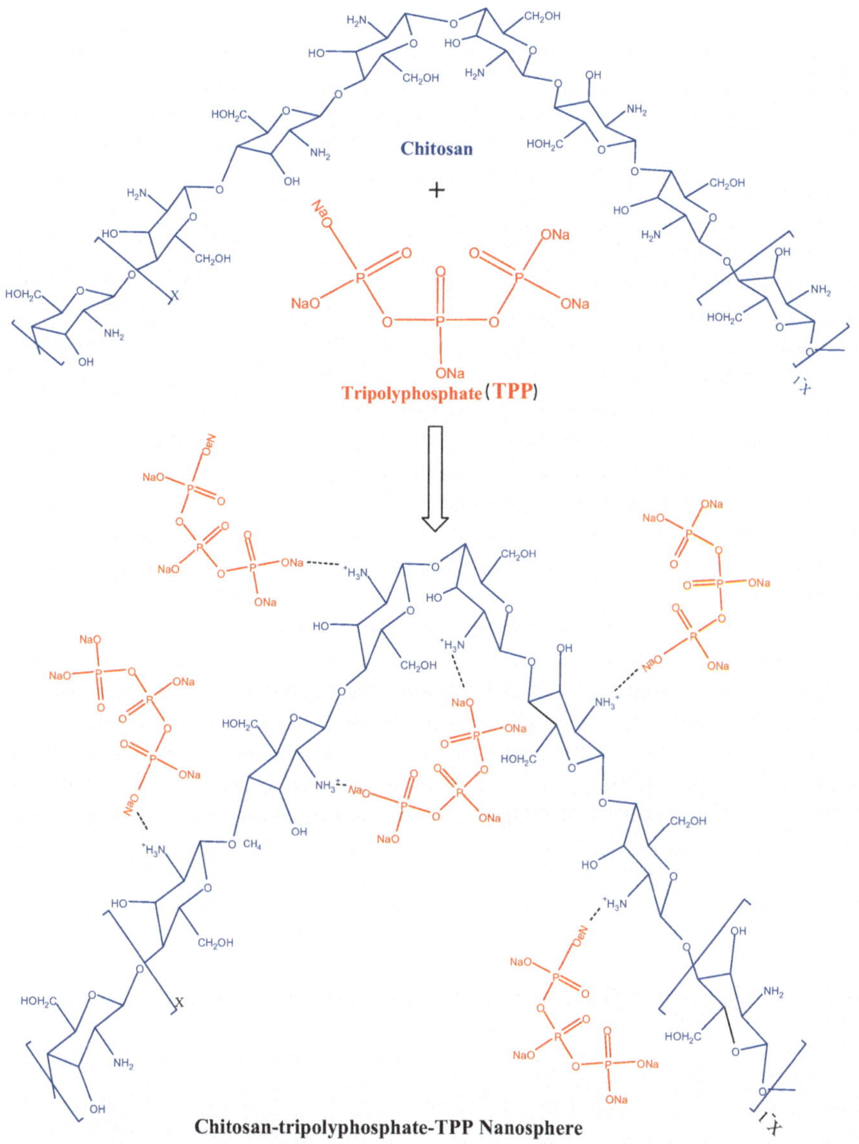

Fig. 6.4 Structural representation of the chitosan-TPP nanosphere [35]

identification by analyzing the functional groups and possible interactions between them.

The compounds of different nature either organic or inorganic in an unknown mixture can be identified in three basic solid, liquid, and gaseous forms. The functionality of the FTIR is successful because most molecules absorb light in the infrared region of the electromagnetic spectrum. During FTIR analysis, frequency is measured as wave numbers that have a typical range of 4000–400 cm^{-1}, the data is recorded in the form of an interference pattern which is later converted into an easily interpretable spectrum [42]. The FTIR spectrum has characteristic spectral lines, typically these lines are narrow and distinct by the virtue of which different bands corresponding to the nature of the bonds present can be identified.

For the preparation of the sample KBr (100:1) is used, after which the sample is compressed by using a hydraulic press machine. Then the sample is forwarded to get scanned against blank KBr background, at 25 °C on absorbance/transmittance mode. The identification peaks of –OH and $–NH_2$ are recognized as a band of 3420 cm^{-1} wavelengths, this band is attributed to the stretching vibration of these groups in chitosan. While, the corresponding to the $–CONH_2$, C–N stretching vibration of type I amine and anhydrous glucosidic bond show at 1649, 1385, and 892 cm^{-1}, respectively. When the same peaks are checked out in Cu-chitosan nanoparticles, these showed sharp in appearance and got shifted toward 3407, 1639, 1316, and 894 cm^{-1}. The shift in wavelength of the concerned peaks is said to be a result of increased interactions. From this pattern of wavelength shifting from higher to lower values, a new leading way emerges to get a conclusion about the formation of the nanoparticles.

In a research work, chitosan/1,2,4-triazole nanoparticles have been produced and the reason for the extreme interest is in achieving the properties that are not usually governed by either ionic cross-linked chitosan–tripolyphosphate (CS-TPP) nanoparticles or 1,2,4-Triazoles alone. The resultant nanoparticles are more effective in terms of having reduced 1,2,4-triazole toxicity and are recommended for long-term applications. The FT-IR spectra is used to analyze the spectral properties of CS, TPP, CS-TPP, CS-TPP-III [4,5-Diphenyl-4H-1,2,4-triazole-3-thiol (III)], CS-TPP-IV [2-(5-mercapto-4-phenyl-4H-1,2,4-triazol-3-yl) phenol (IV)] and CS-TPP-V and 2(5mercapto-4H-1,2,4-triazol-3-yl) phenol (V) nanoparticles (Fig. 6.5).

6.5.1 Chitosan Infrared (CS-IR) Spectrum

In chitosan IR spectrum seven bands observed in the CS spectrum, –OH, $–NH_2$ groups are specified by the band at 3437 cm^{-1}, C–H stretching is confirmed by the band at 2921 cm^{-1}, the band at 1425 cm^{-1} not only defined the stretching of C=O in the amide I but also the stretching of NH in amide II. Further, the presence of the CH_3 in amide group and antisymmetric stretching belonging to the C–O–C bridge are detailed by the bands at 1383 and 1153 cm^{-1}, respectively. The stretching of CO which causes skeletal vibration are attributed to two bands at 1095 and 1010 cm^{-1} [43].

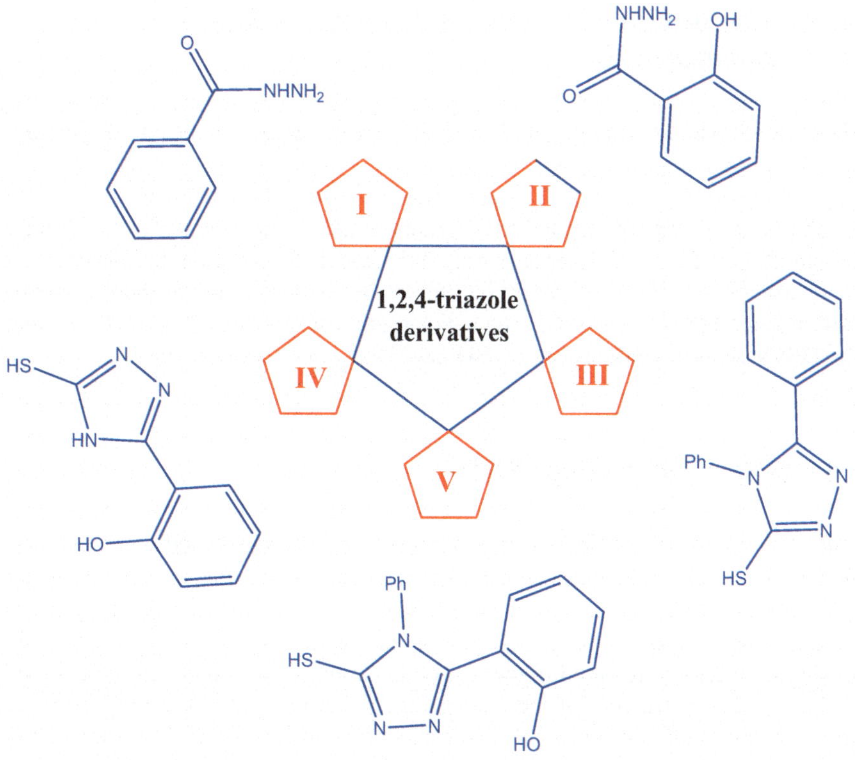

Fig. 6.5 Derivatives of 1,2,4-triazole

6.5.2 *Spectrum of Tripolyphosphate (TPP)*

The band at 1217 cm^{-1} is characteristically assigned to P–O stretching, symmetric and antisymmetric stretching vibrations associated with PO_2 are shown by the bands at 1149 and 898 cm^{-1}, while a single band at 1069 cm^{-1} stated the existence of antisymmetric stretching of the P–O–P Bridge and PO_3 group [44].

6.5.3 *Spectrum of Triazoles (III–V)*

Mono-substituted and di-substituted aromatic rings are identified by the peaks at 660–770 cm^{-1}. The peaks at 2930 cm^{-1} are the pinpoint for the identification of C–H aliphatic, while the C–H aromatic is assigned by the peak that appeared at 3038 cm^{-1}. Similarly, a peak at 3326 cm^{-1} is because of the O–H functional group.

6.5.4 *The Spectrum of Chitosan-Tripolyphosphate (CS-TPP) Nanoparticles*

The absorption bands at 1542, 3447, and 1624 cm^{-1} appeared in the FT-IR spectrum of CS-TTP nanoparticles, and these peaks are linked with the existences of the NH_3^+, OH, and C=O groups. The generation of electrostatic interactions between the phosphoric groups of TPP and NH_3^+ groups of CS are substantiated by the shifting pattern of the OH and C=O groups from 3437 and 1649 cm^{-1} on the formation of nanoparticles [45]. Further, due to the interaction between the phosphoric groups of TPP and NH_3^+ groups of CS, the nanoparticles of CS-TPP-III, CS-TPP-IV and CS-TPP-V showed peaks at 1549, 1543, and 1565 cm^{-1}, respectively.

6.6 X-Rays Diffraction (XRD)

With the increasing demand for high-performance chitosan-based nanomaterials, there is also a need to finalize elemental content and surface composition, and to get to know about the homogeneity of the elemental composition to grade the surface of nanomaterials [46]. Surface properties of nanomaterials play an important role in making the interaction with different materials. In the same context, the elucidation of the molecular structure of the nanomaterials can be made by getting information about the solid-state properties. To get an efficient and accurate functionalizing drug delivery system, the basic structure of nanomaterials is needed to be considered. Understanding the basic properties of the nanomaterial enhances the chances of succor of the extraction method of chitosan nanomaterials as a result improved and appropriate structures containing nanomaterial could be obtained for diverse applications.

X-ray diffraction is an important channel for the determination of solid-state properties. X-ray diffraction (XRD) is a non-destructive technique used to evaluate the structures, phases, crystal orientations, and other structural data of crystalline materials. It is commonly employed for characterizing and determining the crystallinity of synthesized nanoparticles. The basic principle upon which it works is the dual nature of X-rays which helps in studying the structure of crystalline materials. By using Bragg's law ($n\lambda = 2d\sin\theta$), this technique describes the diffraction pattern of different crystalline materials and distinguishes them on the same grounds. A crystalline material can be completely outlined by the knowledge of the size and shape of its unit cell.

There could be different types of atoms and unique arrangements of them in the structure because of which intensities get most often vary. Bruker d8 Advance X-ray Diffractometer (Bruker, USA) configured with a Cu X-ray tube with 1.5418 Å can be used to secure the XRD pattern. Zero diffraction plate (MTI Corporation, USA) is used for the analysis of the sample of Cu-chitosan NPs. An angle range of 10–60° with a step size of 0.02° is used to scan the XRD data while the sample is rotated at a rate of 15 rotations per min and under the 0.4 s per step count time. To process and scan the X-ray diffraction data Eva program is used for evaluation. It is observed in many reports that the peaks with low intensities are characteristic of the chitosan during XRD analysis [47, 48]. Similarly, Cu-chitosan NPs have been identified in XRD diffractogram by crystalline peak intensity ($2\theta = 19.5$ and $21°$) with the help of which it becomes very easy to determine solid-state properties.

6.7 Scanning Electron Microscopy (SEM)

The dispersed and agglomerated chitosan-based nanomaterials can be studied with an important tool called scanning electron microscopy (SEM) which deals with the surface morphology, shape, and functionalization [43]. In SEM, the surface of the nanomaterial is analyzed with a high-energy electron beam. Gold, silver, chromium, or platinum, are used for ultrathin coating (2–20 nm) so that the material can enhance its electrical conductivity and contrast the nanomaterial sample is scanned by a fine beam of a high-energy electron, there occurs an interaction between the sample and the electrons, as a result, electrons are emitted and signals are developed.

Three-dimensional images of nanomaterials are achieved by displaying the collected signals as a pixel on a monitor. There are three types of electron emission from the surface of the sample after the bombardment of electrons on it, low-energy secondary electrons, high-energy backscattered electrons, and X-rays. Composition, the sample is based on, ultimately decides the nature of high-energy backscattered electrons and X-rays scattered and emitted from the lower surface of the sample. The resolution of the SEM can be improved and become advanced under ideal conditions. For example, nanomaterial can be chemically investigated when SEM combines with energy-dispersive X-ray spectroscopy (SEM-EDX).

There is a diary of literature where a large number of reports are available in which SEM analysis is heavily used to completely characterized nanomaterials [49, 50]. The Cu-chitosan nanoparticles when treated by SEM tool, many micro- and nano-sized pores are observed at different magnifications. It has been also reported that chitosan nanoparticles due to their porous surface and uniform sphere are involved in the fair absorption of Cu [43].

The shape as well as the surface morphology of nanoparticles decides how these particles will interact. Scanning electron microscope (SEM) can be employed to report the surface morphology of nanoparticles. Chitosan/DNA nanoparticles investigated by scanning electron microscope (SEM) are found to be spherical in one study, and, toroids, globular and rod-like particles in other studies [51–54]. Modifications of the chitosan backbone along with the complexation conditions are the possible causes of this discrepancy.

In the scanning electron microscope (SEM), a high-energy beam of electrons is used to analyze the surface of the sample. As compared to TEM, in addition to drying the sample, there is also a need to coat it with a thin layer of some metals like platinum or gold before processing. An unmediated picture of the surface of nanoparticles is provided in SEM. The surface morphology is widely investigated by using SEM. Moreover, it also provides information about the size of the nanoparticles. In one study, nanoparticles of chitosan cross-linked with TPP are examined with SEM. These particles which are prepared by ionic gelation method and are also loaded with salicylic acid and gentamicin have been founded to have a spherical shape and 200 nm is the reported average size [55]. The chitosan/siRNA nanoparticles studied by environmental SE microscopy have suggested being in an arrangement of approximately 50 nm concerning the diameter. The preparation of these nanoparticles involved the rapid pipetting of the chitosan solution to get mix it well with the siRNA [56].

6.8 Transmission Electron Microscopy (TEM)

Transmission electron microscopy can be used to confirm the size and surface morphology of nanoparticles when are visualized under it. In TEM, a sample is irradiated by a beam of high-energy electrons as a result of which an image is seen on a fluorescent screen. After passing through the sample, the high-energy electron beam undergoes both transmission and diffraction. These diffracted and transmitted beams interfere with each other and result in a formation of an image. A very high resolution of the order of 2 Å can be achieved by focusing the image on a fluorescent screen, an imaging device, and a charged coupled device (CCD) camera. However, the sample should be flimsy type approximately 500 Å thin, to get preferable results.

The preparation of samples for TEM analysis involves the preparation of dilute suspension and subsequently, its deposition on carbon-coated copper grids or a uranyl acetate which is a negative staining material can be used for this objective characterized the size of fabricated chitosan nanoparticles by TEM and DLS techniques. These glutaraldehydes cross-linked chitosan nanoparticles are achieved by a process taking place in an aqueous core of the reverse micellar droplets. The nanoparticles having an average size of 90 nm within a narrow size distribution are revealed by TEM. The TEM is also used to investigate the chitosan/DNA nanoparticles with an average size of 200 nm. These nanoparticles are formed by the electrostatic interaction between the anionic DNA and cationic chitosan. The nanoparticle formation took place due to electrostatic interactions between anion (DNA) and cation (chitosan), followed by charge neutralization. It has been reported in the literature, chitosan nanoparticles have been investigated to get information about surface morphology. Glutaraldehyde is used to crosslink the chitosan nanoparticles, in the TEM image these particles showed a spherical shape, and the whole sample divulged a uniform distribution of the smooth surface. The methodology by which the sample is prepared consisted of the use of a formvar resin-coated copper grid, on the surface of which the sample is loaded, after which air drying is exercised to get the dry sample. It is pertinent to mention here that formvar refers to any of several thermoplastic resins that are polyvinyl formals, which are polymers formed from polyvinyl alcohol and formaldehyde as copolymers with polyvinyl acetate. They are typically used as coatings, adhesives, and molding materials.

For the coating of the copper grids, 0.5% (w/v) solution of formvar is dropped in chloroform, on the surface of the degassed water. Several copper grids after cleaning are fixed in the thin film that have been formed on the water surface having a matt surface. The researchers made use of a piece of filter paper with forceps and air dried to lift the grids along with the film after 2–3 s of setting up the copper grids. The TEM is also used to study the chitosan/DNA nanoparticles, these particles has a narrow size distribution and are figured out spherically [56]. As compared to SEM, transmission electron microscopy (TEM) because of its higher spatial resolution becomes the most common electron microscopy technique to analyze chitosan-based nanoparticles. Not only size and internal structure of the nanomaterials can be characterized but also it provides a piece of great information about the interactions between different nanomaterials. During TEM analysis, the sample of the nanomaterial interacts with the high-energy beam of accelerated electrons. As a result of this interaction, a cascade of transmitted and reflected particles is emitted from the sample. Magnified images of high resolution are generated due to the detected difference in electron energy [57].

For the TEM analysis, the sample is prepared by coating a carbon film-covered copper grid with the suspension of chitosan-based nanomaterials and the coating must be dried before processing it under TEM. The TEM can resolve down to 0.2 nm under the ideal conditions of processing. Due to the chances of the presence of the aggregation of the nanomaterial with the water vapors during the preparative pathway, there is less certainty in the size measurement by TEM whereas DLS measurement is carried out in aqueous suspension. Chitosan-based nanomaterials showed a spherical shape under the TEM catalogue in many research reports. A comprehensive study showed that Cu-chitosan nanoparticle are spherical with 100–500 nm size under TEM study and the presence of copper is confirmed as darker background. Similarly, another research work done on Cu-chitosan nano materials confirmed the spherical smooth surface with a size range of 100–400 nm [58].

6.8.1 Characterization of CS-TM-PEG Copolymer-Insulin Nanoparticles by TEM

In one of the research works, CS-TM has been formulated through free radical polymerization of chitosan and N-trimethyl aminoethyl methacrylate chloride (TMAEMC). The graft copolymerization of chitosan and TMAEMC has been proposed to proceed by the opening of the vinyl of TMAEMC resulting from the redox action of the $-NH_2$ group of chitosan and APS is used as an initiator. Through one step mechanism quaternized chitosan is prepared. Structural representation of CS-TM-PEG has been presented in Fig. 6.6. To avoid the shortcomings of the first-generation including side reactions and unstable links PEGylation chemistry, the

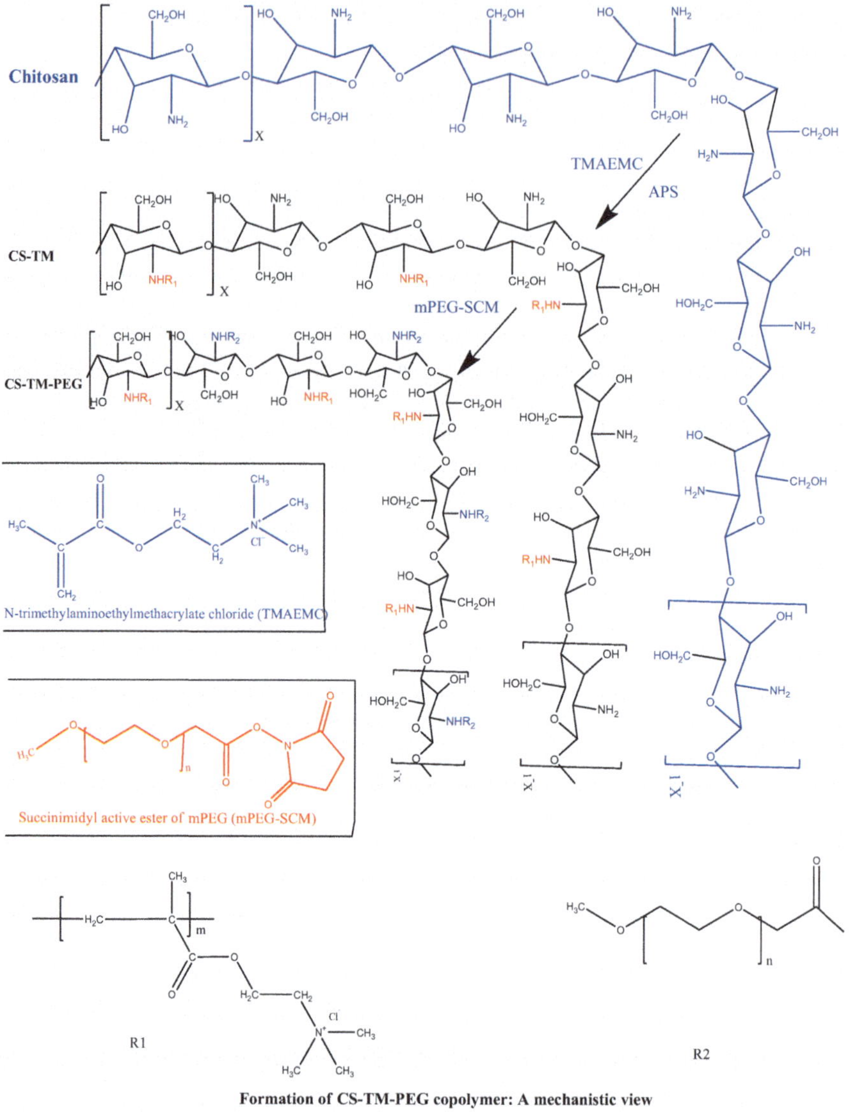

Fig. 6.6 Structural representation of CS-TM-PEG [59]

second-generation PEGylation chemistry is used to design the activation of mPEG. A low molecular weight mPEG causes low toxicity so mPEG of 5000 Da molecular weight is selected for the study. A group of a catalyst such as naphthalene and sodium are used to catalytic conversion of mPEG into carboxylic acid derivatives [59]. The PEG-carboxylic acid is reacted with N-hydroxy succinimide (NHS) to form activated NHS-terminated mPEG (mPEG-SCM). The nucleophile –NH_2 present on CS-TM would have been coupled with electrophilic mPEG and the outcome of this coupling reaction resulted in the formation of CS-TM-PEG graft copolymer.

Electrostatic interaction played a vital role in the formation of the nanoparticles from the negatively charged insulin and passively charged copolymer [60]. When the negative and positive charges on the insulin and copolymer attracted each other, the carriers came close and formed gel-like aggregates in no time, and the pH of the medium is maintained at 7.4 to complete the formation of nanoparticles. The aptness of the spontaneity of gel formation lied in the fact that multivariant polyanion triggered inter- and intramolecular cross-linkages with the CS and CS-TM. At a fixed pH of 7.4, a phosphate buffer solution is chosen to proceed with the ionic gelation reaction between the appropriate added amounts of copolymer and insulin. The same procedure is followed to form the nanoparticles in 0.1 M Tris buffer solution in which 1 mg/ml insulin solution is added in copolymer solution under gentle magnetic stirring and incubation period of 15 min at room temperature [59].

The authors preferred transmission electron microscopy (TEM) to explore the surface morphology and characteristics of both types of CS-TM-PEG copolymer-insulin nanoparticles differed in the method of their preparation one type of which is prepared in PBS and the other one in Tris buffer solution. The information extracted from the transmission electron micrographs (TEM) of both types of nanoparticles is shown in Fig. 6.7. The different shapes and surface morphology are observed due to the nature of the medium in which these particles are formed. The CS-TM-PEG copolymer-insulin nanoparticles which are synthesized in PBS are not close to each other and has spherical shape (left side of Fig. 6.7).

The possible explanation of the compact structure observed could be made on the strength of the unassailable interactivity between the $-N^{+}(CH^{3})^{3}$ of copolymer

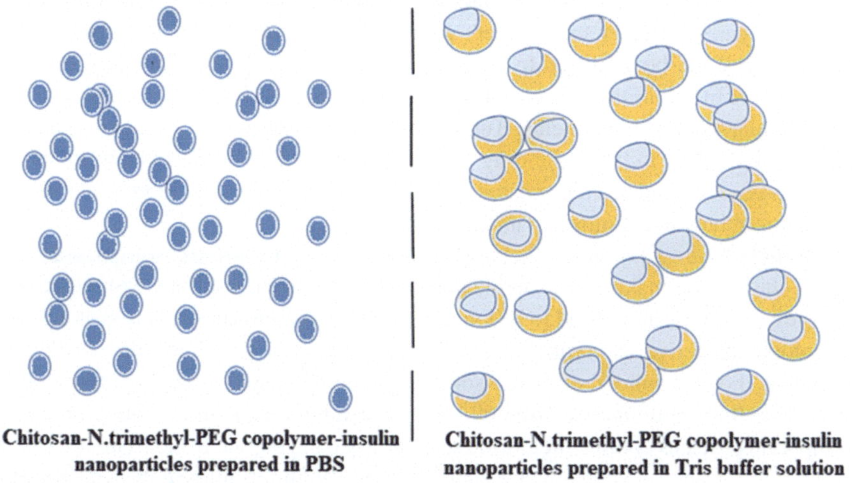

Fig. 6.7 Pictorial representation of chitosan-N.trimethyl-PEG copolymer-insulin nanoparticles prepared in **a** PBS and **b** Tris buffer solution [59]

and PEG which ended in strong inter-chain gelation formation. On the other hand, the nanoparticles formed in Tris buffer solution are shown a specific morphology, these particles are of irregular shape and are examined to have a diffuse and fuzzy coat around themselves (right side of Fig. 6.7). The possible reason for the loose structure's formation is associated with the dominancy of interaction caused by PEG with CS. By linking the outcomes with the previously reported research work, the authors described the morphology of the nanoparticles achieved in Tris buffer solution, according to it, there is evidence of the presence of intermolecular hydrogen bonding that exists between the electro-positive amino hydrogen and the electro-negative oxygen atom whenever PEG and CS derivative react [61]. During the formation of nanoparticles, the interactive quotient of the negative groups of insulin might be overcome by the interactive quotient of the oxygen atom of PEG to make an interaction with $–N^{+}(CH^{3})^{3}$ of the copolymer. Therefore, the likelihood of insulin interaction with the copolymer becomes less in the presence of PEG and has serious consequences on the morphology of the formed particles, a loose structure.

6.9 Nuclear Magnetic Resonance (NMR)

With a deep understanding of chitosan chemistry, drug delivery has become an interesting topic. Chemical modification, competent effects on different tissues, biodegradability, distribution capacity in various organs of the body, mucoadhesion, association ability with inorganic compounds, and advancement in technological transformations are all prominent properties that draw great attention to chitosan. In addition to biocompatibility, another attribute of chitosan is, it does not have an adverse effect when it comes in contact with human cells. In the human body, ubiquitous enzymes function as degraders which can potentially degrade chitosan, and macrophages can be activated by oligomers with the synthesis of hyaluronan into the bargain. In addition, chitosan also acts as a building block to reconstruct the extracellular matrix components. Tumor cells can recognize chitosan and makes it the genial element to bring the drug to the targeted sites. For the human organism chitosan is a safe and affable substance but because of the poor solubility above pH 6.5 chitosan becomes limited in its action [62–65].

Strongly hydrogen bonding network structure restricts various biological activities of chitosan by making it insoluble under physiological conditions. Solubility and working pH range are needed to be knotted to improve chitosan's function as a pharmacologically active agent [66, 67]. Many researchers have prepared chitosan derivatives that showed quite good water solubility who have synthesized carbohydrate-branched chitosan derivatives [68, 69]. The researchers made use of functionalized compounds such as choline dichloride structurally carrying the trimethylammonium group. This group is responsible for yielding highly cationic chitosan. When chitosan reacts with glycidyl trimethylammonium chloride (GTMAC), N-[(2-Hydroxy-3-trimethyl-ammonium) propyl] chitosan chloride (HTCC) is prepared. Polymeric quaternary ammonium compounds are attentively identified as biocidal polymers over the years [66]. Due to its accommodating polycationic nature, chitosan acts as an antimicrobial safe agent against fungi and bacteria [70].

Reaction Mechanism: Chitosan and epoxides undergo nucleophilic substitution reaction which results in the opening of the ring of the epoxide. There are many nucleophilic centers in chitosan including the amino group and hydroxyl groups but the most favorable nucleophilic center is amine group [71]. To investigate the mechanistic way of the reaction. Chitosan is provided with different excesses of GTMAC in the presence of water. In the IR spectrum when these modified chitosan products are analyzed, there appeared two new bonds at one at 1630 and the other ones at 1470 cm^{-1}. The first band is associated with the N–H deformation while the second reported band is assigned to methyl groups of the ammonium [72].

6.10 Dynamic Light Scattering (DLS)

The primacy associated with the chitosan nanoparticles produced by the ionic gelation method lies in the fact that there is zero involvement of the chemical cross-linking in their formation. Due to the preference for reversible physical cross-linking, the reagents formed do not show undesirable effects in addition to toxicity. Although, the ionic gelation process is mild and simple (Fig. 6.8), but TPP/CS microparticles formed by this method are needed to be ameliorated in terms of mechanical strength to make them conductive to drug delivery systems. One of the research works reported ionotropic gelation to produce insulin-loaded CS nanoparticles. The particle size investigated by photon correlation spectroscopy is in the range of 250–400 nm with the insulin association up to 80%. Similarly, CS microparticles prepared by another researcher following the same method showed particle size varied from 500 to 710 μm having more than 90% drug encapsulation efficiencies. It is observed that molecular weight, size distribution, and zeta potential are important to perceive the working efficiency of the CS nanoparticles. Dynamic Light Scattering (DLS) Analysis is used to measure the hydrodynamic diameter of the CSNPs in the nanometer size. In a research study, DLS is used to characterize the chitosan nanoparticles formed by the ionic gelation method [73].

DLS specified nanoparticles by measuring their size by plotting a graph between the intensity (%) and the size (d.nm), by analyzing the size distribution by intensity. The Z-average is found which came up 216 nm and zeta potential is measured from a plot of a zeta potential distribution which is 50 mV. When the size is measured by SEM it ranged from 80 to 100 nm, which is lower than the hydrodynamic diameter detected by the DLS. It is buttressed by the high swelling capacity of CS nanoparticles. To fathom the distinction, it should be mentioned here that these both techniques are different as SEM estimates the projected area diameter whereas DLS is pronounced in measuring the hydrodynamic radius of the particle. During the estimation process by DLS, a very thin electric dipole layer of the solvent tends to get adhere to the surface of the dispersed particles when these are passed through a liquid medium.

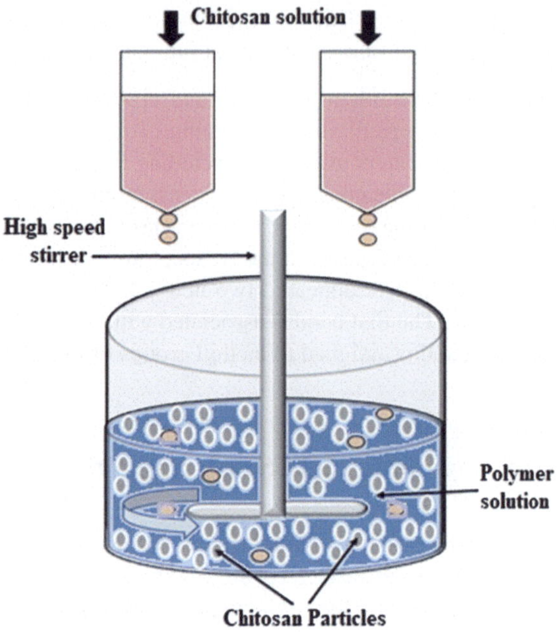

Fig. 6.8 Schematic representation of ionic gelation method

This layer has a consequential effect on the movement of the particle in the design medium. For that reason, hydrodynamic diameter gives information about the inorganic core along with coating material by augmenting the knowledge about the Brownian motion of the particles because of the attached solvent layer with them. By using a statistical approach DLS provides information about the average size (by intensity), and average polydispersity index (PDI), while using mathematical inversion to find out a moderately peak-resolved distribution. While in SEM, both the layers by coating material and solvent are ignored and only the inorganic core is highly focused.

In DLS, the nanoparticles are dispersed in deionized distilled water and the readings become superfluous by the presence of the dust particles in the sample medium as a result of which the greater size of nanoparticles in DLS is analyzed.

Zeta potential on the other hand is evidence of the stability of the nanoparticles in suspension, the high value of the zeta potential 50.3 mV estimated by DLS is construed by the stability particles. Irrespective of this fact, zeta potential is also a tool for measuring in vivo interaction between the nanoparticles and the negatively charged cell membrane of bacteria. It has been reported that, anion adsorption is greatly hindered by the long amino groups which prevented the aggregation by keeping the value of the thickness of the electrical double layer high.

Furthermore, the dynamic light-scattering technique is widely used for the determination of many properties associated with chitosan-based nanomaterials such as surface charge, particle size, and size distribution. This technique is exalted the most because of the accuracy and efficiency grades it has. It is also known as photon correlation spectroscopy (PCS) which measures the size and its distribution throughout that's why it is a well-established and non-invasive technique. The characterization of emulsions, particles, and molecular dispersions (mostly dissolved) substantially come in the prime application zone of DLS. The basis of the functionality of the DLS technique is its capability to measure the Brownian motion at different scattered angles (θ) of the particles dispersed in the suspension.

At molecular level, both the solvent and particles are in a continuous motion and have an appreciable amount of kinetic energy because of which collide with each other. As more solvent molecules are present in the system, it act as bombarding agents on the suspended particles as a result solute particles move randomly with no specific direction as surrounded by a solvent cage around them. This random movement of the suspended particles is referred to as Brownian motion. The measurement that generally involves particles suspended in a liquid medium comes under the concern of DLS. Hence, particle size influences the Brownian motion, as larger-sized suspended particles will move slowly, on the other hand, the particles having small size will have a fast Brownian motion. The measurement involves in DLS uses the Stokes–Einstein equation to calculate the hydrodynamic size after analyzing the velocity of Brownian motion and fluctuations in the intensity.

The hydrodynamic size is a hypothetical size of particles fully surrounded by the solvent molecules. This hypothetical particle has been supposed to have the same diffusional properties as that of the real suspended particle. The hydrodynamic size provides a clue to the apparent size of the particle being measured but in the hydrated/solvated phase. Further, by combining light scattering and other microscopy techniques, a researcher could become enough enable to make a more sophisticated analysis of the size and distribution of nanoparticles [74]. However, suspended nanoparticles due to their enhanced surface properties tend to combine with the other particles and can make aggregates which can reduce the reliability and reproducibility parameters of the DLS, to overcome these unnoticed minor defects, it is compulsory to fully disperse the chitosan nanomaterials. The chitosan nanoparticles when got dispersed in an aqueous medium and have been analyzed by DLS have shown a size range of 40–374 nm with an average of ~ 250 nm. When the light is scattered by different particles in suspension, relative intensity is monitored, based on which intensity distribution is obtained which correlates with the particle size.

While the number proportion of the different types of particles is used to find the number distribution. From the number distribution, a quick conclusion about the particle size can be made, for example, after getting an idea of the number distribution it is found that 91.6% of chitosan nanoparticles has a size lesser than 100 nm, which extended to 43.82 (9.1%), 50.72 (26.8%), 58.77 (30.2%), 68.06 (17.6%), 78.82 (6.3%), and 91.28 nm (1.6%). As the particles in suspension are different in terms of their size and shape, the polydispersity index (PDI) value proves to be an important parameter here.

The polydispersity index (PDI) value gives information about the nature of dispersity of the nanoparticles in the medium under observation, for example, those particles which are monodispersed will have lower PDI values (< 0.5), while the particles having higher PDI value (> 0.5) will show polydisperse distribution. The chitosan nanoparticles generally have a PDI value varying from 0.1 to 0.4, this can be regarded as a lower PDI value (< 0.5) which estimates that there is a uniform distribution of the particles and the suspension is quite stable. The particle surface of nanomaterials has charged groups on it, which causes the generation of the electrostatic repulsion between them that is the point where Zeta potential becomes an important parameter to further characterize the nature of the material.

For the measurement of the Zeta potential, an external electric field is applied, and by using the electrophoretic light scattering, the mobility of particles is determined. To convert the electrophoretic mobility into zeta potential Henry equation is used [75]. The value of zeta potential > 20 mV represents a quite stable colloidal system. When the range of the value of zeta potential for the chitosan nanomaterials has been found, it fell from + 22 to + 88 mV [76]. It is observed that chitosan nanomaterials show a positive zeta potential value, which turns out to be an important characteristic in terms of their applicability where it can show better activity when it will come in contact with the negatively charged biological membranes [77, 78].

6.11 Inductively Coupled Plasma Mass Spectrometry (ICP-MS)

Biocompatibility, non-toxicity, biodegradability, and mucoadhesive are some of the various properties of chitosan nanoparticles (CSNPs) because of which these have been intensively studied and extensively administered in drug delivery systems [79, 80]. Chitosan nanoparticles as a whole are a defendable material because from the core to the surface it has applicability. Different therapeutic agents can be easily encapsulated in the core of the CSNPs [81], and can penetrate the cell membrane of the biological system due to its positively charged surface [76, 82]. Chitosan has also the potential to inhibit the growth of tumor cells in the bladder [83].

6.11.1 Characterization of Gd (TPyP) Conjugated Chitosan Nanoparticles

Chitosan nanoparticles can operate as MRI contrast enhancers when neutral metallated porphyrin Gd (III) meso-tetrakis (4-pyridyl) porphyrin [Gd (TPyP)] are encapsulated in them. In one of the research projects' findings, chitosan nanoparticles are formed by ionic gelation with narrow size distributions are conjugated with Gd (TPyP) through passive method (Fig. 6.9). UV–visible spectroscopy and inductively coupled plasma mass spectrometry (ICP-MS) are used synergistically to probe into the loading of Gd (TPyP) to the nanoparticles (Fig. 6.9).

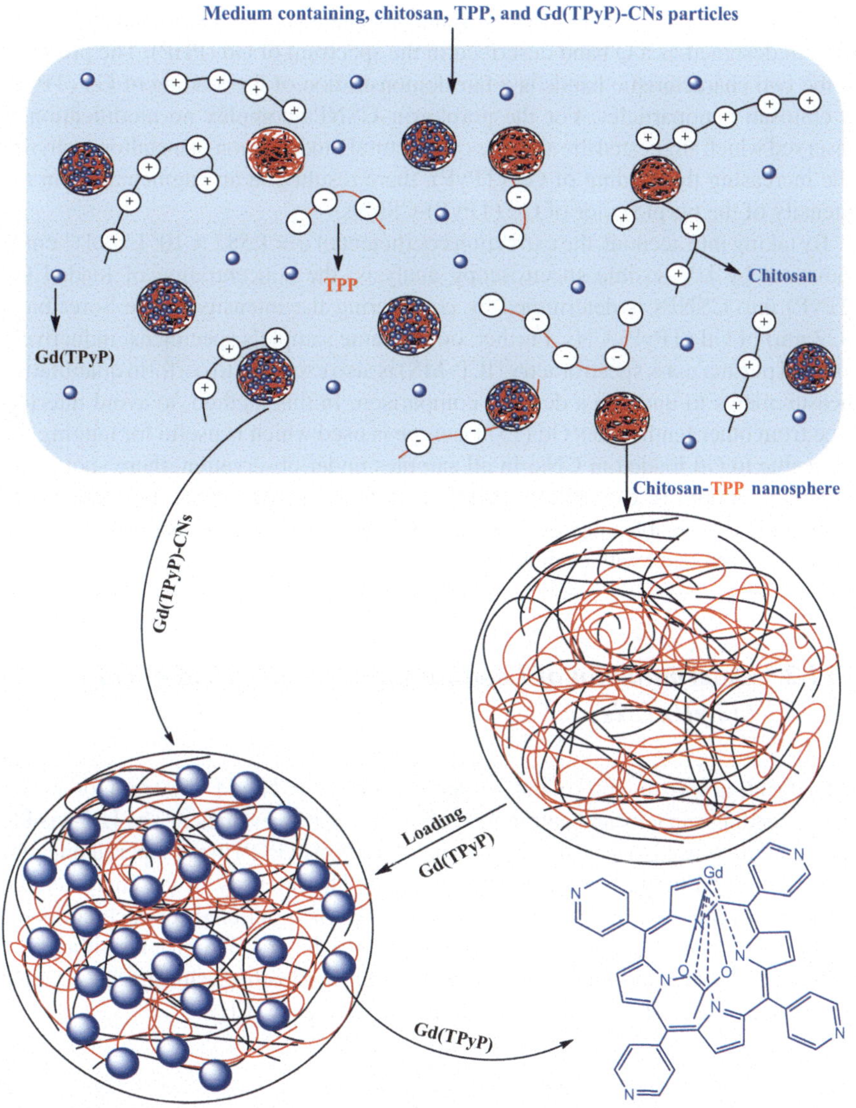

Fig. 6.9 Schematic diagram depicting the formation of chitosan nanoparticles (CSNPs) and Gd (TPyP)-CSNPs particles [83]

It has been reported that chitosan-based nanoparticles are loaded with Gd (TPyP) by exercising a passive absorption process in which copious concentrations of Gd (TPyP) are mixed with CSNPs. The 1 mg of CSNPs loaded with 0.5, 1, 2, and 3 mg of Gd (TPyP) has been tagged as samples 1, 2, 3, and 4. Electronic absorption spectra of the Gd (TPyP)-loaded CSNPs are obtained to compare with the electronic absorption spectra of free Gd (TPyP) in ethanol solvent. An intense band at 422 nm displayed on spectra belonging to Gd (TPyP)-CSNPs designated as the Soret band. A band at 554 nm designed as a Q band described in the spectrum of Gd (PHP). The presence of the two characteristic bands is a fair demonstration of the loading of Gd (TPyP) in chitosan nanoparticles. For the porphyrin–CSNPs complex no modification is observed which suggested the absence of structural modification in metalloporphyrin. The increasing the loading of Gd (TPyP), there resulted in an augmentation in the intensity of the purple color of Gd (TPyP)-CSNPs.

By taking into account, the extinction coefficient of $\varepsilon = 1.582 \times 10^5$ L mol^{-1} cm^{-1} came out by UV–visible spectroscopy analysis, the concentration of loaded Gd (TPyP) into CSNPs is determined by considering the intensity of the Soret band (422 nm) of Gd (TPyP)-CNs. Further, on the same sample's specimens, inductively coupled plasma mass spectrometry (ICP-MS) is also executed to perform quantitative measurements to undergo a detailed comparison. In this method, to avoid interference from other lanthanides Gd (157) isotope is used which is useful for tagging the real value to Gd loaded in CNs. In all samples under observation, there spotted an increased amount of Gd (TPyP) loaded in chitosan nanoparticles. Both analytical techniques UV–visible spectroscopy and ICP-MS gave such measurements which are highly correlated with each other.

6.11.2 Determination of Loading Capacity, the Yield of Gd (TPyP)/CNs

After the ultracentrifugation of the prepared Gd (TPyP)-CNs at 18,500 rpm for 20 min at 4 °C loading capacity and entrapment efficiency can be evaluated. The UV–visible spectrometry is used to measure the amount of free Gd (TPyP) in the supernatant after passing through the ultracentrifugation process. The intensity of the characteristic porphyrin Soret band by UV–visible spectroscopy is monitored for each collected colorless (to the eye) solution to get an insight into the present concentration of Gd (TPyP). Following formulas are used to find out the loading capacity (LC), yield percentage of Gd (TPyP)/CNs, and entrapment efficiency (EE):

$$LC\ (\%) = \frac{(A - B)}{C} \times 100$$

$$EE\ (\%) = \frac{(A - B)}{A} \times 100$$

$$Yield\ (\%) = \left(\frac{actual\ mass}{theoretical\ mass}\right) \times 100$$

where "A" represented the initially added mass of Gd (TPyP), "B" represented the mass of free Gd (TPyP) in the supernatant after centrifuge, and "C" is used as a notation of the "e" mass of nanoparticles. In addition, it is also monitored how the different concentrations of Gd (TPyP) have affected the loading capacity (LC) %, entrapment efficiency (EE) %, and yield percentage of prepared samples. It is found that no change in EE could be considered significant and at a ratio of 1:2 of CNs: Gd (TPyP) a maximum of (87%) is achieved while at a ratio of 1:3 LC and yield (%) achieved an appreciably maximum.

6.12 Zeta Potential

The charge present at the surface of the nanoparticles is reported as zeta potential. The surface potential measures the strength of the repulsive and attractive forces between the particles. An ionic solution is a medium containing positive and negative charges in bulk, any charged particle in it experiences an electrical double layer of ions and counterions. At the hydrodynamic boundary of the charged particle, a potential develops which is regarded as the zeta potential. To determine the zeta potential the velocity of the particle is measured during electrophoresis, for this laser Doppler velocimetry is recommended. The ability of a colloidal system to remain stable is governed by the magnitude of the zeta potential. If the suspended particles have low negative or positive zeta potential values, then there is more probability to form aggregates and the respective system would be unstable. On the other hand, if this same system has high values of zeta potential for its suspended particles, the repulsion will get dominated and these particles would not be able to come close to each other as a result of which there will be less probability of aggregation. In such a manner, zeta potential accounts for the stability of any colloidal system, the particles which have low zeta potential values will mark their good tendency of agglomeration, and vice versa.

In one of the reported studies, Zetasizer, a zeta potential analyzer, has been used to observe the possible effect of pH on the charge pattern present on the surface of the chitosan/DNA nanoparticles [84]. At high pH there observed a low positive charge while by decreasing the pH values the suspension medium bent away to carry a high positive charge. These observations helped to conclude that there has been a noticeable effect of the pH of the medium on the zeta potential. In one of the other studies, zeta potential of the chitosan nanoparticles having a difference in molecular weight but with a fixed N/P ratio has been determined. With increasing the difference in molecular weight, there is an appreciable increase in zeta potential. There is a fair correlation between the molecular weight and zeta potential, zeta potential value turned to + 23 from + 1 as the MW got increased. In the same study, it has been observed that zeta potential tended to decrease from 0 to − 23 mV with an increase of the pH values from 6.9 to 7.5.

6.13 Dark Field Microscopy

The understanding of the interface between living cells and nanomaterials is of great importance because nanomaterials may have toxic effects on tissues [85]. Gold nanoparticles (AuNPs) have proved themselves a set of good nanoprobes which have a versatile role in biomedical applications including cell labeling and imaging [86], cancer diagnostics and therapeutics, and drug delivery [87]. Colloidal gold nanoparticles can be easily prepared by chemical methods, due to this easiness in obtaining these particles are of special interest [88], in addition, their small size (less than 100 nm) has also made them enable them to hold a good position among the other probes that could be manageability imported by cells and tissues [9]. However, these have the potential to cause long-term adverse effects and cytotoxicity when used in vivo delivery or if some preparative conditions are not under control [8].

It is also an essential step to carefully address their surface chemistry before making a conjugation with biological molecules (e.g., proteins and DNA) because most of the properties such as functionality, stability trends, and application are highly dependent on the surface characteristics. It show strong reactivity due to the presence of free electrons on their surface, this factor makes them sensitive to the various environmental fluctuations in which these are probably designed to use such as electrolytes, temperature, pH, and the nature of the solvents. This issue of getting aggregated in different media can be sorted out if ligand-protected AuNPs are synthesized [89]. In this regard, Esumi et al. have used chitosan to stabilize gold nanoparticles [90]. The study conducted by a researcher also demonstrated the process of stabilization of spherical AuNPs by chitosan biopolymer and further investigated the conjugation impact by monitoring the effect on Chinese Hamster Ovary (CHO) cells in vitro. To get an idea about the pattern of the nanoparticles in the cytoplasm dark field imaging and optical scattering measurements both can be used.

6.13.1 Darkfield and Chitosan-Capped Gold Nanoparticles

A Zeiss Axiovert 200 inverted microscope with a hydrogen lamp, a Canon Photo camera, and a condenser under 20× Ph2 magnification is used to perform dark field imaging. A dark field condenser focused the narrow beam of white light emitting from the halogen lamp on the sample, and the scattered light is collected by a 40× objective lens. More specifically, internalization and the particles' location in biological CHO cells incubated with chitosan-conjugated gold nanoparticles are deeply understood by the image obtained by dark field microscopy. Due to the surface plasmon resonance in the visible region, yellow light is strongly scattered by the nanoparticles. From the dark field microscopy image, a guess about the affinity of the particles for the cell is made, as reported, as the nanoprobes has crossed the plasmatic membrane and by endocytosis entered the cell. The presence of chitosan-capped AuNPs inside the CHO cells is additionally supported through the evidence obtained from the light-scattering spectra in dark field microscopy mode.

A single cell is observed to get s a large asymmetric shape, scattering spectrum, there observed two overlapping bands at about 587 and 700 nm. One of the bands with a lower wavelength is assumed to be due to o surface plasmon resonance of individual particles [91]. Particles show an increase in the refractive index when these are inside cells, this factor has been proved when a red shift of the low wavelength band relative to the position of the extinction maximum of chitosan-coated AuNPs in water at 530 nm has been measured by light scattering. A research work depicted the scattering spectra which made a comparison among the different zones of the biological tissue. It became very easy to conclude that no specific signal is detected in case of the zones having no cells or no nanoparticles.

6.14 Aerodynamic Particle Sizer (APS)

An instrument that functions by considering an aerodynamic and optical property such as time-of-fight and light-scattering intensity for counting and size distribution profile of particles is termed as aerodynamic particle sizer (APS). There are two modes on which it is operated including correlated or summing. In the first mode of action, each particle is analyzed to find its aerodynamic diameter and light-scattering intensity simultaneously. During this process, the aerodynamic size of the particle is determined by the quantity of light scattering. In the second mode, aerodynamic sizes, light-scattering sizes, or both are measured [92]. This technique has been used in many research works to find the particle size distribution. One of the researcher developed nanomediated isoniazid (INH)-loaded dry powder for inhalation (Nano-DPI) to deliver macrophage-target in pulmonary tuberculosis (TB). Ionic gelation method is used to prepare hybrid nanosuspension by the conjugation of CD44 receptors of macrophages, hyaluronic acid (HA), and mannosylated chitosan (MC) using cross-linker, sodium tripolyphosphate (TPP) followed by freeze-drying to obtain a dry powder. For the aerodynamic characteristics, an Aerodynamic Particle Sizer (APS-TSI 3321, Shoreview, MN, USA) is used, and time-of-flight measurements in an accelerated flow are called for analyzing the aerodynamic size.

6.15 Scanning Mobility Particle Sizer (SMPS)

Chitosan is a natural polymer with remarkable characteristics including biocompatibility, biodegradability, and non-toxicity, due to which this polymer won a good position among the gene transfection reagents [93]. In addition to this, chitosan as a polymer is widely used as a drug delivery substance [94], as a coating material for nanoparticles [95], and in scaffolds for tissue engineering [96]. Irrespectively, the reticuloendothelial system limits the drug delivery by chitosan to the target tissues because of the cationic properties of the chitosan-based material [97]. From this point of view, it becomes compulsory to mask the cationic surface with the reduction in the opsonization [98], so that modified chitosan substance could find a way to biomedical applications such as intracellular delivery of genetic material, and condensation [99–101].

No doubt, nanoparticle–drug/gene delivery platforms can be easily prepared due to chitosan's ability to show little solubility in a medium set at neutral pH [102], but this ability switches to an obstacle if the same platforms are intended to be synthesized under physiological conditions [103]. When primary amines of chitosan are reacted with succinic anhydride, a new unique, and modified chitosan is achieved named Zwitterionic chitosan (ZCS), one of the characteristics of being solubilized at neutral pH makes them a prominent macromolecular drug carrier [104–106]. By the virtue of this unique pH-dependent charge profile of the modified chitosan, a pH-sensitive coating on the cationic carrier surface can be created [104]. Chitosan-based systems in the form of formulation exist as colloidal liquids that are generally stable only for short periods and are prepared by batch wet chemical processes which themselves a group of time-consuming methods.

Also, liquid suspensions become unstable because in many formulations the carrier/active substances degrade, or insoluble aggregates form, most probably due to the unexpected loss of bioactivity [105, 106]. One approach to overcome these stability limitations is to formulate and store dry powders. Aerosol processing when put in comparison with the classical wet chemical methods, is an easy way because it involves a few numbers of preparation steps. This method generates low wastage by allowing direct collection of powder due to the continuous production of the material [107]. By atomization of liquid solutions in an appropriate carrier followed by solvent removal, gas dry particles in the aerosol state can be formulated [108]. Processing provides an opportunity to get uniform particle-size distributions, with a high percentage of reproducibility, and have a relatively easy-to-scale-up attribute [109]. The study conducted by researchers used a one-step aerosol method to fabricate ZC nanoparticles (ZCNPs) and determined the effects of UV irradiation on the surface charge.

6.15.1 Size Distribution and Scanning Mobility Particle Sizer (SMPS)

In the same study, to get a profile of the size distribution of the aerosol particles measurements are done by a scanning mobility particle sizer (SMPS). It consisted of a condensation particle counter, a soft X-ray charger, a differential mobility analyzer, and an electrostatic classifier. The operating condition of the SMPS system to measure the mobility equivalent diameter included a sample flow of 0.3 L min^{-1}, a scan time of 135 s (measurement range: 15.1–661.2 nm), and a sheath flow of 3.0 L min^{-1}. The size distribution measurements of ZCSNPs are investigated by measuring the

Table 6.1 Aerosol size distributions of CSNPs and ZCSNPs [92]

Case	Geometric mean diameter (nm)	Geometric standard deviation (–)	Total number concentration (particles cm^{-1})
CSNPs	164.6	1.71	1.23×10^6
ZCSNPs (An/Am = 0.3)	176.5	1.67	1.06×10^6
UV-ZCSNPs (An/Am = 0.3)	180.4	1.68	1.16×10^6
ZCSNPs (An/Am = 0.7)	189.7	1.69	1.12×10^6
UV-ZCSNPs (An/Am = 0.7)	195.2	1.69	1.22×10^6

geometric mean diameter (GMD), geometric standard deviation (GSD), and total number concentration (TNC) of the CNPs (Table 6.1). From the analogous data for ZCSNPs (An/Am = 0.3) and ZCNPs (An/Am = 0.7), a culmination is formed that ZCNPs are smaller in size than ZCNPs. Additionality, a higher An/Am ratio indicated that aerosol size increased, and UV irradiation slightly increased the size distribution.

6.16 Matrix-Assisted Laser Desorption/Ionization Mass Spectrometry (MALDI-MS)

Magnetite nanoparticles (e.g., iron oxide (Fe_3O_4) MNPs) possess cubic inverse spinel structure and have face-centered cubic (fcc) close packing formed with oxygen. In addition, these particles also contain interstitial tetrahedral sites and octahedral sites which are occupied by Fe cations [110]. Magnetic nanoparticles have shown atypical applications in biomedicine [111], in biotechnology helpful for the purification of protein. Similarly, these particles are also bidding to become useful in other fields including environmental applications [112], catalytical applications [113], medical imaging [114], and have many detection and separation applications [102, 115]. Magnetic nanoparticles and their modification are also used to remove methyl orange (MO), using chitosan/Al_2O_3/magnetic iron oxide nanoparticle [116], Hg(II) [117], antidepressant drugs [118], aflatoxins [119], and endocrine disrupting phenols. Magnetite nanoparticles can also be observed to be helpful in the administration of enrichment and separation of analytes such as endotoxins [120], pathogenic bacteria [121], biothiols [122], glycoproteins [123], phosphopeptides [124], and many others [125–127].

Chitosan (CS) modified magnetic nanoparticles (CS@Fe_3O_4 MNPs) are prepared, characterized, and applied to preconcentrate the surfactants. There are myriad fields where surfactants are widely used such as consumer products, and hospital and household detergents, these are also used for the preparation of nanoparticles, as natural dispersion to emulsify oil in water. Every year surfactants are produced in tons and it is reported that more than one million tons are disposed of annually by the US. Surfactants can react with biomolecules of every kind such as humic acids [128], DNA [129], and the outer membrane of bacterial cells [130]. It has also been highlighted that hospital wastewater consists of such detergents and disinfectants that could cause eco-toxicity very easily [131]. Surfactants can affect the cyto-toxicities of the nanoparticles acting as capping agents, which can ultimately harm blood T lymphocytes (TIB-152) and human skin keratinocytes (HaCaT) [132].

In keeping these facts in view, there is a clear need to determine the presence of these species. Surfactants have been followed by tracing their concentration or the way of distribution by several techniques including methylene blue active substances (MBAS) [133], liquid chromatography-mass spectrometry (LC/MS/MS), capillary electrophoresis [134], and many others [135]. Irrespective of the availability, these all methods involve tedious procedures where a large number of chlorinated solvents are required which is not an ecofriendly step as these solvents are not biodegradable. When it comes to the analysis of non-volatile compounds, matrix-assisted laser desorption/ionization mass spectrometry (MALDI-MS) proves to be a good handful of analytical technical tools [136–139]. A number of large analytes such as bacterial cells [140–142], metals [143, 144], fungi [145], proteins [146, 147], and others [148, 149], have been easily analyzed by this technique.

To process the synthesis of CS@Fe_3O_4-MNPs basic medium is utilized through the hydrothermal method. Matrix-assisted laser desorption/ionization mass spectrometry (MALDI-MS) is used in many research work for the characterization of the nanoparticles. For the preparation of magnetic nanoparticles modified with chitosan (CS@Fe_3O_4-MNPs), the backbone of the chitosan is coordinated with Fe^{+2} and Fe^{+3} ions before co-precipitation in the basic medium (Fig. 6.10).

Transition metal ions can be easily coordinated to the hydroxyl functional group of the chitosan under hypothermic conditions, black magnetic nanoparticles are formed. The magnetic field is applied to separate the free chitosan and, in this way, chitosan-coated nanoparticles can be achieved. In one of the reported research works, in the standard solutions of surfactants CS@Fe_3O_4 magnetic nanoparticles is added in each solution of different surfactants. The external magnet is used to separate magnetic nanoparticles containing surfactants.

The analytic species must be able to absorb laser energy to get analyzed by laser desorption/ionization mass spectrometry (LDI-MS). Surfactants do not show laser absorption and do not apply to LDI-MS. Therefore, to compensate for this impecunious impairment, 2,5-DHB, an organic matrix is used. However, the conventional matrix undergoes self-ionization which not only causes ion suppression but also complicates the spectrum by giving additional peaks. The MALDI-MS analysis of CTAB using 2,5-DHB is studied. A plot is given out as an output which is made between intensity and *m*/*z*. Different peaks is interpreted to detect the CTAB, hence it has been mentioned as a useful method for the internalization of the nanoparticles in any biological system [149].

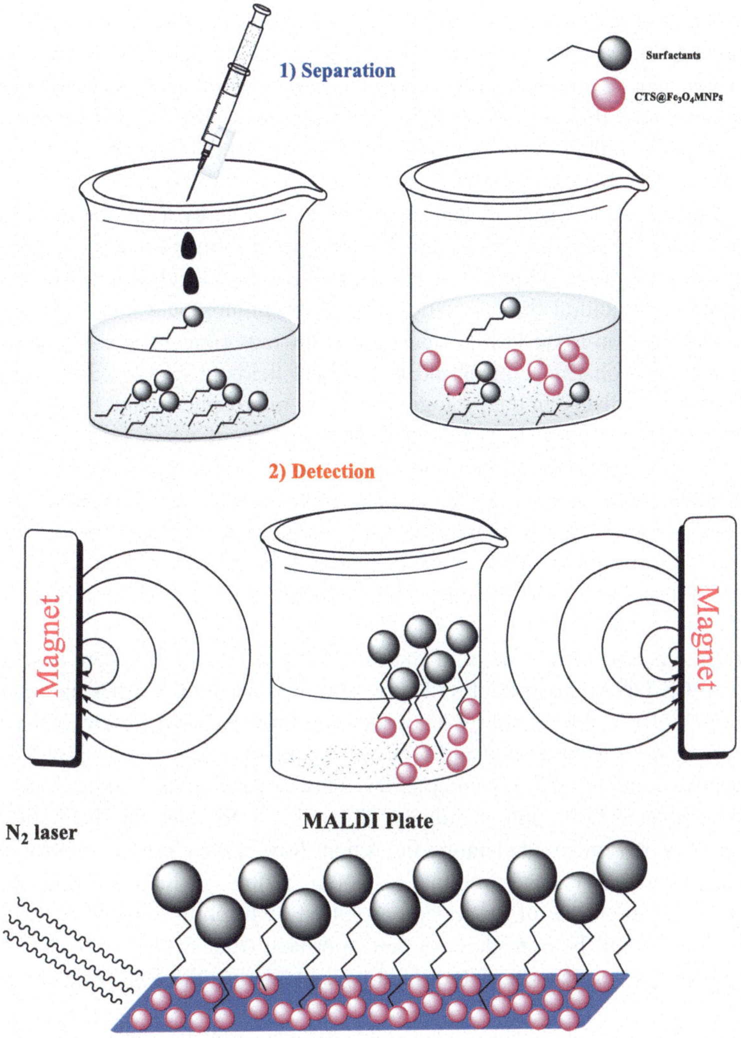

Fig. 6.10 Schematic presentation of pre-concentration of aqueous surfactants using CTS@Fe_3O_4-MNPs [150]

6.17 Size Exclusion Chromatography

In drug delivery polymer nanoparticles have been widely used as a carrier [150]. Synthetic biodegradable polymers have been found efficacious for the production of nanoparticles including polylactide (PLA) [132], poly ε-caprolactone (PCL) [151, 152], and their copolymers [153, 154], due to their high biocompatibility, the good extent of biodegradability, and novel drug release system. However, the strong hydrophobic property of these nanoparticles does not allow them to act as an ideal carrier for a drug of hydrophilic nature. Therefore, there is a dire need to improve the hydrophilic property of the hydrophobic nanoparticles so that effective hydrophilic drug carriers can be obtained and relied upon. In this regard, polyester nanoparticles modified by poly(ethylene glycol) (PEG) not only due to their hydrophilic property but also because of the physio-chemical and biological remarkable properties are the promising system for the hydrophilic drugs delivery [153–157].

The procedure involves in the preparation of the hydrophobic–hydrophilic nanoparticles have several limitations as sonication or homogenization, surfactants, and organic solvents are required. A natural polysaccharide, chitosan, which shares some common structural characteristics with glycosaminoglycans, and is biodegradable as well as non-toxic [158], has made it an important bridge to achieve the improvements in pharmaceutical and biomedical fields [159–161]. For anticancer drug or protein delivery in the biological system, microspheres and beads prepared from chitosan have been heavily relied upon in recent years [162–164]. Many approaches have used the water-in-oil method spray drying process and emulsion-droplet coalescence technique, others to form chitosan beads. These all-reported methods also bank on the employment of some organic solvents and surfactants, and beads prepared by these methods do not meet the requirement of a good administration system, like vein injection due to their large size (larger than 2 μm) [165–167].

To overcome these drawbacks, in an acidic solution CS and DNA are coacervated to prepare CS-DNA nanoparticles. The size range that the CS-DNA nanoparticles hold is 100–250 nm, the nanoparticles of such size limit can easily intercept the degradation of the encapsulated plasmid DNA by nuclease. Alonso et al. [168] utilized the captative aptness of ionic interactions between negatively charged polymer-tripolyphosphate (TPP) and positively charged CS to develop the hydrophilic chitosan–polyethylene oxide nanoparticles that showed the ability to control release with a great capacity to load proteins. By polymerizing acrylic acid into a chitosan template in the presence of an aqueous solution hydrophilic nanoparticles can be prepared, this could be considered a new approach (Fig. 6.11) [169]. The significance of this system can be determined by considering the following interesting features.

(1) There is no need to use organic solvents, surfactants, and the applicability of other special advanced experimental technologies. In other words, mild conditions are enough to get the full protocol.
(2) These positively charged nanoparticles with small particle size have shown improved stability even in the presence of biological cations [170], and the specific interactions that these nanoparticles show with the negatively charged biological membranes and various targeting sites is the foremost requirement for many drugs [7, 171, 172].
(3) The other key factor that proves the dominance of the nanoparticles is their pH-dependent dissolution behavior.

Various analytical techniques can be used to characterize these chitosan-based nanoparticles, one of them is size exclusion chromatography (SEC) [7, 173–176].

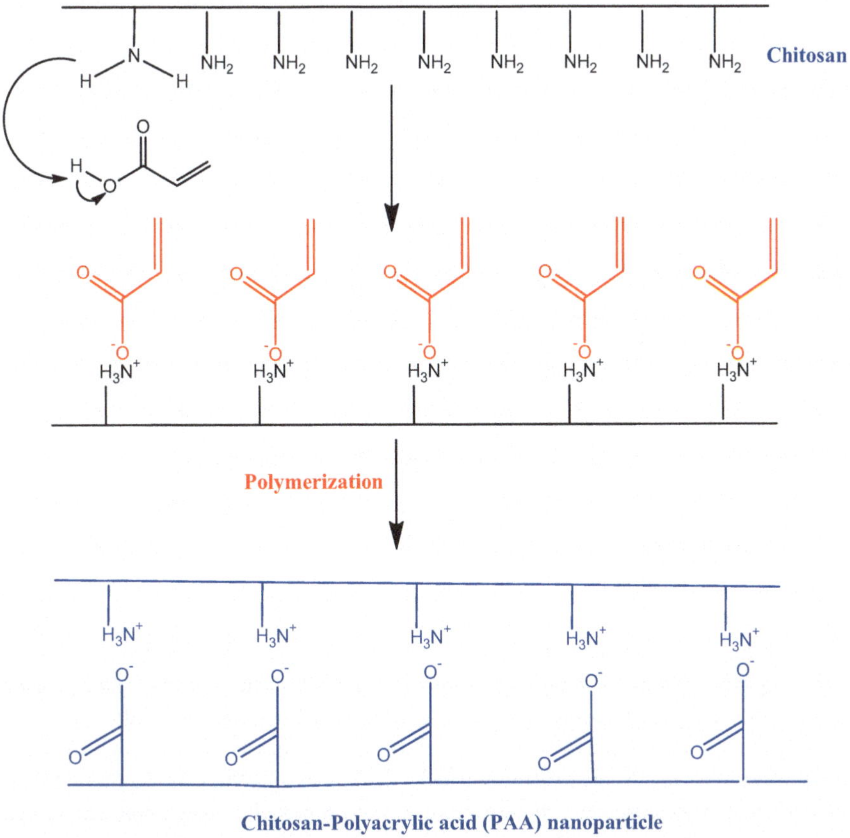

Fig. 6.11 A mechanistic overview of the CS–PAA nanoparticles preparation [169]

One research work has applied SEC to analyze the characteristics of these types of CS–PAA nanoparticles Size exclusion chromatography (SEC) measured the molecular weight of PAA in nanoparticles. That information can be used to find out the yields of the CS–PAA nanoparticles given in the following equation.

$$Nanoparticle\ yield\ (\%) = \frac{Weight\ of\ nanoparticles}{Weight\ of\ chitosan\ and\ monomer\ fed\ initially} \times 100$$

Further, it is found that with the increase of the molecular weight of CS, the number average molecular weight ($\overline{M}n$) of PAA in the CS–PAA nanoparticles also increased (Fig. 6.12) while the yield followed the range of 60–70%. Additionally, the CS template restricted the propagation of PAA chains when the critical molecular weight of the PAA is reached. Therefore, it is concluded that AA followed a template polymerization in the CS solution [177].

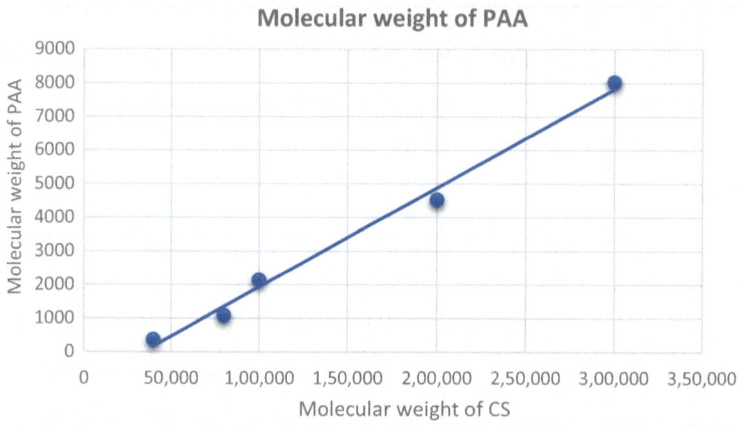

Fig. 6.12 Correlation between the molecular weight of CS and PAA [177]

6.18 Summary

This chapter reviews various techniques for structural confirmation and property analysis of chitosan derivatives and nanoparticles. This chapter provides an overview of various characterization techniques employed to confirm the structure and properties of chitosan derivatives and nanoparticles. The spectroscopic techniques involved to study the materials are UV–Visible spectroscopy, Fourier Transform Infra-Red (FTIR) spectroscopy, X-ray Diffraction (XRD) analysis and Nuclear Magnetic Resonance (NMR) spectroscopy. The structure of the proposed material was elucidated by the researchers by using the mentioned techniques. The microscopic techniques employed to study chitosan nanoparticles and their derivatives are Scanning Electron Microscopy (SEM), Transmission Electron Microscopy (TEM) and Darkfield microscopy. The Particle Size and Distribution Analysis is usually carried out by Dynamic Light Scattering (DLS), Aerodynamic Particle Sizer (APS), Scanning Mobility Particle Sizer (SMPS) and Zeta potential measurements. Mass Spectrometry and chromatographic techniques involved are Inductively Coupled Plasma Mass Spectrometry (ICP-MS), Matrix-Assisted Laser Desorption/Ionization Mass Spectrometry (MALDI-MS) and Size Exclusion Chromatography (SEC). The results and discussion demonstrate that various characterization techniques have proven instrumental in understanding the properties of chitosan nanoparticles. These techniques facilitate development in biomedical devices, drug delivery systems, tissue engineering, biosensors etc., Optimizing chitosan nanoparticle properties enhances their potential in these applications.

References

1. Wang W, Meng Q, Li Q, Liu J, Zhou M, Jin Z, Zhao K (2020) Chitosan derivatives and their application in biomedicine. https://doi.org/10.3390/ijms21020487
2. Huang W, Xu H, Xue Y, Huang R, Deng H, Pan S (2012) Layer-by-layer immobilization of lysozyme-chitosan-organic rectorite composites on electrospun nanofibrous mats for pork preservation. Food Res Int 48. https://doi.org/10.1016/j.foodres.2012.06.026
3. Affes S, Maalej H, Aranaz I, Kchaou H, Acosta N, Heras Á, Nasri M (2020) Controlled size green synthesis of bioactive silver nanoparticles assisted by chitosan and its derivatives and their application in biofilm preparation. Carbohydr Polym 236. https://doi.org/10.1016/j.carbpol.2020.116063
4. Andonegi M, Las Heras K, Santos-Vizcaíno E, Igartua M, Hernandez RM, de la Caba K, Guerrero P (2020) Structure-properties relationship of chitosan/collagen films with potential for biomedical applications. Carbohydr Polym 237. https://doi.org/10.1016/j.carbpol.2020.116159
5. Nematidil N, Sadeghi M, Nezami S, Sadeghi H (2019) Synthesis and characterization of Schiff-base based chitosan-g-glutaraldehyde/NaMMTNPs-APTES for removal Pb^{2+} and Hg^{2+} ions. Carbohydr Polym 222. https://doi.org/10.1016/j.carbpol.2019.114971
6. Lv Y, Li K, Li Y (2013) Surface modification of quantum dots and magnetic nanoparticles with PEG-conjugated chitosan derivatives for biological applications. Chem Pap 67. https://doi.org/10.2478/s11696-013-0401-1
7. Calvo P, Remuñán-López C, Vila-Jato JL, Alonso MJ (1997) Novel hydrophilic chitosan-polyethylene oxide nanoparticles as protein carriers. J Appl Polym Sci 63. https://doi.org/10.1002/(SICI)1097-4628(19970103)63:1<125::AID-APP13>3.0.CO;2-4
8. Giunchedi P, Genta I, Conti B, Muzzarelli RAA, Conte U (1998) Preparation and characterization of ampicillin loaded methylpyrrolidinone chitosan and chitosan microspheres. Biomaterials 19. https://doi.org/10.1016/S0142-9612(97)00181-6
9. Bratskaya SY, Azarova YA, Matochkina EG, Kodess MI, Yatluk YG, Pestov AV (2012) N-(2-(2-pyridyl)ethyl)chitosan: synthesis, characterization and sorption properties. Carbohydr Polym 87. https://doi.org/10.1016/j.carbpol.2011.08.081
10. Bratskaya SY, Azarova YA, Portnyagin AS, Mechaev AV, Voit AV, Pestov AV (2013) Synthesis and properties of isomeric pyridyl-containing chitosan derivatives. Int J Biol Macromol 62. https://doi.org/10.1016/j.ijbiomac.2013.09.017
11. Nikonorov VV, Ivanov RV, Kil'Deeva NR, Bulatnikova LN, Lozinskii VI (2010) Synthesis and characteristics of cryogels of chitosan crosslinked by glutaric aldehyde. Polym Sci Ser A 52. https://doi.org/10.1134/S0965545X10080092
12. Zou P, Yang X, Wang J, Li Y, Yu H, Zhang Y, Liu G (2016) Advances in characterisation and biological activities of chitosan and chitosan oligosaccharides. https://doi.org/10.1016/j.foodchem.2015.06.076
13. Mujtaba M, Morsi RE, Kerch G, Elsabee MZ, Kaya M, Labidi J, Khawar KM (2019) Current advancements in chitosan-based film production for food technology; a review. https://doi.org/10.1016/j.ijbiomac.2018.10.109
14. Salah R, Michaud P, Mati F, Harrat Z, Lounici H, Abdi N, Drouiche N, Mameri N (2013) Anticancer activity of chemically prepared shrimp low molecular weight chitin evaluation with the human monocyte leukaemia cell line, THP-1. Int J Biol Macromol 52. https://doi.org/10.1016/j.ijbiomac.2012.10.009
15. Rinaudo M (2006) Chitin and chitosan: properties and applications. https://doi.org/10.1016/j.progpolymsci.2006.06.001
16. Hedström M, Plieva F, Galaev IY, Mattiasson B (2008) Monolithic macroporous albumin/chitosan cryogel structure: a new matrix for enzyme immobilization. Anal Bioanal Chem 390. https://doi.org/10.1007/s00216-007-1745-6
17. Yang J, Zhang YS, Yue K, Khademhosseini A (2017) Cell-laden hydrogels for osteochondral and cartilage tissue engineering. https://doi.org/10.1016/j.actbio.2017.01.036

18. Prabaharan M (2008) Review paper: chitosan derivatives as promising materials for controlled drug delivery. https://doi.org/10.1177/0885328208091562
19. Colangelo F, Roviello G, Ricciotti L, Ferone C, Cioffi R (2013) Preparation and characterization of new geopolymer-epoxy resin hybrid mortars. Materials 6. https://doi.org/10.3390/ma6072989
20. Mansouri S, Cuie Y, Winnik F, Shi Q, Lavigne P, Benderdour M, Beaumont E, Fernandes JC (2006) Characterization of folate-chitosan-DNA nanoparticles for gene therapy. Biomaterials 27. https://doi.org/10.1016/j.biomaterials.2005.09.020
21. Park JH, Saravanakumar G, Kim K, Kwon IC (2010) Targeted delivery of low molecular drugs using chitosan and its derivatives. https://doi.org/10.1016/j.addr.2009.10.003
22. Singh R, Lillard JW (2009) Nanoparticle-based targeted drug delivery. https://doi.org/10.1016/j.yexmp.2008.12.004
23. Gan Q, Wang T, Cochrane C, McCarron P (2005) Modulation of surface charge, particle size and morphological properties of chitosan-TPP nanoparticles intended for gene delivery. Colloids Surf B Biointerfaces 44. https://doi.org/10.1016/j.colsurfb.2005.06.001
24. Chuc-Gamboa MG, Vargas-Coronado RF, Cervantes-Uc JM, Cauich-Rodríguez JV, Escobar-García DM, Pozos-Guillén A, del Barrio JSR (2019) The effect of PEGDE concentration and temperature on physicochemical and biological properties of chitosan. Polymers (Basel) 11. https://doi.org/10.3390/polym11111830
25. Husain S, Al-Samadani KH, Najeeb S, Zafar MS, Khurshid Z, Zohaib S, Qasim SB (2017) Chitosan biomaterials for current and potential dental applications. https://doi.org/10.3390/ma10060602
26. Bajić M, Jalšovec H, Travan A, Novak U, Likozar B (2019) Chitosan-based films with incorporated supercritical CO_2 hop extract: structural, physicochemical, and antibacterial properties. Carbohydr Polym 219. https://doi.org/10.1016/j.carbpol.2019.05.003
27. Chandrasekaran M, Kim KD, Chun SC (2020) Antibacterial activity of chitosan nanoparticles: a review. https://doi.org/10.3390/PR8091173
28. Richardson SCW, Kolbe HVJ, Duncan R (1999) Potential of low molecular mass chitosan as a DNA delivery system: biocompatibility, body distribution and ability to complex and protect DNA. Int J Pharm 178. https://doi.org/10.1016/S0378-5173(98)00378-0
29. Choksakulnimitr S, Masuda S, Tokuda H, Takakura Y, Hashida M (1995) In vitro cytotoxicity of macromolecules in different cell culture systems. J Controlled Release 34. https://doi.org/10.1016/0168-3659(95)00007-U
30. Thanou MM, Kotzé AF, Scharringhausen T, Lueßen HL, De Boer AG, Verhoef JC, Junginger HE (2000) Effect of degree of quaternization of N-trimethyl chitosan chloride for enhanced transport of hydrophilic compounds across intestinal Caco-2 cell monolayers. J Controlled Release. https://doi.org/10.1016/S0168-3659(99)00131-5
31. Janes KA, Calvo P, Alonso MJ (2001) Polysaccharide colloidal particles as delivery systems for macromolecules. Adv Drug Deliv Rev 47. https://doi.org/10.1016/S0169-409X(00)00123-X
32. Quellec P, Gref R, Perrin L, Dellacherie E, Sommer F, Verbavatz JM, Alonso MJ (1998) Protein encapsulation within polyethylene glycol-coated nanospheres. I. Physicochemical characterization. J Biomed Mater Res 42. https://doi.org/10.1002/(SICI)1097-4636(199810)42:1<45::AID-JBM7>3.0.CO;2-O
33. Fischer D, Li Y, Ahlemeyer B, Krieglstein J, Kissel T (2003) In vitro cytotoxicity testing of polycations: influence of polymer structure on cell viability and hemolysis. Biomaterials 24. https://doi.org/10.1016/S0142-9612(02)00445-3
34. Amiji MM (1997) Synthesis of anionic poly(ethylene glycol) derivative for chitosan surface modificarion in blood-contacting applications. Carbohydr Polym 32. https://doi.org/10.1016/S0144-8617(97)00006-4
35. Gref R, Minamitake Y, Peracchia MT, Trubetskoy V, Torchilin V, Langer R (1994) Biodegradable long-circulating polymeric nanospheres. Science (1979) 263. https://doi.org/10.1126/science.8128245

36. Otsuka H, Nagasaki Y, Kataoka K (2003) PEGylated nanoparticles for biological and pharmaceutical applications. https://doi.org/10.1016/S0169-409X(02)00226-0
37. Hu Y, Jiang X, Ding Y, Zhang L, Yang C, Zhang J, Chen J, Yang Y (2003) Preparation and drug release behaviors of nimodipine-loaded poly(caprolactone)-poly(ethylene oxide)-polylactide amphiphilic copolymer nanoparticles. Biomaterials 24. https://doi.org/10.1016/S0142-9612(03)00021-8
38. Xie W, Xu P, Wang W, Liu Q (2002) Preparation and antibacterial activity of a water-soluble chitosan derivative. Carbohydr Polym 50. https://doi.org/10.1016/S0144-8617(01)00370-8
39. Saharan V, Pal A (2016) Current and future prospects of chitosan-based nanomaterials in plant protection and growth. https://doi.org/10.1007/978-81-322-3601-6_5
40. Saharan V, Mehrotra A, Khatik R, Rawal P, Sharma SS, Pal A (2013) Synthesis of chitosan based nanoparticles and their in vitro evaluation against phytopathogenic fungi. Int J Biol Macromol 62. https://doi.org/10.1016/j.ijbiomac.2013.10.012
41. Fan W, Yan W, Xu Z, Ni H (2012) Formation mechanism of monodisperse, low molecular weight chitosan nanoparticles by ionic gelation technique. Colloids Surf B Biointerfaces 90. https://doi.org/10.1016/j.colsurfb.2011.09.042
42. Choudhary RC, Kumaraswamy RV, Kumari S, Pal A, Raliya R, Biswas P, Saharan V (2017) Synthesis, characterization, and application of chitosan nanomaterials loaded with zinc and copper for plant growth and protection. In: Nanotechnology: an agricultural paradigm. https://doi.org/10.1007/978-981-10-4573-8_10
43. Lawrie G, Keen I, Drew B, Chandler-Temple A, Rintoul L, Fredericks P, Grøndahl L (2007) Interactions between alginate and chitosan biopolymers characterized using FTIR and XPS. Biomacromolecules 8. https://doi.org/10.1021/bm070014y
44. Zamora-Mora V, Fernández-Gutiérrez M, Román JS, Goya G, Hernández R, Mijangos C (2014) Magnetic core-shell chitosan nanoparticles: rheological characterization and hyperthermia application. Carbohydr Polym 102. https://doi.org/10.1016/j.carbpol.2013.10.101
45. Jingou J, Shilei H, Weiqi L, Danjun W, Tengfei W, Yi X (2011) Preparation, characterization of hydrophilic and hydrophobic drug in combine loaded chitosan/cyclodextrin nanoparticles and in vitro release study. Colloids Surf B Biointerfaces 83. https://doi.org/10.1016/j.colsurfb.2010.11.005
46. Baer DR, Gaspar DJ, Nachimuthu P, Techane SD, Castner DG (2010) Application of surface chemical analysis tools for characterization of nanoparticles. https://doi.org/10.1007/s00216-009-3360-1
47. Liu H, Gao C (2009) Preparation and properties of ionically cross-linked chitosan nanoparticles. Polym Adv Technol 20. https://doi.org/10.1002/pat.1306
48. Li LH, Deng JC, Deng HR, Liu ZL, Xin L (2010) Synthesis and characterization of chitosan/ZnO nanoparticle composite membranes. Carbohydr Res 345. https://doi.org/10.1016/j.carres.2010.03.019
49. Jaiswal M, Chauhan D, Sankararamakrishnan N (2012) Copper chitosan nanocomposite: synthesis, characterization, and application in removal of organophosphorous pesticide from agricultural runoff. Environ Sci Pollut Res 19. https://doi.org/10.1007/s11356-011-0699-6
50. Nguyen Van S, Dinh Minh H, Nguyen Anh D (2013) Study on chitosan nanoparticles on biophysical characteristics and growth of Robusta coffee in green house. Biocatal Agric Biotechnol 2. https://doi.org/10.1016/j.bcab.2013.06.001
51. Danielsen S, Vårum KM, Stokke BT (2004) Structural analysis of chitosan mediated DNA condensation by AFM: influence of chitosan molecular parameters. Biomacromolecules 5. https://doi.org/10.1021/bm034502r
52. MacLaughlin FC, Mumper RJ, Wang J, Tagliaferri JM, Gill I, Hinchcliffe M, Rolland AP (1998) Chitosan and depolymerized chitosan oligomers as condensing carriers for in vivo plasmid delivery. J Controlled Release 56. https://doi.org/10.1016/S0168-3659(98)00097-2
53. Köping-Höggård M, Tubulekas I, Guan H, Edwards K, Nilsson M, Vårum KM, Artursson P (2001) Chitosan as a nonviral gene delivery system. Structure-property relationships and characteristics compared with polyethylenimine in vitro and after lung administration in vivo. Gene Ther 8. https://doi.org/10.1038/sj.gt.3301492

54. Huang M, Fong CW, Khor E, Lim LY (2005) Transfection efficiency of chitosan vectors: effect of polymer molecular weight and degree of deacetylation. J Controlled Release 106. https://doi.org/10.1016/j.jconrel.2005.05.004
55. Alameh M, Jean M, Dejesus D, Buschmann MD, Merzouki A (2010) Chitosanase-based method for RNA isolation from cells transfected with chitosan/siRNA nanocomplexes for real-time RT-PCR in gene silencing. Int J Nanomedicine 5. https://doi.org/10.2147/ijn.s10879
56. Manchanda R, Nimesh S (2010) Controlled size chitosan nanoparticles as an efficient, biocompatible oligonucleotides delivery system. J Appl Polym Sci 118. https://doi.org/10.1002/app.32508
57. Carter CB, Williams DB (2016) Transmission electron microscopy: diffraction, imaging, and spectrometry. https://doi.org/10.1007/978-3-319-26651-0
58. Choudhary RC, Kumaraswamy RV, Kumari S, Sharma SS, Pal A, Raliya R, Biswas P, Saharan V (2017) Cu-chitosan nanoparticle boost defense responses and plant growth in maize (*Zea mays* L.). Sci Rep 7. https://doi.org/10.1038/s41598-017-08571-0
59. Zhu S, Qian F, Zhang Y, Tang C, Yin C (2007) Synthesis and characterization of PEG modified N-trimethylaminoethylmethacrylate chitosan nanoparticles. Eur Polym J 43. https://doi.org/10.1016/j.eurpolymj.2007.03.042
60. Mao S, Germershaus O, Fischer D, Linn T, Schnepf R, Kissel T (2005) Uptake and transport of PEG-graft-trimethyl-chitosan copolymer-insulin nanocomplexes by epithelial cells. Pharm Res 22. https://doi.org/10.1007/s11095-005-8175-y
61. Kim SS, Lee YM, Cho CS (1995) Synthesis and properties of semi-interpenetrating polymer networks composed of β-chitin and poly(ethylene glycol) macromer. Polymer (Guildf) 36. https://doi.org/10.1016/0032-3861(95)96859-7
62. Collins JJ, Lineker GA (2004) A review and meta-analysis of formaldehyde exposure and leukemia. Regul Toxicol Pharmacol 40. https://doi.org/10.1016/j.yrtph.2004.04.006
63. Boran S, Usta M, Gümükaya E (2011) Decreasing formaldehyde emission from medium density fiberboard panels produced by adding different amine compounds to urea formaldehyde resin. Int J Adhes Adhes 31. https://doi.org/10.1016/j.ijadhadh.2011.06.011
64. Wu Z, Lei H, Du G, Cao M, Xi X, Liang J (2016) Urea–formaldehyde resin prepared with concentrated formaldehyde. J Adhes Sci Technol 30. https://doi.org/10.1080/01694243.2016.1193963
65. Elbert AA (1995) Influence of hardener systems and wood on the formaldehyde emission from urea-formaldehyde resin and particleboards. Holzforschung 49. https://doi.org/10.1515/hfsg.1995.49.4.358
66. Costa NA, Pereira J, Ferra J, Cruz P, Martins J, Magalhães FD, Mendes A, Carvalho LH (2013) Scavengers for achieving zero formaldehyde emission of wood-based panels. Wood Sci Technol 47. https://doi.org/10.1007/s00226-013-0573-4
67. Shalbafan A, Welling J, Hasch J (2016) Geopolymers as potential new binder class for the wood based composite industry. Holzforschung 70. https://doi.org/10.1515/hf-2015-0206
68. Ji X, Guo M (2018) Preparation and properties of a chitosan-lignin wood adhesive. Int J Adhes Adhes 82. https://doi.org/10.1016/j.ijadhadh.2017.12.005
69. Park BD, Ayrilmis N, Kwon JH, Han TH (2017) Effect of microfibrillated cellulose addition on thermal properties of three grades of urea-formaldehyde resin. Int J Adhes Adhes 72. https://doi.org/10.1016/j.ijadhadh.2016.10.003
70. Hassannejad H, Shalbafan A, Rahmaninia M (2020) Reduction of formaldehyde emission from medium density fiberboard by chitosan as scavenger. J Adhes 96. https://doi.org/10.1080/00218464.2018.1515631
71. Loubaki E, Sicsic S, Le Goffic F (1989) Modification chimique du chitosane avec la δ-gluconolactone, la β-propiolactone et le glycidol. Eur Polym J 25. https://doi.org/10.1016/0014-3057(89)90154-7
72. O-T WJ (1970) K. Nakanishi, infrared absorption spectroscopy—practical. J Mol Struct 5. https://doi.org/10.1016/0022-2860(70)80011-4
73. Agarwal M, Agarwal MK, Shrivastav N, Pandey S, Das R, Gaur P (2018) Preparation of chitosan nanoparticles and their in-vitro characterization. Int J Life-Sci Sci Res 4. https://doi.org/10.21276/ijlssr.2018.4.2.17

74. Boluk Y, Danumah C (2014) Analysis of cellulose nanocrystal rod lengths by dynamic light scattering and electron microscopy. J Nanopart Res 16. https://doi.org/10.1007/s11051-013-2174-4
75. Israelachvili J (2011) Intermolecular and surface forces. https://doi.org/10.1016/C2009-0-21560-1
76. Hasegawa M, Yagi K, Iwakawa S, Hirai M (2001) Chitosan induces apoptosis via caspase-3 activation in bladder tumor cells. Jpn J Cancer Res 92. https://doi.org/10.1111/j.1349-7006.2001.tb01116.x
77. Du WL, Niu SS, Xu YL, Xu ZR, Fan CL (2009) Antibacterial activity of chitosan tripolyphosphate nanoparticles loaded with various metal ions. Carbohydr Polym 75. https://doi.org/10.1016/j.carbpol.2008.07.039
78. Saharan V, Sharma G, Yadav M, Choudhary MK, Sharma SS, Pal A, Raliya R, Biswas P (2015) Synthesis and in vitro antifungal efficacy of Cu-chitosan nanoparticles against pathogenic fungi of tomato. Int J Biol Macromol 75. https://doi.org/10.1016/j.ijbiomac.2015.01.027
79. Bowman K, Leong KW (2006) Chitosan nanoparticles for oral drug and gene delivery. https://doi.org/10.2147/nano.2006.1.2.117
80. Schipper NGM, Vårum KM, Stenberg P, Ocklind G, Lennernäs H, Artursson P (1999) Chitosans as absorption enhancers of poorly absorbable drugs. 3: influence of mucus on absorption enhancement. Eur J Pharm Sci 8. https://doi.org/10.1016/S0928-0987(99)00032-9
81. Trapani A, Sitterberg J, Bakowsky U, Kissel T (2009) The potential of glycol chitosan nanoparticles as carrier for low water soluble drugs. Int J Pharm 375. https://doi.org/10.1016/j.ijpharm.2009.03.041
82. Nimesh S, Thibault MM, Lavertu M, Buschmann MD (2010) Enhanced gene delivery mediated by low molecular weight chitosan/DNA complexes: effect of pH and serum. Mol Biotechnol 46. https://doi.org/10.1007/s12033-010-9286-1
83. Farrer RA, Butterfiefd FL, Chen VW, Fourkas JT (2005) Highly efficient multiphoton-absorption-induced luminescence from gold nanoparticles. Nano Lett 5. https://doi.org/10.1021/nl050687r
84. Paciotti GF, Kingston DGI, Tamarkin L (2006) Colloidal gold nanoparticles: a novel nanoparticle platform for developing multifunctional tumor-targeted drug delivery vectors. https://doi.org/10.1002/ddr.20066
85. Huang X, Qian W, El-Sayed IH, El-Sayed MA (2007) The potential use of the enhanced nonlinear properties of gold nanospheres in photothermal cancer therapy. Lasers Surg Med 39. https://doi.org/10.1002/lsm.20577
86. Yelin D, Oron D, Thiberge S, Moses E, Silberberg Y (2003) Multiphoton plasmon-resonance microscopy. Opt Express 11. https://doi.org/10.1364/oe.11.001385
87. Hauck TS, Ghazani AA, Chan WCW (2008) Assessing the effect of surface chemistry on gold nanorod uptake, toxicity, and gene expression in mammalian cells. Small 4. https://doi.org/10.1002/smll.200700217
88. Brunel F, Véron L, David L, Domard A, Delair T (2008) A novel synthesis of chitosan nanoparticles in reverse emulsion. Langmuir 24. https://doi.org/10.1021/la801917a
89. Bhattarai SR, Remant RB, Aryal S, Bhattarai N, Kim SY, Yi HK, Hwang PH, Kim HY (2008) Hydrophobically modified chitosan/gold nanoparticles for DNA delivery. J Nanopart Res 10. https://doi.org/10.1007/s11051-007-9233-7
90. Esumi K, Takei N, Yoshimura T (2003) Antioxidant-potentiality of gold-chitosan nanocomposites. Colloids Surf B Biointerfaces 32. https://doi.org/10.1016/S0927-7765(03)00151-6
91. Miller MM, Lazarides AA (2006) Sensitivity of metal nanoparticle plasmon resonance band position to the dielectric environment as observed in scattering. J Opt A Pure Appl Opt 8. https://doi.org/10.1088/1464-4258/8/4/S26
92. Armendariz AJ, Leith D (2002) Concentration measurement and counting efficiency for the aerodynamic particle sizer 3320. J Aerosol Sci 33. https://doi.org/10.1016/S0021-8502(01)00152-5
93. Duceppe N, Tabrizian M (2010) Advances in using chitosan-based nanoparticles for in vitro and in vivo drug and gene delivery. https://doi.org/10.1517/17425247.2010.514604

94. Liang XF, Wang HJ, Luo H, Tian H, Zhang BB, Hao LJ, Teng JI, Chang J (2008) Characterization of novel multifunctional cationic polymeric liposomes formed from octadecyl quaternized carboxymethyl chitosan/cholesterol and drug encapsulation. Langmuir 24. https://doi.org/10.1021/la703775a
95. Nandanan E, Jana NR, Ying JY (2008) Functionalization of gold nanospheres and nanorods by chitosan oligosaccharide derivatives. Adv Mater 20. https://doi.org/10.1002/adma.200702193
96. Madihally SV, Matthew HWT (1999) Porous chitosan scaffolds for tissue engineering. Biomaterials 20. https://doi.org/10.1016/S0142-9612(99)00011-3
97. Satija J, Gupta U, Jain NK (2007) Pharmaceutical and biomedical potential of surface engineered dendrimers. https://doi.org/10.1615/CritRevTherDrugCarrierSyst.v24.i3.20
98. Owens DE, Peppas NA (2006) Opsonization, biodistribution, and pharmacokinetics of polymeric nanoparticles. https://doi.org/10.1016/j.ijpharm.2005.10.010
99. Saranya N, Moorthi A, Saravanan S, Devi MP, Selvamurugan N (2011) Chitosan and its derivatives for gene delivery. https://doi.org/10.1016/j.ijbiomac.2010.11.013
100. Kim TH, Jiang HL, Jere D, Park IK, Cho MH, Nah JW, Choi YJ, Akaike T, Cho CS (2007) Chemical modification of chitosan as a gene carrier in vitro and in vivo. https://doi.org/10.1016/j.progpolymsci.2007.05.001
101. Thanou M, Nihot MT, Jansen M, Verhoef JC, Junginger HE (2001) Mono-N-carboxymethyl chitosan (MCC), a polyampholytic chitosan derivative, enhances the intestinal absorption of low molecular weight heparin across intestinal epithelia in vitro and in vivo. J Pharm Sci 90. https://doi.org/10.1002/1520-6017(200101)90:1<38::AID-JPS5>3.0.CO;2-3
102. Gao J, Gu H, Xu B (2009) Multifunctional magnetic nanoparticles: design, synthesis, and biomedical applications. Acc Chem Res 42. https://doi.org/10.1021/ar9000026
103. Van Der Merwe SM, Verhoef JC, Verheijden JHM, Kotzé AF, Junginger HE (2004) Trimethylated chitosan as polymeric absorption enhancer for improved peroral delivery of peptide drugs. https://doi.org/10.1016/j.ejpb.2004.03.023
104. Xu P, Bajaj G, Shugg T, Van Alstine WG, Yeo Y (2010) Zwitterionic chitosan derivatives for pH-sensitive stealth coating. Biomacromolecules 11. https://doi.org/10.1021/bm100481r
105. Kato Y, Onishi H, Machida Y (2000) Biological fate of highly-succinylated N-succinyl-chitosan and antitumor characteristics of its water-soluble conjugate with mitomycin C at i.v. and i.p. administration into tumor-bearing mice. Biol Pharm Bull 23. https://doi.org/10.1248/bpb.23.1497
106. Le-Tien C, Millette M, Mateescu M-A, Lacroix M (2004) Modified alginate and chitosan for lactic acid bacteria immobilization. Biotechnol Appl Biochem 39. https://doi.org/10.1042/BA20030158
107. Flores-Fernández GM, Solá RJ, Griebenow K (2009) The relation between moisture-induced aggregation and structural changes in lyophilized insulin. J Pharm Pharmacol 61. https://doi.org/10.1211/jpp/61.11.0016
108. Ré MI (2006) Formulating drug delivery systems by spray drying. https://doi.org/10.1080/07373930600611877
109. Rizi K, Green RJ, Donaldson M, Williams AC (2011) Production of pH-responsive microparticles by spray drying: investigation of experimental parameter effects on morphological and release properties. J Pharm Sci 100. https://doi.org/10.1002/jps.22291
110. Sun S, Zeng H (2002) Size-controlled synthesis of magnetite nanoparticles. J Am Chem Soc 124. https://doi.org/10.1021/ja026501x
111. Pankhurst QA, Connolly J, Jones SK, Dobson J (2003) Applications of magnetic nanoparticles in biomedicine. https://doi.org/10.1088/0022-3727/36/13/201
112. Tang SCN, Lo IMC (2013) Magnetic nanoparticles: essential factors for sustainable environmental applications. https://doi.org/10.1016/j.watres.2013.02.039
113. Yin R, Guo W, Zhou X, Zheng H, Du J, Wu Q, Chang J, Ren N (2016) Enhanced sulfamethoxazole ozonation by noble metal-free catalysis based on magnetic Fe_3O_4 nanoparticles: catalytic performance and degradation mechanism. RSC Adv 6. https://doi.org/10.1039/c5ra25994k
114. Atar N, Eren T, Yola ML, Karimi-Maleh H, Demirdögen B (2015) Magnetic iron oxide and iron oxide@gold nanoparticle anchored nitrogen and sulfur-functionalized reduced graphene oxide electrocatalyst for methanol oxidation. RSC Adv 5. https://doi.org/10.1039/c5ra03735b

115. Jung JH, Lee JH, Shinkai S (2011) Functionalized magnetic nanoparticles as chemosensors and adsorbents for toxic metal ions in environmental and biological fields. Chem Soc Rev 40. https://doi.org/10.1039/c1cs15051k
116. Abdelhamid HN (2016) Laser assisted synthesis, imaging and cancer therapy of magnetic nanoparticles. Mater Focus 5. https://doi.org/10.1166/mat.2016.1336
117. Tanhaei B, Ayati A, Lahtinen M, Sillanpää M (2015) Preparation and characterization of a novel chitosan/Al_2O_3/magnetite nanoparticles composite adsorbent for kinetic, thermodynamic and isotherm studies of methyl orange adsorption. Chem Eng J 259. https://doi.org/10.1016/j.cej.2014.07.109
118. Nasirimoghaddam S, Zeinali S, Sabbaghi S (2015) Chitosan coated magnetic nanoparticles as nano-adsorbent for efficient removal of mercury contents from industrial aqueous and oily samples. J Ind Eng Chem 27. https://doi.org/10.1016/j.jiec.2014.12.020
119. Zare F, Ghaedi M, Daneshfar A (2015) Solid phase extraction of antidepressant drugs amitriptyline and nortriptyline from plasma samples using core-shell nanoparticles of the type Fe_3O_4@ZrO_2@N-cetylpyridinium, and their subsequent determination by HPLC with UV detection. Microchimica Acta 182. https://doi.org/10.1007/s00604-015-1499-3
120. Jiang X, Cheng J, Zhou H, Li F, Wu W, Ding K (2015) Polyaniline-coated chitosan-functionalized magnetic nanoparticles: preparation for the extraction and analysis of endocrine-disrupting phenols in environmental water and juice samples. Talanta 141. https://doi.org/10.1016/j.talanta.2015.04.017
121. Gopal J, Abdelhamid HN, Hua PY, Wu HF (2013) Chitosan nanomagnets for effective extraction and sensitive mass spectrometric detection of pathogenic bacterial endotoxin from human urine. J Mater Chem B 1. https://doi.org/10.1039/c3tb20079e
122. Bhaisare ML, Abdelhamid HN, Wu BS, Wu HF (2014) Rapid and direct MALDI-MS identification of pathogenic bacteria from blood using ionic liquid-modified magnetic nanoparticles (Fe_3O_4@SiO_2). J Mater Chem B 2. https://doi.org/10.1039/c4tb00528g
123. Abdelhamid HN, Wu HF (2014) Facile synthesis of nano silver ferrite ($AgFeO_2$) modified with chitosan applied for biothiol separation. Mater Sci Eng C 45. https://doi.org/10.1016/j.msec.2014.08.071
124. Zhang X, Wang J, He X, Chen L, Zhang Y (2015) Tailor-made boronic acid functionalized magnetic nanoparticles with a tunable polymer shell-assisted for the selective enrichment of glycoproteins/glycopeptides. ACS Appl Mater Interfaces 7. https://doi.org/10.1021/acsami.5b06445
125. Chen Y, Xiong Z, Peng L, Gan Y, Zhao Y, Shen J, Qian J, Zhang L, Zhang W (2015) Facile preparation of core-shell magnetic metal-organic framework nanoparticles for the selective capture of phosphopeptides. ACS Appl Mater Interfaces 7. https://doi.org/10.1021/acsami.5b03335
126. Abdelhamid HN, Talib A, Wu HF (2015) Erratum: facile synthesis of water soluble silver ferrite ($AgFeO_2$) nanoparticles and their biological application as antibacterial agents. RSC Adv 5:34594–34602. https://doi.org/10.1039/c5ra90041g
127. Abdelhamid HN (2015) Delafossite nanoparticle as new functional materials: advances in energy, nanomedicine and environmental applications. https://doi.org/10.4028/www.scientific.net/MSF.832.28
128. Deese RD, Leblanc MR, Cook RL (2016) Surfactant toxicity to *Artemia franciscana* and the influence of humic acid and chemical composition. Environ Chem 13. https://doi.org/10.1071/EN15108
129. Guo Q, Zhang Z, Song Y, Liu S, Gao W, Qiao H, Guo L, Wang J (2017) Investigation on interaction of DNA and several cationic surfactants with different head groups by spectroscopy, gel electrophoresis and viscosity technologies. Chemosphere 168. https://doi.org/10.1016/j.chemosphere.2016.11.019
130. Zhou C, Wang F, Chen H, Li M, Qiao F, Liu Z, Hou Y, Wu C, Fan Y, Liu L, Wang S, Wang Y (2016) Selective antimicrobial activities and action mechanism of micelles self-assembled by cationic oligomeric surfactants. ACS Appl Mater Interfaces 8. https://doi.org/10.1021/acsami.5b12688

131. Boillot C, Perrodin Y (2008) Joint-action ecotoxicity of binary mixtures of glutaraldehyde and surfactants used in hospitals: use of the toxicity index model and isoblogram representation. Ecotoxicol Environ Saf 71. https://doi.org/10.1016/j.ecoenv.2007.08.010
132. García I, Henriksen-Lacey M, Sánchez-Iglesias A, Grzelczak M, Penadés S, Liz-Marzán LM (2015) Residual CTAB ligands as mass spectrometry labels to monitor cellular uptake of Au nanorods. J Phys Chem Lett 6. https://doi.org/10.1021/acs.jpclett.5b00816
133. APHA (2005) Standard methods for the examination of water and wastewater. Standard Methods. ISBN 9780875532356
134. Heinlg K, Vogt C (1999) Determination of surfactants by capillary electrophoresis. https://doi.org/10.1002/(sici)1522-2683(19991001)20:15/16<3311::aid-elps3311>3.0.co;2-i
135. Hussain S, Malik AH, Iyer PK (2015) Highly precise detection, discrimination, and removal of anionic surfactants over the full pH range via cationic conjugated polymer: an efficient strategy to facilitate illicit-drug analysis. ACS Appl Mater Interfaces 7. https://doi.org/10.1021/am507731t
136. Abdelhamid HN, Lin YC, Wu HF (2017) Magnetic nanoparticle modified chitosan for surface enhanced laser desorption/ionization mass spectrometry of surfactants. RSC Adv 7. https://doi.org/10.1039/c7ra05982e
137. Abdelhamid HN, Chen ZY, Wu HF (2017) Surface tuning laser desorption/ionization mass spectrometry (STLDI-MS) for the analysis of small molecules using quantum dots. Anal Bioanal Chem 409. https://doi.org/10.1007/s00216-017-0433-4
138. Wu HF, Gopal J, Abdelhamid HN, Hasan N (2012) Quantum dot applications endowing novelty to analytical proteomics. https://doi.org/10.1002/pmic.201200295
139. Abdelhamid HN (2018) Nanoparticle assisted laser desorption/ionization mass spectrometry for small molecule analytes. https://doi.org/10.1007/s00604-018-2687-8
140. Abdelhamid HN, Bhaisare ML, Wu HF (2014) Ceria nanocubic-ultrasonication assisted dispersive liquid-liquid microextraction coupled with matrix assisted laser desorption/ionization mass spectrometry for pathogenic bacteria analysis. Talanta 120. https://doi.org/10.1016/j.talanta.2013.11.078
141. Abdelhamid HN, Gopal J, Wu HF (2013) Synthesis and application of ionic liquid matrices (ILMs) for effective pathogenic bacteria analysis in matrix assisted laser desorption/ionization (MALDI-MS). Anal Chim Acta 767. https://doi.org/10.1016/j.aca.2012.12.054
142. Wu BS, Abdelhamid HN, Wu HF (2014) Synthesis and antibacterial activities of graphene decorated with stannous dioxide. RSC Adv 4. https://doi.org/10.1039/c3ra43992e
143. Li B, Comi TJ, Si T, Dunham SJB, Sweedler JV (2016) A one-step matrix application method for MALDI mass spectrometry imaging of bacterial colony biofilms. J Mass Spectrom 51. https://doi.org/10.1002/jms.3827
144. Abdelhamid HN, Wu HF (2015) Synthesis and characterization of quantum dots for application in laser soft desorption/ionization mass spectrometry to detect labile metal-drug interactions and their antibacterial activity. RSC Adv 5. https://doi.org/10.1039/c5ra11301f
145. Abdelhamid HN, Wu HF (2014) Ultrasensitive, rapid, and selective detection of mercury using graphene assisted laser desorption/ionization mass spectrometry. J Am Soc Mass Spectrom 25. https://doi.org/10.1007/s13361-014-0825-z
146. Abdelhamid HN, Kumaran S, Wu HF (2016) One-pot synthesis of $CuFeO_2$ nanoparticles capped with glycerol and proteomic analysis of their nanocytotoxicity against fungi. RSC Adv 6. https://doi.org/10.1039/c6ra13396g
147. Abdelhamid HN, Wu HF (2016) Gold nanoparticles assisted laser desorption/ionization mass spectrometry and applications: from simple molecules to intact cells. https://doi.org/10.1007/s00216-016-9374-6
148. Abdelhamid HN, Wu HF (2015) Synthesis of a highly dispersive sinapinic acid@graphene oxide (SA@GO) and its applications as a novel surface assisted laser desorption/ionization mass spectrometry for proteomics and pathogenic bacteria biosensing. Analyst 140. https://doi.org/10.1039/c4an02158d
149. Abdelhamid HN, Talib A, Wu HF (2017) One pot synthesis of gold—carbon dots nanocomposite and its application for cytosensing of metals for cancer cells. Talanta 166. https://doi.org/10.1016/j.talanta.2016.11.030

150. Abdelhamid HN, Lin YC, Wu HF (2017) Thymine chitosan nanomagnets for specific preconcentration of mercury(II) prior to analysis using SELDI-MS. Microchimica Acta 184. https://doi.org/10.1007/s00604-017-2125-3
151. Couvreur P, Grislain L, Lenaerts V, Brasseur F, Guiot P, Biernacki A (2018) Biodegradable polymeric nanoparticles as drug carrier for antitumor agents. In: Polymeric nanoparticles and microspheres. https://doi.org/10.1201/9781351075893-2
152. Legrand P, Barratt G, Mosqueira V, Fessi H, Devissaguet JP (1999) Polymeric nanocapsules as drug delivery systems: a review
153. Wehrle P, Magenheim B, Benita S (1995) The influence of process parameters on the PLA nanoparticle size distribution, evaluated by means of factorial design. Eur J Pharm Biopharm 41
154. Guzmán M, Aberturas MR, Rodríguez-Puyol M, Molpeceres J (2000) Effect of nanoparticles on digitoxin uptake and pharmacologic activity in rat glomerular mesangial cell cultures. Drug Deliv J Deliv Target Therapeut Agents 7. https://doi.org/10.1080/107175400455146
155. Lamprecht A, Ubrich N, Hombreiro Pérez M, Lehr CM, Hoffman M, Maincent P (2000) Influences of process parameters on nanoparticle preparation performed by a double emulsion pressure homogenization technique. Int J Pharm 196. https://doi.org/10.1016/S0378-5173(99)00422-6
156. Ge H, Hu Y, Yang S, Jiang X, Yang C (2000) Preparation, characterization, and drug release behaviors of drug-loaded ε-caprolactone/L-lactide copolymer nanoparticles. J Appl Polym Sci 75. https://doi.org/10.1002/(SICI)1097-4628(20000214)75:7<874::AID-APP3>3.0.CO;2-G
157. Wang N, Wu XS, Li JK (1999) A heterogeneously structured composite based on poly(lactic-co-glycolic acid) microspheres and poly(vinyl alcohol) hydrogel nanoparticles for long-term protein drug delivery. Pharm Res 16. https://doi.org/10.1023/A:1018911411381
158. Chacón M, Molpeceres J, Berges L, Guzmán M, Aberturas MR (1999) Stability and freeze-drying of cyclosporine loaded poly(D,L-lactide-glycolide) carriers. Eur J Pharm Sci 8. https://doi.org/10.1016/S0928-0987(98)00066-9
159. Ryu JG, Il Jeong Y, Kim IS, Lee JH, Nah JW, Kim SH (2000) Clonazepam release from core-shell type nanoparticles of poly(ε-caprolactone)/poly(ethylene glycol)/poly(ε-caprolactone) triblock copolymers. Int J Pharm 200. https://doi.org/10.1016/S0378-5173(00)00392-6
160. De Jaeghere F, Allémann E, Leroux JC, Stevels W, Feijen J, Doelker E, Gurny R (1999) Formulation and lyoprotection of poly(lactic acid-co-ethylene oxide) nanoparticles: influence on physical stability and in vitro cell uptake. Pharm Res 16. https://doi.org/10.1023/A:1018826103261
161. Heald CR, Stolnik S, De Matteis C, Garnett MC, Illum L, Davis SS, Leermakers FAM (2001) Self-consistent field modelling of poly(lactic acid)-poly(ethylene glycol) particles. Colloids Surf A Physicochem Eng Asp 179. https://doi.org/10.1016/S0927-7757(00)00729-9
162. Muzzarelli R, Baldassarre V, Conto F, Ferrara P, Biagini G, Gazzanelli G, Vasi V (1988) Biological activity of chitosan: ultrastructural study. Biomaterials 9. https://doi.org/10.1016/0142-9612(88)90092-0
163. Hirano S, Seino H, Akiyama Y, Nonaka I (1990) Chitosan: a biocompatible material for oral and intravenous administrations. Prog Biomed Polym. https://doi.org/10.1007/978-1-4899-0768-4_28
164. Muzzarelli RAA (1993) Biochemical significance of exogenous chitins and chitosans in animals and patients. Carbohydr Polym 20. https://doi.org/10.1016/0144-8617(93)90027-2
165. Lorenzo-Lamosa ML, Remuñán-López C, Vila-Jato JL, Alonso MJ (1998) Design of microencapsulated chitosan microspheres for colonic drug delivery. J Controlled Release 52. https://doi.org/10.1016/S0168-3659(97)00203-4
166. Hari PR, Chandy T, Sharma CP (1996) Chitosan/calcium-alginate beads for oral delivery of insulin. J Appl Polym Sci 59. https://doi.org/10.1002/(sici)1097-4628(19960314)59:11<1795::aid-app16>3.0.co;2-t
167. Gupta KC, Ravi Kumar MNV (2000) Drug release behavior of beads and microgranules of chitosan. Biomaterials 21. https://doi.org/10.1016/S0142-9612(99)00263-X

168. Lim ST, Martin GP, Berry DJ, Brown MB (2000) Preparation and evaluation of the in vitro drug release properties and mucoadhesion of novel microspheres of hyaluronic acid and chitosan. J Controlled Release 66. https://doi.org/10.1016/S0168-3659(99)00285-0
169. Onishi H, Shimoda J, Machida Y (1996) Chitosan-drug conjugate microspheres: preparation and drug release properties of microspheres composed of the conjugate of 2′- or 3′-(4-carboxybutyryl)-5-fluorouridine with chitosan. Drug Dev Ind Pharm 22. https://doi.org/10.3109/03639049609069355
170. Mi FL, Kuan CY, Shyu SS, Lee ST, Chang SF (2000) Study of gelation kinetics and chain-relaxation properties of glutaraldehyde-cross-linked chitosan gel and their effects on microspheres preparation and drug release. Carbohydr Polym 41. https://doi.org/10.1016/S0144-8617(99)00104-6
171. Mao HQ, Roy K, Troung-Le VL, Janes KA, Lin KY, Wang Y, August JT, Leong KW (2001) Chitosan-DNA nanoparticles as gene carriers: synthesis, characterization and transfection efficiency. J Controlled Release 70. https://doi.org/10.1016/S0168-3659(00)00361-8
172. Tokumitsu H, Ichikawa H, Fukumori Y (1999) Chitosan-gadopentetic acid complex nanoparticles for gadolinium neutron-capture therapy of cancer: preparation by novel emulsion-droplet coalescence technique and characterization. Pharm Res 16. https://doi.org/10.1023/A:1018995124527
173. Ahn JS, Choi HK, Cho CS (2001) A novel mucoadhesive polymer prepared by template polymerization of acrylic acid in the presence of chitosan. Biomaterials 22. https://doi.org/10.1016/S0142-9612(00)00256-8
174. Calvo P, Remuñán-López C, Vila-Jato JL, Alonso MJ (1997) Development of positively charged colloidal drug carriers: chitosan-coated polyester nanocapsules and submicron-emulsions. Colloid Polym Sci 275. https://doi.org/10.1007/s003960050050
175. Meisner D, Pringle J, Mezei M (1989) Liposomal ophthalmic drug delivery. III. Pharmacodynamic and biodisposition studies of atropine. Int J Pharm 55. https://doi.org/10.1016/0378-5173(89)90030-6
176. Meisner D, Pringle J, Mezei M (1989) Liposomal pulmonary drug delivery I. In vivo disposition of atropine base in solution and liposomal form following endotracheal instillation to the rabbit lung. J Microencapsul 6. https://doi.org/10.3109/02652048909019920
177. Hu Y, Jiang X, Ding Y, Ge H, Yuan Y, Yang C (2002) Synthesis and characterization of chitosan-poly(acrylic acid) nanoparticles. Biomaterials 23. https://doi.org/10.1016/S0142-9612(02)00071-6

Chapter 7
Biomedical Applications of Modified Chitosan Nanoparticles

7.1 Pickering Emulsions-Chitosan Tripolyphophate Nanoparticles in Drug Delivery System

Chitosan possesses outstanding biological characteristics that has contributed to its huge relevance in biomaterials formulation and substantial study in drug delivery systems. Chitosan is very biodegradable and biocompatible. Other essential qualities include hemostatic, bacteriostatic, anticholesteremic, anticarcinogenic, and fungistatic et al. [1]. Chitosan may be employed in several ways, based on the supplier's function and uses. Figure 7.1 displays the numerous compositions of chitosan materials in drug carriers and other biomedical applications, and their manufacturing processes [2].

The creation of extremely porous water-insoluble substances with huge surface areas and linked porous structures is increasing appeal in a variety of applications, include biomedical devices, packaging, sorption and separation research, electronics, and sensor systems [3, 4]. Due to their surfactant-free features and high microspores, biodegradable and biocompatible fully integrated porous structures formed from extremely high internal phase emulsions (HIPEs) have aroused the interest of scientists. The manufacture of such porous materials for biomedical applications (e.g., scaffolds) based on HIPEs includes the polymerization or crosslinking of the HIPEs' dispersion medium, accompanied by separation of the dispersed phase [5]. In a recent study, Zhou et al. created a very porous material that is straightforward and non-toxic for the first time by freeze-drying. Pickering gliadin-chitosan complex particle-stabilized high internal phase emulsions (HIPEs) [4]. When used as oil absorbent materials, the porous materials created utilizing HIPEs (PMHIPEs) shown a strong capacity for oil absorption [4]. Due to their great permeability, that supplies them with such a strong absorption capability, these substances are particularly significant in the abatement of oil spillage [6].

Even with its anti-oxidant and antibacterial qualities, chitosan is particularly enticing to the cosmetic sector. And for its great "water-holding" capability, it has

K. M. Zia, *Chitosan*, Smart Nanomaterials Technology,
https://doi.org/10.1007/978-981-96-2629-8_7

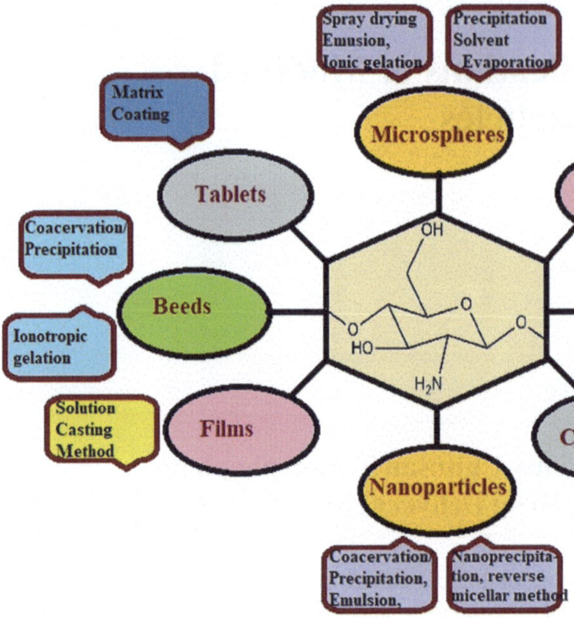

Fig. 7.1 Chitosan uses in drug delivery [2]

been claimed to offer better skin hydrating advantages. Chitosan is indeed employed as a thickness regulating ingredient in cosmetic formulations since polymer content and molecular weight may be altered to vary product fluidity. It's also been utilized to retain aroma sensory characteristics and lengthen their discharge [6]. Emulsion-containing creams, gels, pastes, ointments, and lotions are widely used in cosmetic, food, and medicinal applications. Colloidal dispersions comprised of two distinct stages—oil and aqueous—are the basis of straightforward conventional emulsions. Phase separation, coalescence, and Ostwald ripening; all contribute to the fact that emulsions are thermodynamically unstable systems [7]. To prevent these unwanted events that conflict with stabilization, stabilizers such as emulsifying agent and texture modifier are incorporated in emulsion compositions. Pickering emulsions seem to be emulsions that are stabilized by solid particles instead of standard emulsifying agents. Since these are surfactant-free, therefore are possible targets for mitigating the detrimental impacts of existing emulsifiers on the health and environment. It's have been named for S. U. Pickering, who found them in 1907. Pickering emulsions are retained by lowering the interfacial energy between the two immiscible phases by solid particle adsorption at the emulsion droplet surface. The hydrophilicity of the particles impacts whether an oil-in-water (o/w) or a water-in-oil (w/o) emulsion is formed [8]. Pickering emulsions based on chitosan are useful for food and cosmetic use owing to their biocompatibility and excellent stability [6].

Traditional injectable drugs are cumbersome to carry and are susceptible to induce inflammatory reaction as well as a whole-body allergic response when administered, among other reasons. Whereas the sustained-release performance of bio-microcapsules may sustain locally high medication levels and prevent drug loss at other sites, really improving therapy is achievable. To examine the sustained-release performance as a drug delivery carrier, CS/TPP nanoparticles (EE of 78.78%) are used. The release of BSA in CS/TPP nanoparticles are tested at 0.5, 1, 2, 4, 6, and 8 days, and the CS/TPP nanoparticles BSA discharge are reported [9]. The BSA release curve of BSA-loading CS/TPP nanoparticles in PBS (pH = 7.4) (BSA-loading CS/TPP nanoparticles are prepared at 0.5% CS, 0.5% TPP, pH = 5.7. The EE of BSA is about 78.78%) [9].

The studies has shown that the BSA released from the CS/TPP nanoparticles appeared sudden in the first 0.5 day, and subsequently the curve gradually moderate from first 0.5 day to the eighth day. Actually, BSA might release regularly and smoothly for up to one week. The transition in the curve seems to be due to two reasons: first, some BSA soaked up in the shell of the microcapsules during prep work is quickly released into the PBS solution; second, from 0.5 to 8 days, the BSA discharge is mainly reliant on the micro-porous exterior of the CS/TPP nanoparticles, which might control the releasing of BSA. The bulk of the BSA is discharged in 8 days because the CS/TPP shell progressively crumbled and the BSA contained within the CS/TPP nanoparticles released visually [9].

7.2 Delivery of Salazosulfapyridine

Salazosulfapyridine (SASP), a 1930s-discovered sulfa antibiotic, is frequently used to diagnose rheumatoid as well as inflammatory bowel diseases such as ulcerative colitis and Crohn's disease [10]. However, due to its low solubility (0.9 g/mL), this drug's bioavailability is significantly reduced [11]. As a result, to compensate for its low exposure, an oral dose of 1–3 g day 1 has been proposed [12]. The SASP's poor water solubility presents substantial hurdles in terms of generating high exposure following intravenous administration, which is limited by the administrable dosage volume. Several methods for increasing SASP solubility or dissolution rate have been developed. Shadid and colleagues [13] investigated SASP ionic liquid, and SASP absorption and dispersion in an ionic liquid salt have been assessed utilizing in vitro and in vivo models. da Costa and his colleagues [14, 15] described the characterization and in vitro dissolving behavior of sulfasalazine suspensions for the treatment of chronic intestinal inflammatory disorders.

Nanoparticles (NPs) have revolutionized drug delivery by Solubilizing pharmaceuticals, protecting against degradation, modulating tissue distribution and enhancing absorption and bioavailability. Chitosan, derived from chitin, is a naturally occurring; abundant polysaccharide composed of (1–4)-linked D-glucosamine and N-acetylated derivatives. Its unique properties make chitosan an attractive material for NP-based drug delivery systems [16]. The CS is rapidly being employed

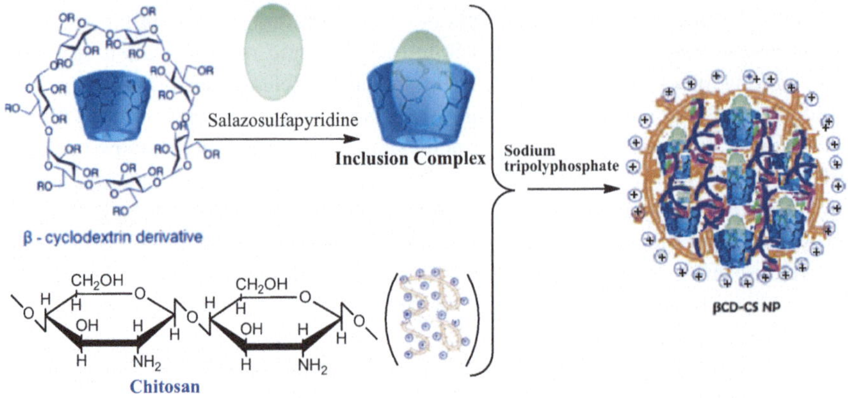

Fig. 7.2 Scheme of the formation of DMβCD/SASP inclusion complex-loaded CS nanoparticles [15]

for bio-medical reasons, including nano-sized drug delivery systems, because to its great features like as muco-adhesiveness, biodegradability, ease of chemical modification, and pH sensitivity. The integration of CDs into CS NPs for hydrophobic drug delivery has piqued the interest of researchers [17]. These NPs also offer many benefits, such as high drug loading, prolonged release capacity, and biocompatibility, which makes them excellent for drug delivery systems [18]. The CS may also produce NPs with polyanionic sources through ionotropic gelation [19]. When non-ionic CDs are coupled with sodium tripolyphosphate (TPP), it may create NPs with CS [18]. As a consequence, the DMCD/SASP inclusion complex is suggested to be included into CS to produce a new sustained-release NP system for SASP delivery. To produce and characterize the SASP inclusion complex with DMCD, several analytical approaches have been applied. Ionic gelation is used to produce complex-loaded NPs from the DMCD/SASP inclusion complex. The building strategy of DMCD/SASP inclusion complex-loaded CS nanoparticles is shown in Fig. 7.2. Several critical influencing factors, including the CS/TPP mass ratio, CS concentration, pH, and DMCD/SASP inclusion complex concentration, are optimized for the complex-loaded NPs. Tang et al. studied the particle size distribution, zeta potential, shape, in vitro drug release, and in vitro cytotoxicity of the resultant DMCD/SASP/CS NPs [15].

7.3 Polyelectrolyte Complex CSNPs and Poly(2-Acrylomido-2-Methylpropane Sulfonic Acid)

A novel manufacturing approach has been developed for creating theophylline granules coated with a polyelectrolyte complex comprising chitosan and sodium tripolyphosphate. Following the standard procedure, theophylline granules containing sodium tripolyphosphate are agitated in a chitosan HCl solution and sodium tripolyphosphate migrates to the granule surface, reacting with chitosan to form a polyelectrolyte complex film. Parameters influencing drug content, grain size, and coating-film thickness were investigated by increasing sodium tripolyphosphate/theophylline ratio (decreased chitosan in coating solution and increased theophylline in coated granule) or by rising chitosan content in coating solution (increased coated granule size and reduced reaction rate). It can be further optimized by enhancing chitosan concentration, coating solution pH, and sodium tripolyphosphate/theophylline ratio with increased coating-film thickness. Pharmacological evaluation involved drug release followed zero-order kinetics and extraction efficiency significantly reduced compared to uncoated granules. This coated granule formulation demonstrates potential for controlled drug release and improved pharmacokinetics [20].

Acyclovir (ACV) is really a purine nucleoside analogue antiviral medication used to treat herpes viruses [21]. It is presently accessible as a topical ointment, capsules with 200 mg, and tablets with 200, 400, and 800 mg [22]. The typical plasma half-life of the medication is 3 h [23]. The ACV should be delivered 5 times per day in the standard medication delivery mechanism, which leads in negative side effects especially with major dosages. Furthermore, the existing commercial formulations have a variety of drawbacks after orally administered. Since ACV adsorption from the gastrointestinal tract is facilitated by carriers, raising the ACV dosage led in carrier system saturate, dose-related adverse effects, and reduced bioavailability. Because of the deficiencies of commercially available products, novel strategies like polymeric drug delivery systems [24] have just been developed. Because of its biocompatibility, biodegradability, and low toxicity, chitosan (CS) has acquired appeal as a polymeric carrier for drug delivery systems, recently [25]. This is a copolymer of glucosamine and N-acetylglucosamine connected by (1–4) linkages. It is the sole polymer having cationic characteristics, enabling it to attach to negatively charged molecules such as anionic polymers, nucleic acids, and enzymes. By exploiting its $-NH_2$ and –OH functional groups in the graft polymerization procedure, it may be employed to easily manufacture hydrogels with good physicochemical parameters [26]. The bacteria *Xanthomonas campestris* secretes the anionic extracellular polysaccharide Xanthan gum (XG). The inclusion of glucuronic acid and pyruvic acid groups in the side chain confers anionicity to the chemical [27]. Because of its higher efficacy, particularly in harsh environments such as acid, high salt, and increased shear stress, it has attracted a lot of attention as among the most effective hydrocolloids. It can also conjugate with other polymers, proteins, peptides, and non-peptides, and these conjugates have high stability against proteolytic enzymes, inertness, biocompatibility, and efficient solubility. The water affinity of Xanthan gum (XG) increases the solubility of hydrophobic medicines or carriers [28]. Xanthan gum can be employed in modest doses to delay release in vitro and achieve zero-order release kinetics [29]. AMPS (2-Acrylamido-2-Methylpropanesulfonic acid) is a versatile hydrophilic monomer with salient features like ionic and non-ionic groups, strong salt resistance, enhanced hydrolytic stability due to sulfonic functional group and pH-sensitive behavior via ionizable sulfonate groups [30]. Cross-linking AMPS with biopolymers enables pH-responsive polymer systems, tailored release profiles and improved formulation stability. AMPS's unique properties make it an attractive component for designing advanced controlled release systems [31]. The goal of this study is to develop a new approach for producing pH-sensitive ACV hydrogels through non-covalent grafting of AMPS onto a polymer system comprising XG and CS. The hydrogels have been created through crosslinking and ionic interactions between the amino groups of chitosan and the carboxyl groups of xanthan gum. Graft copolymerization of monomer AMPS with the AMPS sulfonate group on the CS-XG backbone. Because of its high biocompatibility and non-toxicity, CS, a cationic polymer, is best chosen as a polymeric carrier. The XG, an anionic hydrophilic polymer, boosts the solubility of hydrophobic medications and transporters. Furthermore, monomer AMPS exhibit the previously described hydrogel pH-sensitive swelling properties. This allows the hydrogel to preferentially target the absorption region in the small intestine, increasing medicine bioavailability. As a result, Malik et al. established a novel method for producing the required hydrogel properties by mixing a cationic polymer CS with an anionic polymer XG [24]. A xanthan-chitosan hydrogel system with the needed structural stiffness, pH-sensitive swelling behavior, and drug release characteristics is produced by changing the formulation factors and content [24].

7.4 Chitosan-5-Flurouracil NPs

Cancer is defined as unchecked cell growth and differentiation that outcomes in tumor development; tumors can be cancerous (malignant) or non-cancerous (benign). In 2020, worldwide cancer statistics estimated nearly 10 million deaths, with up to 18.1 million new cancer cases discovered [32]. The traditional unanswered questions are low stamina rate, non-specific cytotoxicity of malignant cells, and in vitro development of multidrug conflict toxicity [33]. These adverse effects of traditional chemotherapy drugs, alone or in combined effect, indicate that greater emphasis is placed instantly on helping to improve effectiveness, which has a negative socioeconomic impact. Although there are numerous anticancer agents capable of treating various types of cancer, inadequate drug release due to an unspecified targeted site of action can bring injury to other tissues and organs. Chemotherapeutic agents can be supplied to specific areas of action using a silent targeted drug delivery system without the need of a ligand (active targeting) [33]. The leaky vasculature characteristic of cancerous cells enables nanoparticles (NPs) to enhance capillary permeability and prolong residence time at the tumor site. This phenomenon, known as the Enhanced Permeability and Retention (EPR) effect, facilitates targeted drug delivery. Key benefits of EPR-mediated nano-carrier systems involve improved drug accumulation at tumor sites, enhanced therapeutic efficacy, reduced systemic toxicity and prolonged drug release. Ideal characteristics of NPs for EPR affect 10–100 nm size, targeting ligands, stealth coatings surface modification and biocompatible, biodegradable polymers material. By leveraging the EPR effect, nano-carrier drug delivery systems can selectively target cancer cells, improving treatment outcomes [34]. The nanoparticles (NPs) with added features, such as stable and compact behavior in reaction to external, have been developed by modern researchers and are known as clever, sophisticated, ecologically sensitive, or responsive nanoparticles [35]. Drug delivery systems (DDSs) based on nanotechnology are novel methodologies for the pharmaceutical industry. Nanotechnology involves nanoparticles (NPs) of nanoscale size. The primary goals of developing nanoparticles for oral drug delivery systems (ODDSs) are discovered to be working to improve drug(s) solubility, dissolution rate, drug release percentage control, and target specificity, resulting in increased bioavailability [36]. These tiny drug delivery systems are built of nanostructures that are hydrophilic, hydrophobic, biodegradable, biocompatible, and non-toxic. These have been developed and put to use using a range of physical and chemically based nanoparticle formulation techniques, such as foremost and multiple emulsion solvent evaporation approaches, ionic gelation, aerosol, supercritical fluid technology, and antisolvent precipitation using compact fluid tools. Through the chemical production of silica nanoparticles with uneven structural components, medicinal agent solubility, permeability, and bioavailability(s) have now been modified [37]. Chitin deacetylation produces the biopolymer chitosan, an amino polysaccharide that is non-toxic and biocompatible (poly-1,4-D-glucosamine). Chitosan derived from crustacean shells, insect cuticles, and fungal cell walls. Applications involved oral drug delivery systems (stabilizing/suspending agent), chelating agent for toxic metal removal, nuclear waste management and ion exchange handling (various industries) having biodegradable, biocompatible, non-toxic and high molecular weight properties. Chitosan's unique properties make it an ideal material for diverse applications across pharmaceutical, environmental remediation, biotechnology and cosmetics. Its versatility and sustainability have led to increasing demand in various sectors. Chitosan-NPs, however, are subject to a variety of external factors that affect rapid drug dispersion throughout the meshwork [35]. A recent research by Sethi et al. showed that a cross-linking matrix containing a di-aldehyde such as glutaraldehyde may successfully limit or limit drug diffusion from a chitosan network. The schematic preparation and TEM micrograph of 5-FU/GA-LMWCH-NPs is shown in Fig. 7.3 [32].

5-Fluorouracil (5-FU), an anticancer medication developed in 1958, exhibits high therapeutic effectiveness against solid tumors, including colon, rectum, and breast malignancies [38]. However, its oral formulation faces limitations due to irregular absorption via the gastrointestinal tract (GIT), rapid peak blood levels (15–60 min), and non-specific toxicity in healthy cells, quick renal clearance and metabolism by dihydropyrimidine dehydrogenase (DPD) enzyme and severe digestive discomfort. Regarding its metabolic fate, the 80% of orally administered 5-FU is metabolized by the liver and kidney [39], F-β-alanine is detoxified and excreted in urine with short half-life (8–20 min) due to rapid metabolism in blood [40]. To overcome these challenges, researchers have developed orally administered 5-FU formulations using biocompatible and biodegradable polysaccharides, pH-sensitive polymers for prolonged and targeted release and cross-linked chitosan nanoparticles for improved drug loading and entrapment efficiency [41, 42]. The primary goal of this study is

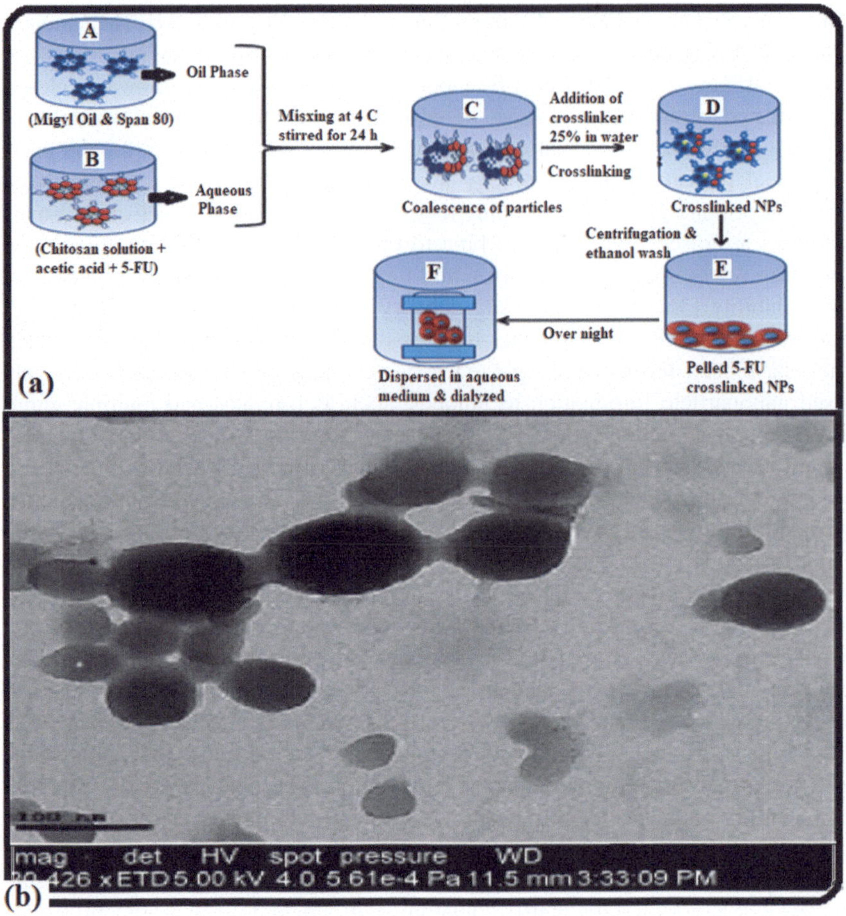

Fig. 7.3 **a** Schematic preparation of 5-FU/GA-LMWCH-NPs; **b** TEM micrograph of 5-FU/GA-co-LMWCH-NPs [32]

to design and evaluate low-molecular-weight cross-linked chitosan nanoparticles as novel drug delivery carriers, aiming to enhance 5-FU dissolution, solubility, and bioavailability and improve efficacy and reduce toxicity in solid cancer treatment [32].

7.5 Propranolol CSNPs

Nanoparticles have garnered significant attention in pharmaceutical research due to their controlled drug release capabilities, protective properties against harmful environments and enhanced drug absorption and permeation through mucosal membranes

[43, 44]. The buccal mucosa offers an advantageous route for systemic drug delivery via nanoparticles, bypassing first-pass metabolism and gastrointestinal degradation. The benefits involved improved bioavailability, enhanced therapeutic efficacy and reduced systemic side effects. Nanoparticle-mediated buccal delivery holds promise for optimizing drug therapy, particularly for macromolecules (proteins, peptides), hydrophobic compounds and vulnerable molecules (nucleic acids) [45]. Gums, chitosan, alginate, and agarose are examples of natural polymers that have been used to improve the cell penetration of buccal medication delivery [46]. Chitosan is an attractive polymer because to its inherent qualities, which include low toxicity, biocompatibility, biodegradability, mucoadhesion, and membrane permeability. Chitosan facilitates medication penetration through the mucus barrier by promoting both transcellular and paracellular transport [47]. Transcellular transport is performed by the transcytosis of epithelial cells. The positive charges on chitosan engage with negatively charged mucus components, producing tight junction reconfiguration and epithelial junction release [47]. On the other hand, chitosan interacts with negatively charged mucus via hydrogen and hydrophobic bonds. Chitosan has furthermore been investigated as a potential carrier for targeted therapeutic treatment in addition to improving pharmaceutical penetration. When given directly to particular cells using chitosan nanoparticles, an anticancer preparation (a mix of fluorouracil and quercetin) has increased efficacy against pancreatic cancer cells. Both medications are quickly absorbed into cancer cells thanks to the surface amine groups of chitosan nanoparticles [48]. Additionally, the cationic chitosan molecules used in the creation of the nanoparticles may interact with the anionically charged small molecules. Chitosan has been used to make nanoparticles utilizing renin–angiotensin–aldosterone gelation, co-precipitation, microemulsion, emulsification solvent diffusion, solvent evaporation, and reverse micellar methods [49]. Ionotropic gelation is among the most frequent methods for creating drug delivery nanoparticles, and it includes an ionic connection between the positive charges of chitosan and the negative charges of polyanion molecules like tripolyphosphate (TPP) [50, 51]. Ion channel gelation is a straightforward process with various benefits, including gentle circumstances, aqueous formulation, and poor toxicity [34]. Propranolol, a non-selective adrenergic antagonist, has also been used to treat cardiovascular illnesses like hypertension, arrhythmia, and angina pectoris [52]. The fundamental concern with standard propranolol pill administration is first-pass metabolism. Propranolol is rapidly and virtually fully absorbed via the gastrointestinal system following ingestion. However, limited bioavailability (25%) has indeed been found owing to significant first-pass metabolism [53]. Nanoparticle-based propranolol techniques have been created to overcome this problem [43]. Few research has looked on propranolol-loaded nanoparticles. The impacts of chitosan molecular weight and propranolol concentrations on the physical characteristics of nanoparticles have been reported by Duangjit et al. [54]. However, no thorough information on the impacts of propranolol concentrations on nanoparticle characteristics encompassing the spectrum of dosages presently accessible in commercial tablet formulations is presently available, and the effects of the initial chitosan solution pH have not previously been described. To improve nanoparticle formulations, the distinctive characteristics of nanoparticles must be

established. The pH, zeta potential, and nanoparticle size of propranolol-loaded chitosan nanoparticles are well studied. The effects of altering the pH of the chitosan solution on the nanoparticle characteristics are also studied [55]. The concentration and pH of the chitosan solution both influenced the nanoparticle pH, but the zeta potential is solely changed by the quantity of propranolol. Moreover, the pH of the chitosan solution influenced the size of the nanoparticles [55].

7.6 Desvenlafaxine-Loaded PLGA CSNPs

Poor mood, despair, melancholy, low self-esteem, sleep disturbances, and, in certain instances, suicidal thoughts are also symptoms of depression [56]. Depression is among the primary physical illnesses, impacting approximately 350 million people globally, according to the most current WHO data [57]. One out of every twenty individuals has suffered depression, and per the World Mental Health Study, which has been done in 17 nations (WHO 2015). Suicides, which take 1 million lives each year, have also been linked to depression [57]. Interruption in monoaminergic transmission in the brain, i.e., norepinephrine, serotonin, and dopamine, are the major cause of depression, and are induced by a complex combination of various social, behavioral, and biological variables [56]. As a consequence, most antidepressants boost monoamine availability at the junction via a variety of ways. Nonetheless, such antidepressants' therapeutic efficiency is contingent on their protracted availability at the brain's action site [56]. Antidepressants are presently largely taken orally due to the chronic character of therapy spanning many days. The blood–brain barrier (BBB) and blood-cerebrospinal fluid (BSF) barriers restrict oral antidepressants' entry into the central nervous system (CNS), limiting their efficacy. The key challenges involved restricted brain penetration, fluctuating plasma drug concentrations, side effects and intolerance and low remission rates (30% after first trial) [58]. The common antidepressants involve tricyclic antidepressants (TCAs), selective serotonin reuptake inhibitors (SSRIs) and serotonin and norepinephrine reuptake inhibitors (SNRIs) having some limitations like reduced effectiveness over time, high malfunction rate after initial trial. The targeted brain delivery, sustained release formulations, nanoparticle-mediated transport and blood–brain barrier-permeable analogs are considered as novel delivery systems. Advancing antidepressant delivery systems is crucial to improve treatment outcomes and enhance patient quality of life [56]. As a consequence, scientists and pharmaceutical corporations have been striving to find efficient antidepressant medications with immediate and long-term recovery. This is evident in the fact that the FDA has authorized a plethora of new antidepressant medications in the previous three years, including vilazodone, levomilnacipran, vortioxetine, and 20 more that are now in clinical testing. These results support the development of intranasal nano versions that can transport antidepressant medications to the brain non-invasively and boost brain focus while lowering dose, side effects, and intolerance. In recent years, intranasal (i.n.) nose to brain administration has developed as a unique method for delivering therapeutic drugs to the CNS [59]. The olfactory and/or trigeminal nerve systems offer a unique link between the olfactory epithelium and the CNS, circumventing the BBB, allowing enabling nose-to-brain medication transport [59]. Because of its apparent advantages, like avoiding BBB and hepatic first-pass metabolism, non-invasiveness, and convenience of administration, nose-to-brain drug delivery has lately been described in a number of papers [59]. Desvenlafaxine succinate (DVF), a second-generation SNRI identified as a pharmacological option for nose-to-brain administration, as examined by Tong et al. [60]. Desvenlafaxine is indeed an effective venlafaxine metabolite with a plasma half-life of 11 h as well as an oral bioavailability of 80%. Despite having a superior serotonin: norepinephrine ratio (10:1) than Venlafaxine (30:1), Desvenlafaxine is linked to a variety of negative impacts, including elevated blood pressure and heart rate, constipation, agitation, tremor, sweating, nausea, headache, and sleep difficulties [61]. For the delivery of therapeutics, biodegradable polymeric nanoparticles made of polylactide-co-glycolide (PLGA) and chitosan (CS) are optimized. The biodegradable, biocompatible, and bio-adhesive nature of PLGA allows the drug to spend a lot more time inside the nasal cavity and limit nasal mucus secretion, leading to better absorption throughout the nasal epithelium, including the olfactory regions, for nose-to-brain delivery [61]. Tong et al. designed a nano-sized drug delivery system employing PLGA and CS for intranasal nose to brain administration of DVF and used pharmacodynamic and biochemical tests to assess its antidepressant effectiveness in animals. Tong et al. looked examined the pharmacokinetics of DVF-loaded PLGA-CS nanoparticles in the brain and plasma to see whether these might predict pharmacological and biochemical outcomes [60].

7.7 Quetiapine Fumarate-Loaded CSNPs

Two primary categories of antipsychotic medications are conventional antipsychotics and atypical antipsychotics. Atypical antipsychotics are preferred as first-line treatment due to reduced risk of neuropsychiatric symptoms and improved cognitive performance [62]. Quetiapine Fumarate (QF): A Second-Generation Atypical Antipsychotic has demonstrated enhanced efficacy against positive and negative symptoms, improved neurocognition features and superiority over olanzapine and risperidone in schizophrenia treatment. The key benefits of QF involved are effective symptom management, improved cognitive function and better tolerability profile. QF's unique pharmacological profile makes it a valuable option for schizophrenia management [63]. Because of its 6 h plasma half-life, QF must be taken on a daily basis to maintain therapeutic concentrations. Because of considerable first-pass metabolism, QF has a poor oral bioavailability of 9%, resulting in inadequate absorption and concentration at the target location. Little absorption is predicted since QF is a weakly water soluble medication with diminishing solubility as pH rises. The blood–brain barrier (BBB) poses significant obstacles to systemic therapy for schizophrenia due to tight junctions, P-glycoprotein (P-gp), acting as a gatekeeper and restricted permeability. Consequently reduced medication efficacy, variable drug concentrations and increased risk of side effects occur. The strategies to overcome BBB limitations involved targeted drug delivery systems, nanoparticle-mediated transport, BBB-permeable analogs, inhibition of P-gp activity, blood–brain barrier disruption techniques. It is pertinent to mention here that advancing BBB-penetrant therapies is crucial for improving schizophrenia treatment outcomes. The QF has a low concentration in the brain after oral treatment because it is a P-glycoprotein substrate [64]. For improved bioavailability, a well-defined route of administration with an adequate dose form would be necessary to overcome BBB constraints and restricted absorption at the target location. Invasive and non-invasive technologies may both be used to deliver medicines throughout the brain while overcoming hurdles. Invasive drug delivery methods (intracerebral or intraventricular) need a temporary breakdown of barrier integrity to open tight junctions, which is typically dangerous and unfriendly to patients, resulting in more harmful molecular access into the brain. Non-invasive procedures allow for direct medication administration throughout the brain without the need for surgery [65, 66]. Intranasal drug delivery to the brain has attracted a lot of attention in recent decades as a feasible alternative to oral and parenteral administration procedures. Direct transport via olfactory neurons and trigeminal nerves, which bypasses the BBB, and/or absorption into systemic circulation, which depends on molecular weight and lipophilicity, may both traverse the BBB [67, 68]. The intranasal route is favored over the oral route for enhancing drug input at the BBB because of its high absorptive surface area, avoidance of first-pass metabolism, and rich vascular structure [69]. Because of their capacity to target active areas, colloidal carriers with diameters ranging from 10 to 400 nm are replacing traditional medication therapy in the treatment of a variety of brain illnesses. By decreasing drug dosage, these new nanocarriers enable greater solubility, better brain targeting, and less toxicity [70]. Because of their biodegradability, biocompatibility, and simplicity of manufacture, polymeric nanoparticles (NP) are increasingly being employed in nasal nanotherapeutics. The nanoparticle (NP) allows for simple access across the BBB by effectively encapsulating drug molecules and enhancing their diffusion across biological membranes. It may shield encapsulated drugs from biological, chemical, and enzymatic degradation, as well as an extracellular mechanism known as the P-glycoprotein efflux system, allowing them to reach their target location more quickly [71–73]. The nanoparticle (NP) derived from natural polymers such as chitosan has been shown to increase nasal absorption and brain targeting of a variety of neurotherapeutics [71, 74–76]. Chitosan is a natural carbohydrate polymer having bio-adhesion, non-toxicity, and biodegradability among its benefits. Due to the muco-adhesion phenomenon, it has the benefit of avoiding the natural defense mechanism known as mucociliary clearance while simultaneously demonstrating penetration boosting characteristics. Chitosan's positively charged amino groups interact with the nasal mucosal surface's negatively charged surface, enabling it to stay in the nasal cavity for longer. Through brief tight junction opening, this leads in enhanced para-cellular drug trafficking across the membrane [72, 77]. Shah et al. created a Quetiapine fumarate nanoparticles (QF-NP) system to see whether using a non-invasive intranasal (IN) route instead of an intravenous (IV) route may improve brain delivery. The QF-loaded nanoparticles

(QF-NP) are created using the ionic gelation process, which are subsequently optimized using the Box–Behnken design (BBD). The QF-NP has been investigated in terms of physicochemical properties, solid state characterization, morphology, nasal diffusion, and histology. Pharmacokinetic investigations are undertaken to measure brain targeting efficiency and direct transport % after IN and IV doses of QF-NP and ex-vivo nasal diffusion of QF-NP and QF solution (QF-DS) in PBS pH 6.4 [78].

7.8 Rutin-Encapsulated CSNPs

Stroke is the world's third most frequent neurological illness, leading to long-term impairment and increased morbidity and death [79]. Stroke is produced by cerebral ischemia, which permanently affects the central nervous system (CNS) [80]. The determining factor (produced as a consequence of inflammatory process activation) is assumed to be the cause of severe brain damage, morbidity, and death after cerebral ischemia occurs [81, 82]. In addition to the aforementioned factors, the study presented below shows that oxidative stress plays an important role in brain ischemia reperfusion damage. As a consequence of the requirement to consume a significant quantity of oxygen in reaction to oxidative stress, free radicals and reactive oxygen species (ROS) are produced, which may be damaging [79]. Because the brain's antioxidant defense system is immature, the oxidants or free radicals produced by inflammatory cells endanger tissue viability in the ischemic core, making ischemia reperfusion a key source of pathogenesis [79, 83]. As a consequence, pharmacological modification for oxidative damage therapy might become a crucial component of treatment. Neuroglobin (Ngb), a recently discovered globin, has been associated to protection against hypoxic/ischemic cell damage, indicating that it may play a role in brain protection [84, 85]. Hypoxic/ischemic damage, according to the study, results in the formation of reactive oxygen species (ROS) and reactive nitrogen species (RNS), which successfully combats and so promotes neuronal cell survival [86]. After middle cerebral artery occlusion (MCAO) injury to the brain, ubiquilin-1 (a biomarker utilized in RNS and ROS) may oxidize multiple proteins, producing misfolding and alterations in three-dimensional protein structures [87]. In animal models of cerebral ischemia, antioxidant medicines such as thymoquinone, curcumin, ropinirole, and thioperamide have been demonstrated to diminish ROS-mediated responses and preserve neurons from reperfusion-induced neuronal death [79, 88]. Rutin [88, 89], a lipophilic molecule found in *Ruta graveolens* and *Carpobrotus edulis*, has recently been proven to be an effective therapy for cerebral ischemia. Rutin, on the other hand, has several drawbacks, including low water solubility and thus lower bioavailability, chemical and enzymatic degradation in the gastrointestinal tract as a result of its lipophilic nature, which can cause permeability issues, and extensive hepatic first-pass metabolism [90]. The blood brain barrier (BBB), a protective natural barrier around the central nervous system (CNS) that inhibits medication delivery to the brain, is another major challenge in drug discovery and development of an effective formulation to treat and control cerebral ischemia. The BBB is unaffected by even the most powerful drug delivery systems (DDS), such as intravenous, buccal, and transdermal routes [91, 92]. Several magnetic drug targeting and drug carrier system techniques, including as liposomes, antibodies, nanoemulsions, nanoparticles (NPs), or nanomicelles, have been developed to overcome the aforementioned problems. Nanoparticle production, among the aforementioned methods, may be the most effective for establishing regulated and targeted delivery systems for a variety of medications, including natural and synthetic pharmaceuticals, vaccines, proteins, and biological macromolecules [93]. Drug transfer from blood to cerebral spinal fluid (CSF) occurs across the choroid plexus, which is leaky compared to the BBB. Virtually all small and large molecules in blood penetrate into CSF, at a rate inversely related to the molecular weight of the substance. Similarly, the intranasal route demonstrated a number of benefits, including significant drug transportation into CSF and the olfactory bulb via olfactory sensory neurons, non-invasiveness, effectiveness, safety, and painlessness, less skill required for drug delivery, and the property of localized therapeutic effect with fewer side effects [79]. Ahmad et al. used the same in-drug delivery system and brain targeting method to achieve therapeutic goals by avoiding variables that could result in reduced Rutin access in the brain, such as first-pass metabolism and distribution to non-targeted locations, and thus reducing peripheral adverse effects [90]. Rutin-loaded muco-adhesive polymeric nanoparticles (NPs) developed as a result to increase nasal residency and delay medicine release to the brain [94]. To assess the direct nose-to-brain transport route, Ahmad et al. investigated the pharmacokinetics of Rutin-loaded NPs in the brain and plasma following intravenous and intramuscular injection in Wistar rats [90].

7.9 Piperine-Loaded CSNPs

Herbal therapy has lately been suggested as an alternate technique for developing drugs for a number of neurological illnesses, such as Alzheimer's disease and epilepsy [95]. Piperine, the major alkaloid constituent in black pepper fruits, has received a lot of attention in the treatment of epilepsy among all natural products [96]. Antimicrobial activity [97], antidepressant activity [98], anticancer activity [99], anti-inflammatory activity [100], cognitive enhancement [101], and antioxidant activity [102] are some of the pharmacological benefits of this medicine. Piperine or its derivatives have been shown in a rising number of trials to considerably decrease seizure behavior [103]. Piperine's anticonvulsant impact has lately been investigated in terms of cellular processes. According to Chen et al. [104], piperine's antiseizure action is mediated via the activation of transient receptor potential cation channel subfamily V member 1 (TRPV1) receptors. Piperine has been proven to lessen seizure behavior through the hippocampus, cortex, and other brain regions by Mishra et al. the TRPV1 receptors have been shown to be abundantly expressed in the hippocampus, cortex, and other brain locations [104]. The TRPV1 overexpression has also been seen in the temporal lobe epileptic mice's dentate gyrus. The addition inhibits sodium and calcium channels [103]. Despite its anticonvulsant properties, piperine's medicinal applicability is restricted owing to its weak water solubility and, as a consequence, low bioavailability [104]. Piperine water solubility has been increased using a variety of methods, including liposomes [105], hydrogel [106], and others. Chitosan (CS), a biodegradable and biocompatible polymer, has been suggested as a potential drug carrier lately [107]. The CS is a cost-effective, non-toxic, and biodegradable material. The CS may interact favorably with cell membranes due to its positive charge [108]. Ionic gelation has been found to be a viable approach for producing CSNPs in many investigations. A CS solution is combined with an adequate amount of STPP as a cross linker in this method [109], giving CS-STTP NPs of the required size. Piperine may be encapsulated with a number of molecules, including CS-STPP, PLGA, and solid lipid NPs. Piperine-loaded NPs have been researched for their impact on breast cancer, Alzheimer's disease, and Rheumatoid arthritis, and experimental data shows that it dramatically lessen the severity of these illnesses. The pathophysiology of epilepsy has been studied using a variety of animal models. The GABAA antagonist pentylenetetrazol (PTZ) is often employed in animals to generate the chemical kindling paradigm. One of the brain areas most vulnerable to kindling stimulation has been identified as the dentate gyrus. In the hippocampus of fully ignited mice, neuronal loss and glial activation have been extensively described [109]. Using a PTZ-induced kindling paradigm, Anissian et al. studied the antiseizure and neuroprotective characteristics of piperine-loaded NPs. The amount of activation of astrocytes in completely kindled rats given piperine-loaded NPs has also been studied [109].

7.10 Hyaluronic Acid-Coated CSNPs

Cancer is without a doubt one of the most common diseases on the planet. Every year, we encounter new cases that are becoming more prevalent. The survival rate for this disease has not improved in the last 30 years, despite new research procedures and therapies [110]. Cancer is without a doubt one of the leading causes of death on a national and global scale. When we look back at the history of cancer, we can see that over 15.5 million children, youths, and adults in the United States has survived cancer as of January 1, 2016. The number of cancer survivors is expected to reach 20.3 million by January 1, 2026, with roughly 10 million men and 10.3 million women [111]. Prior to the previous several decades, there have been standard treatment improvements such as chemotherapy and radiation that saw numerous breakthroughs in cancer-cure, but has limitations and side effects. Because it acts on non-target areas, a powerful anticancer bioactive has significant side effects when injected. Because of its non-specific activity, the drug has a low impact at the tumor site. However, when the medicine enters a healthy tissue, it causes toxicity, rendering the effective dosage useless.

To overcome limitations in cancer treatment, researchers have focused on developing tumor-specific drugs and delivery systems. Recent nanotechnology advancements hold promise for improved targeted delivery [112]. Hyaluronic Acid (HA)

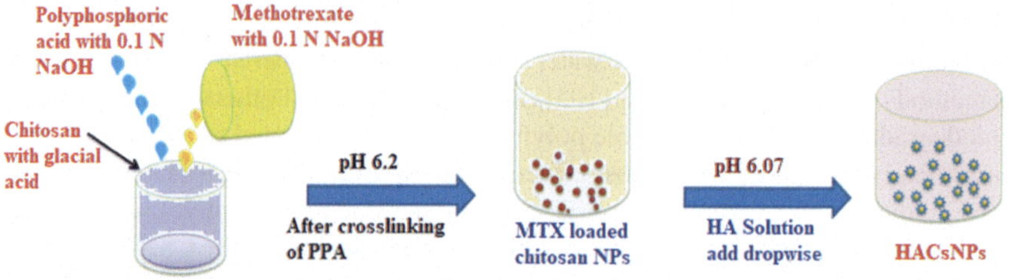

Fig. 7.4 Preparation of MTX-loaded HA anchored CSNPs containing polyphosphoric acid (HACSNPs) [112]

is a naturally occurring, high-molecular-weight glycosaminoglycan composed of repeating disaccharide units [113] with molecular weight range: 100–5000 kDa [114] derived from vitreous fluid of cow eyes. Biological functions include extracellular matrix development and organization, nutrient transport, cell adhesion regulation, morphogenesis and inflammatory modulation. HA receptors in cancer includes cluster of differentiation 44 (CD44), receptor for hyaluronic-acid-motility (RHAMM) and associated with cancer initiation and progression [112]. Preparation of MTX-loaded HA anchored CSNPs containing polyphosphoric acid (HACSNPs) is shown in Fig. 7.4. The HA is involved in the development and organization of the extracellular matrix (ECM), as well as nutrient transport, cell adhesion regulation, morphogenesis, and inflammatory modulation. The HA-binding proteins on cell surfaces had been discovered in the late 1980s. Two HA receptors, cluster of differentiation 44 (CD44) and the receptor for hyaluronic-acid-motility motility (RHAMM), have been associated to cancer initiation and progression RHAMM [112]. HA's unique properties and receptor specificity make it an attractive component for targeted cancer therapy.

Chitosan-based nanoparticles for methotrexate delivery have been reported by many researchers. Chitosan, a cationic amino-saccharide polymer having cellulose-like carbohydrate backbone structure with N-acetyl-D-glucosamine and D-glucosamine repeating units with (1–4)-glycosidic linkage. The advantages involved muco-adhesive properties, easy cross-linking, and polymeric particle formation without harmful solvents. On the other hand polyphosphoric acid (PPA): an inorganic cross-linker forms intermolecular and intramolecular cross-links with chitosan creates nanosized polymeric network. It is worth to mention that chitosan/PPA nanoparticle are soft, rubbery stiffness with high water content, particle size, shape, and surface charge controlled by chitosan/PPA ratio. Methotrexate (MTX), a folate analogue inhibits dihydrofolate reductase (DHFR) and treats cancer (lymphoblastic leukemia, breast cancer) and autoimmune diseases. The pharmacokinetics of the materials involved peak plasma concentrations: 1–10 μM (25–100 mg/m^2), renal clearance: 2–3 h half-life and triphasic elimination: 8–10 h half-life [112, 115, 116]. Chitosan/PPA nanoparticles offer a promising platform for controlled MTX release and improved cancer therapy.

7.11 Naringenin-Loaded Sulfobutylether-β-Cyclodextrin CSNPs

The most common cause of severe and permanent vision loss in senior people in industrialized nations is age-related macular degeneration (AMD), a form of senile spontaneous disorder [117]. In this condition, the retinal pigment epithelium and macular degenerate over time, resulting in irreversible central vision loss. Because of the relatively complex and diverse pathogenic variables, such as age, inheritance, smoking, a lack of nutrition, obesity, chronic illnesses, and so on, the etiology of AMD is unknown. Photodynamic therapy, irradiation, sub macular surgery, laser photocoagulation, trans pupillary thermotherapy, and medication treatment have all been studied in recent decades for various types of AMD, including neovascular AMD. Anti-VEGF medications have largely replaced earlier treatments for wet AMD, such as photodynamic therapy (PDT) and laser photocoagulation, in terms of visual prognosis [118]. The utilization of local ocular drug administration to treat AMD is the focus of our research. A potent flavanone found in nature called naringenin (Nag) is widely taken as a dietary supplement. Nag has recently been demonstrated to have a variety of pharmacologic effects for oculopathy, including choroidal neovascularization, degeneration of the retinal pigment epithelium, restoration of retinal function, and age-related macular degeneration [119, 120]. These effects may be due to enhance in ocular blood circulation and antioxidant activity. Because ocular blood flow is linked to age-related muscle degeneration, Nag was chosen as the model medication for AMD therapy. Nag, on the other hand, has a very limited application due to its low solubility. As a result, for Nag's therapeutic potential, finding a safe and effective carrier is critical. Some researchers used SBE—CD to build an inclusion complex to increase Nag's water solubility. The SBE—CD is a well-known modified cyclodextrin molecule that has been widely used in tablet, injectable, and capsule dosage forms [121]. It is highly soluble and has a low toxicity. Negatively charged CD's charged side chain can also cause ionic gelation, resulting in nanoparticles containing positively charged chitosan [122]. The key benefits of chitosan (CS) involved biodegradable, biocompatible, low toxicity, enhanced membrane permeability, abundant and cost-effective. CS is an ideal material for sustained release, improved corneal penetration and reduced side effects and is widely applicable for glaucoma, cataracts, and macular degeneration treatment and antibiotic, anti-inflammatory, and anti-angiogenic delivery. On the basis of these applications it is recommended for Ocular Drug Delivery [123]. Due to the issue of quick and broad pre-corneal loss induced by drainage and high tear fluid turnover in normal topical ophthalmic drug administration, increased medication corneal penetration provides superior therapeutic benefits for many ocular disorders. The cornea and conjunctiva may interact with positively charged CS, increasing drug concentration and length of stay [124]. Zhang et al. used the SBE—CD and CS ionic gelation technique to generate Nag-loaded nanoparticles in a mild environment, which has been shown to increase intraocular drug penetration and lengthen corneal residency in recent years. Because of the charge attraction of cationic CS, this formulation not only increases Nag solubility but also extends the residence time for local ocular mucosa. Physicochemical parameters such as particle size, zeta potential, shape, and structural characterization are used to assess the nanoparticles. The release rates of Nag nanoparticles in rabbits in vivo are compared to the release rates of Nag suspension in rabbits; drug was released at a faster rate from Nag-CD as compared with those from Nag-suspension and Nag-CD/CS-NPs [123].

7.12 Primaquine Phosphate-Loaded CSNPs

The drug delivery systems are made to deliver a large amount of the active pharmaceutical component to the appropriate area at the right time in order to elicit the desired therapeutic response. Smaller delivery technologies (microparticles and nanoparticles) have been developed to improve therapeutic performance [124]. Nanoparticles (NPs) with sizes ranging from 10 to 1000 nm can be used to dissolve, entrap, encapsulate, or connect active ingredients to the outer surface of polymers [125]. When the medicine is encased by polymers in NPs, the formulation is known as nano capsule. Nanospheres (NS) are defined as drugs that are evenly scattered throughout the NPs matrix [126]. The NS are typically used to deliver medications in a controlled and consistent manner, reducing drug administration frequency and systemic toxicity. Several polymers are used in the synthesis of NPs for medicinal and side effect purposes. Distribution studies show that nanoparticles (NPs) are rapidly taken up by reticuloendothelial cells and preferentially accumulated in the liver. The successful applications involved malaria (primaquine, PQ), amoebiasis (metronidazole), liver fibrosis (sodium ferulate), hepatic carcinoma (5-fluorouracil, 5-FU) [127]. The ideal carriers involved chitosan: non-toxic, biodegradable, and non-accumulative and bovine serum albumin (BSA): biocompatible and adaptable [128]. The intravenous colloidal delivery systems involve microspheres, liposomes and emulsions with enable targeted delivery to specific organs or sites, enhancing therapeutic efficacy and reducing systemic side effects. The key benefits involve are improved bioavailability, enhanced liver targeting, reduced toxicity and increased treatment effectiveness. The nanoparticulate drug delivery systems hold promise for optimizing liver-directed therapies. In a colloidal system, particle size is critical for medication distribution in the human body. The liver and spleen are normally responsible for removing large particles. In drug delivery systems, smaller particles are more stable than larger particles. Smaller colloidal particle carriers, such as those in the 100–200 nm range, may be more stable and have a greater chance of exiting the circulatory system via holes in blood vessel linings. It may result in a lower drug concentration in the blood [129]. Chills, fever, discomfort, and sweating are all signs of malaria. Plasmodium parasites have been found in over 200 species of reptiles, birds, and mammals. The five Plasmodium species that cause human malaria are *Plasmodium vivax*, *Plasmodium falciparum*, *Plasmodium ovale*, *Plasmodium malariae*, and *Plasmodium knowlesi*. Primaquine is the sole treatment for parasitic liver schizonts. Hemolytic anemia, methemoglobinemia, leucocytopenia, leukocytosis, gastrointestinal problems, and stomach pain have all been linked to Primaquine (PQ) [127]. Since a consequence, minimizing the amount of PQ in the blood compartment by targeting the medicine to the liver tissue is crucial, as doing so may improve therapeutic efficiency while lowering toxicity and dosage. Rajera et al. synthesized the PQ-CS-NPs that are intended to promote NP entrapment in response to these issues [130].

7.13 Urocanic Acid Modified CSNPs

The CD98, a cell-surface amino acid transporter, plays a crucial role in cell homeostasis, epithelial adhesion/polarity and immunological responses [131, 132]. It has been reported in the established literature that CD98 is highly expressed in activated macrophages, contributing to inflammation initiation and progression [133]. Blocking CD98 expression may mitigate inflammation and RNA interference (RNAi) is mediated by siRNA which shown effective in reducing CD98 expression in macrophages [134] having limitation by poor stability, low cellular absorption, and rapid enzymatic degradation [135]. The delivery systems involve viral carriers with high transfection efficiency, but associated with immunogenicity and carcinogenic risks [136, 137] and non-viral carriers (liposomes, dendrimers, bioconjugates) with safer, simpler to synthesize, and more stable characteristics [138]. Chitosan-based siRNA delivery forms nanoparticles through interpolyelectrolyte complexation [139] with limiting step releasing from acidic endosomes/lysosomes [140]. The future directions involve optimized chitosan-based siRNA delivery systems, enhanced endosomal escape and cytoplasmic release and explored alternative non-viral carriers for CD98-targeting siRNA. Exogenous substances (such as chloroquine) have been added to NPs in order to allow them to escape endosomes/lysosomes. These exogenous agent-containing NPs, on the other hand, are cytotoxic and immunogenic, making them unsuitable for siRNA delivery. Because it is an amino acid building component with a pKa of roughly 6.0, imidazole is especially biocompatible. It may attach to negatively charged endosomal membranes, causing ion and water influx, endosome destabilization, and NP release. As a consequence, it's often used with polymers to boost the effectiveness of siRNA transfection. Urocanic acid-modified chitosan (UAC) with an imidazole group is recently created and proven to enhance UAC/DNA NPs release into the cytoplasm during endosome/lysosome rupture (via the proton-sponge mechanism). No prior research has looked at the effectiveness of UAC-mediated siRNA delivery. Xiao et al. [141] explained the preparation of UAC/siRNA NPs shown in Fig. 7.5 as well as their physicochemical features, such as their capacity to bind with CD98 siRNA (siCD98), hydrodynamic particle size, zeta potential, and shape. The RNAi is used to examine their cytotoxicity and capacity to treat inflammation.

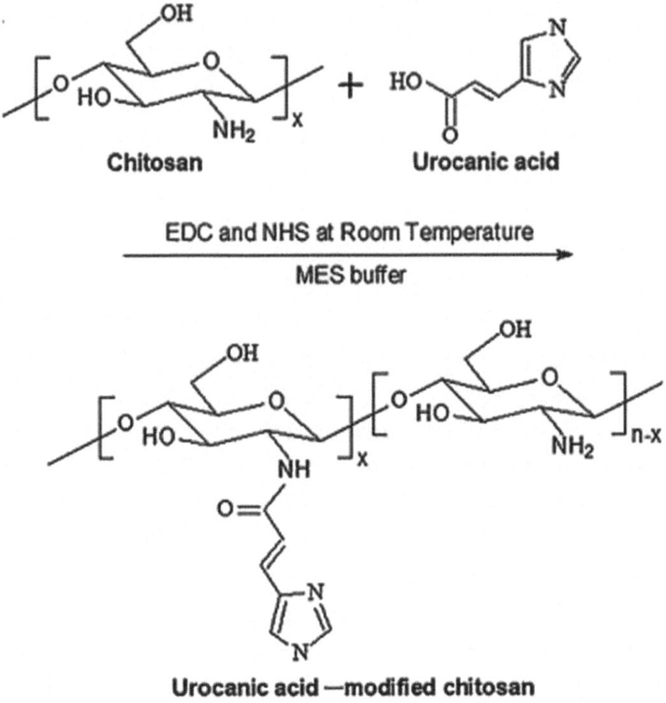

Fig. 7.5 Synthetic scheme of UAC using EDC and NHS as catalysts in MES buffer at room temperature [141]

7.14 PVA Film Containing Tetracycline Hydrochloride-Loaded Quaternized CSNPs for Wound Treatment

For biomedical uses including implant and wound dressing, biopolymer scaffolds have undergone substantial research [142]. The key biopolymers polylactic acid (PLA), polycaprolactone (PCL), chitosan, polyvinyl alcohol (PVA) and blends of these materials have advantages like low cost, biocompatibility, biodegradability, non-toxicity, tunable physical and chemical properties [143]. Scaffold requirements include optimal porosity, high surface area and elongation, controlled roughness [144]. Extracellular Matrix (ECM) Mimicry includes cell adhesion promotion, proliferation enhancement and successful biopolymer-biological system interaction [145]. Ideal scaffold characteristics includes biodegradability, matching tissue regeneration rates, mechanical strength, matching tissue requirements and bioactive surfaces, promoting cell-scaffold interactions applicable for tissue engineering, regenerative medicine, wound healing, implantable devices. By mimicking ECM properties and optimizing physical and chemical characteristics, biopolymer scaffolds can effectively support cell growth and tissue regeneration. Scaffolds made of bio-polymeric nonwoven nanofibers appear to have a three-dimensional structure with interconnect holes that are similar to those found in the extracellular matrix, assisting tissue regeneration [146]. Biopolymer composites, which contain bio-ceramic or/and metallic nanoparticles, as well as biomolecules and pharmaceuticals, to boost the scaffold qualities for tissue regeneration in a short period of time and with excellent repair quality, might be used to create this structure. Depending on the holes' size and density, as well as the biopolymer breakdown rate, drug integration in the nanofibers scaffold might result in a highly efficient drug delivery system with a vast surface area for controlled release. Electrospinning is a simple and common method for producing polymer nanofibers [147]. This technique has seen a surge of attention and significant improvement in recent decades because to its relative simplicity and low cost of creating a wide range of porous structures with long sub-micron fibers. A viscous polymer solution or melt held on a blunt needle is exposed to a strong electric field in the electrospinning procedure. The electrostatic force overcomes the surface tension of the polymer droplets over a particular threshold voltage, and a jet is blasted from the needle tip toward a collecting grounded metallic screen. The solvent in the polymer solution evaporates during the short jet flight, resulting in randomly arranged fibers on the screen. When compared to standard salted leaching membranes, the nanostructured architecture created by such a method may yield in a scaffold with superior physical, chemical, and biological characteristics. Electrospinning parameters that influence fiber characteristics are polymer solution properties, electric field settings, and operational parameters, and ambient conditions. The effects on fiber morphology include fiber diameter, porosity, and continuous fibrillary mat structure. The optimizing electrospinning parameters enable control over fiber uniformity thickness, mechanical properties, and surface area applicable for tissue engineering scaffolds, biomedical implants, drug delivery systems and filtration membranes. The precise control of electrospinning parameters is crucial for producing fibers with tailored properties for specific applications [148]. Electrospun mats may provide a biodegradation or bio absorption rate in tissue engineering applications that is proportional to the time required for the full regeneration of new tissue. This implies that the scaffold's mechanical attributes must be preserved throughout the project. In most cases, enzymatic or hydrolytic breakdown of polymers occurs. Numerous variables, including molecular weight, crystallinity, and cross-linking, may have an impact on the rate of polymer biodegradation. Collagen, alginate, and other biopolymers that are present in nature may be employed to create electrospun scaffolds [149]. However, the absence of mechanical characteristics is one of these materials' most obvious weaknesses. In order to obtain acceptable mechanical and biodegradation rates, blends or composites of natural and synthetic polymers are also required since the bulk of them are difficult to electrospin. There have been several recent research on electrospun scaffolds composed of biopolymers and polymers, both natural and manufactured [150]. Chitosan and poly(vinyl alcohol) blends have been shown to have better mechanical characteristics for biomedical applications because of intermolecular and intramolecular hydrogen bonding between the side chains of chitosan and PVA. Furthermore, even at low PVA concentrations, chitosan thickens, increases viscosity, and generates homogeneous nanofibers in a PVA solution [151].

Chitin, which may be found in the shells of crustaceans, insects, and fungi, is the source of the polysaccharide biopolymer known as chitosan. Chitosan is an interesting biopolymer to research in technology and biology due to its wide availability and low cost. Because of its great biocompatibility, low toxicity, and hemostatic and antibacterial properties, chitosan is an excellent choice for scaffolding in tissue regeneration [152]. Despite their good mechanical qualities, electrospun poly(vinyl alcohol) mats' utility as a tissue development scaffold is restricted owing to their high hydrophilicity, which allows for rapid biodegradation. When compared to other vinyl polymers, this feature is mostly owing to the presence of hydroxyl groups. Crosslinking is a straightforward strategy to modify the rate of biodegradation and mechanical qualities of a polymer. When the concentration of crosslinking agents is carefully managed, it may connect polymer chains to form a tridimensional structure with superior mechanical characteristics and water resistance [153]. In tissue engineering, the use of biodegradable polymeric scaffolds as drug delivery devices is becoming more common. Indeed, healing, antibacterial, and anti-inflammatory compounds might be included into a scaffold to allow for controlled drug release in wounded tissue or implant sites. The use of infection-control drugs or wound-healing promoters within scaffolds is gaining popularity because to the wide range of biopolymers and various physical architectures that enable drug release to be managed from a burst to a long-term maintenance level. Scaffolds, rather than nanoparticles, enable for a controlled release with a wide surface area at the injury site, avoiding excessive systemic drug levels and unfavorable side effects. Tetracycline hydrochloride (TCH) is an antibiotic that is used to treat acne, periodontal disease, and urinary tract infections, among other things. The TCH is shown to have low minimum inhibitory concentrations in both gram-positive and gram-negative bacteria when tested for antibacterial activity against a variety of pathogens [154]. The TCH release kinetic is crucial for tissue engineering applications since large dosages of TCH often result in significant cytotoxicity. Tissue engineering necessitates extensive study into the creation of biopolymeric electrospun scaffolds that include antibacterial medications. Alavarse et al. [154] reviewed at the TCH release profile in electrospun PVA/chitosan scaffolds. The literature on electro-spun mats manufactured from PVA and chitosan blended have been reported. The important feature is the inclusion of the commonly used antibiotic TCH for wound dressing applications. The scaffolds' morphological, thermal, and molecular structures are examined. The efficacy of antibacterial activity and concomitant cell proliferation has also been investigated. The investigations revealed that the TCH release profile from the PVA/CS/TCH scaffold, together with the fitting curves based on the first order and follow Ritger–Peppas models [154].

7.15 Spanish Broom (*Spartium junceum* L.) Fiber with Vancomycin-Loaded CSNPs

In the worldwide medical and pharmaceutical wound care industry, wound dressings and devices play a significant role. Cotton gauze is one of the most effective wound dressings because to the intrinsic characteristics of cotton fibers. Cotton fibers are very hydrophilic, absorbent, and cost-effective. Cotton growing is hazardous to the environment because pesticides are widely used, polluting rivers, and groundwater. Furthermore, extensive cotton farming needs a considerable quantity of water for irrigation, leading in soil desalinization and, as a consequence, soil fertility loss. The researchers examined the possibility of using Spanish Broom fibers for wound care, taking these disadvantages into mind. Several factors influence the use of Spanish Broom fibers as a wound dressing, including cost, availability, and hydrophilic properties. In reality, both Spanish Broom and cotton fibers are made up of cellulose, which can be extracted using a simple, efficient, easy, and quick physical–chemical method, allowing these fibers to be used in a wide range of sectors, including medicines [155]. The Spanish Broom is a tiny plant that grows wild in the Mediterranean. Spanish Broom, unlike flax and hemp, thrives in the most adverse limestone soil and may be utilized for up to twenty years after planting, but hemp and flax need high-quality soil every year. Cellulose-based dressings exist in a range of shapes and sizes, but are not antimicrobial. Antibiotics like neomycin, bacitracin, streptomycin, gentamycin, and polymyxin, as well as combinations of these antibiotics, are used to treat chronic wounds and provide cotton or other fibers some healing activity [156]. Among the antibiotics studied, vancomycin (VM) has been chosen as the model medication for a novel antibacterial dressing made from Spanish Broom fibers. The VM is a water-soluble glycopeptide antibiotic that works against gram-positive bacteria. The therapeutic effectiveness of topical VM treatment on meticillin-resistant S. aureus (MRSA) conjunctivitis, chronic suppurative otitis media, and osteomyelitis has been documented in the literature [157]. Furthermore, local administration of antibiotics is attractive for wound infection prevention because large concentrations of antibiotics are produced locally at the wound site and systemic toxicity is reduced [158]. On the other hand, the delayed release of antimicrobial medicines from wound dressings offers the benefit of gently treating infected wounds. Nanoparticles impregnated into fibers provide good prospects for enhancing wound treatments due to their ease of application and more uniform dispersion over the wound surface [159]. The biomaterial for the creation of nanoparticles is chitosan (CS), a biodegradable, non-toxic, and biocompatible polymer. The CS has been demonstrated to exhibit muco-adhesive, cationic, healing-stimulating, antibacterial, and hemostatic characteristics in wound dressings [160]. The CS is a natural linear polycationic polymer that is formed when chitin is partially N-deacetylated. Cerchiara et al. [160] developed a novel antibacterial wound dressing for the treatment of infected wounds based on Spanish Broom fibers loaded with CS nanoparticles as shown in Fig. 7.6. First, while the experimental settings have been altered, the chemical composition, shape, and tensile characteristics of Spanish Broom extracted via the DiCoDe technique have been studied. The

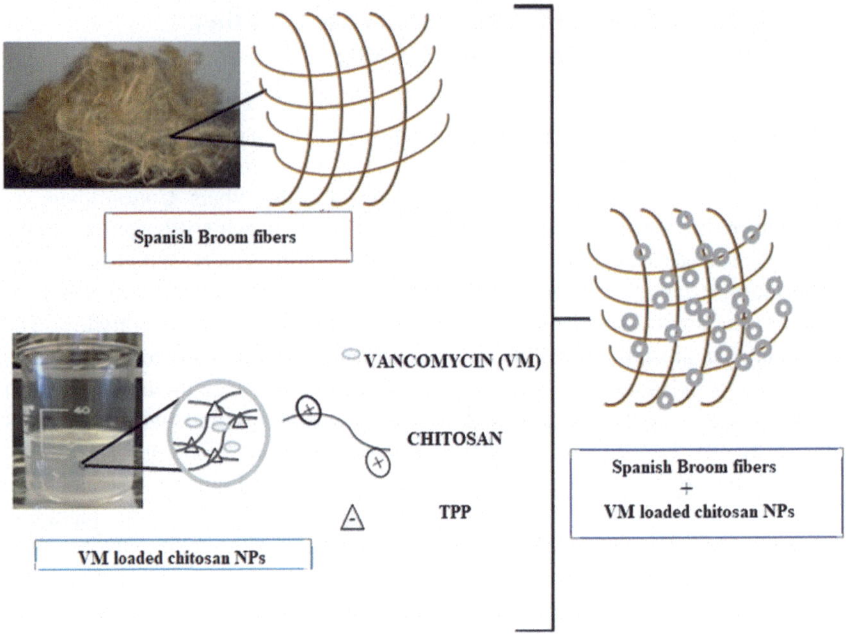

Fig. 7.6 Development of a novel antibacterial wound dressing for the treatment of infected wounds based on Spanish Broom fibers loaded with CS nanoparticles [160]

CS nanoparticles encapsulating VM are then produced utilizing the ionic gelation process and varied weight ratios of CS and tripolyphosphate (TPP). Nanoparticles are measured for their size, zeta potential, yield, encapsulation efficiency, stability, and VM release. Finally, antibacterial activity against *Staphylococcus aureus* has been tested, as well as in vitro cytotoxicity on HaCaT cells [160].

7.16 Silver Sulfadiazine-Loaded CSNPs

Wounds, especially burn wounds, typically offer a perfect habitat for the colonization of microorganisms that cause infection (bacterial or fungal), delaying wound healing. As a consequence, it's vital to establish settings that are unfavorable to microbes but favorable to the host's repair processes in order to boost wound healing prospects. The bacterial load in the wound is reduced when an infection is treated, which enhances the dynamics of oxygen transport and use inside the wound. It also lowers chronic inflammation and improves the tissue's ability to react to cell signaling and produce long-term growth [161]. Topical antibacterial medicines are considered to help this process. Infectious phase, necrosis and agglutination period, proliferation period, and epidermal development period are common wound stages in bedsores and burn wounds. In general, therapy formulations are determined according on the stage of the illness, with the application time being a key consideration. During the infectious phase, silver sulfadiazine (SSD) is the medicine of choice for avoiding wound sepsis and countering the possibility of bacterial infection [162]. It is worth to mention that sepsis is a life-threatening reaction to an infection. It happens when your immune system overreacts to an infection and starts to damage your body's own tissues and organs. Both sides of the SSD have antimicrobial characteristics. Sulfadiazine may inhibit DNA synthesis by interfering with folate formation, and free silver can react with both the sulfhydryl groups of bacterial enzymes and the sulfhydryl groups of DNA [163]. The antimicrobial impact of SSD, on the other hand, is severely limited by its poor aqueous solubility. The SSD has also been shown to be cytotoxic to fibroblasts and keratinocytes in vitro, slowing wound healing in vivo [164]. As addition, SSD is generally found in a 1% cream. This cream dosage form has a variety of common adverse effects, including the difficulty to maintain effective drug concentrations at wet wound surfaces for a long length of time owing to its short residence duration, and its messiness, which causes patients annoyance [165]. Furthermore, according to the SSD cream's producers, the SSD cream creates discoloration of the wound bed, which makes it difficult to judge wound condition after numerous administrations. It also has low silver release levels, which has an effect on the drug's effectiveness since silver ions' antibacterial efficiency is directly related to concentration, which should not decrease below the minimum inhibitory value. Growing demand for silver dressings with precise medication dosages has driven innovation in wound care. Ideal dressings offer protection from physical harm and secondary infection, prevention of contamination and bacterial channeling, assistance in debridement, thermal insulation and easy removal without trauma [166]. Challenges with traditional silver-based dressings involve unpredictable drug release and concentration, cytotoxic effects due to uncontrolled silver release. The possible solution involve encapsulating silver sulfadiazine (SSD) in hydrophilic nano-carriers, coating cotton dressings with SSD-loaded chitosan nanoparticles [167] with following enhanced benefits including controlled SSD release, reduced cytotoxicity, improved wound adherence and extended contact duration. Chitosan an ideal biomaterial with excellent biocompatibility, absorptivity, hypoallergenicity, biodegradability and good wound healing properties [168]. El-Feky et al. [169] created an antimicrobial silver sulfadiazine dressing combining SSD's antibacterial/antifungal efficacy with chitosan nanoparticles' drug delivery and coating capabilities. This innovative approach enables; targeted SSD release, enhanced wound care outcomes, reduced risk of infection and complications with future directions including optimized SSD-loaded chitosan nanoparticle formulations, investigated alternative biomaterials and nano-carriers with clinical trials to validate efficacy and safety.

7.17 Summary

This chapter provides an in-depth examination of the biomedical applications of modified chitosan nanoparticles. Various studies demonstrate their potential in Pickering emulsions and chitosan tripolyphosphate nanoparticles for drug delivery; Salazosulfapyridine delivery for rheumatoid arthritis and inflammatory bowel diseases, Theophylline granules with chitosan-sodium tripolyphosphate complexes and chitosan-5-fluorouracil nanoparticles for cancer treatment. Modified chitosan nanoparticles have also shown promise in cardiovascular treatments (propranolol nanoparticles), neurological disorders (desvenlafaxine-loaded PLGA CSNPs), and antipsychotic medications (quetiapine fumarate-loaded chitosan nanoparticles) and stroke treatment (rutin-encapsulated CSNPs). Herbal therapies and cancer treatments utilizing modified chitosan nanoparticles include piperine-loaded CSNPs for Alzheimer's and epilepsy, hyaluronic acid-coated CSNPs for cancer, naringenin-loaded sulfobutylether-β-cyclodextrin CSNPs. Other applications include Spanish Broom fiber with vancomycin-loaded CSNPs.

References

1. Saikia C, Gogoi P (2015) Chitosan: a promising biopolymer in drug delivery applications. J Mol Genet Med s4. https://doi.org/10.4172/1747-0862.s4-006
2. Ali A, Ahmed S (2018) A review on chitosan and its nanocomposites in drug delivery. https://doi.org/10.1016/j.ijbiomac.2017.12.078
3. Stein A (2003) Advances in microporous and mesoporous solids—highlights of recent progress. Adv Mater 15. https://doi.org/10.1002/adma.200300007
4. Zhou FZ, Yu XH, Zeng T, Yin SW, Tang CH, Yang XQ (2019) Fabrication and characterization of novel water-insoluble protein porous materials derived from pickering high internal-phase emulsions stabilized by gliadin-chitosan-complex particles. J Agric Food Chem 67. https://doi.org/10.1021/acs.jafc.9b00221
5. Tan H, Wei J, Sun G, Mu C, Lin W, Ngai T (2017) Interconnected macroporous 3D scaffolds templated from gelatin nanoparticle-stabilized high internal phase emulsions for biomedical applications. Soft Matter 13. https://doi.org/10.1039/c7sm00706j
6. Sharkawy A, Barreiro MF, Rodrigues AE (2020) Chitosan-based pickering emulsions and their applications: a review. https://doi.org/10.1016/j.carbpol.2020.116885
7. McClements DJ, Jafari SM (2018) Improving emulsion formation, stability and performance using mixed emulsifiers: a review. https://doi.org/10.1016/j.cis.2017.12.001
8. Binks BP, Yin D (2016) Pickering emulsions stabilized by hydrophilic nanoparticles: in situ surface modification by oil. Soft Matter 12. https://doi.org/10.1039/c6sm01214k
9. Yang W, Fu J, Wang T, He N (2009) Chitosan/sodium tripolyphosphate nanoparticles: preparation, characterization and application as drug carrier. J Biomed Nanotechnol 5. https://doi.org/10.1166/jbn.2009.1067
10. Brooks CD (2001) Sulfasalazine for the management of juvenile rheumatoid arthritis. J Rheumatol 28
11. Lindenberg M, Kopp S, Dressman JB (2004) Classification of orally administered drugs on the World Health Organization model list of essential medicines according to the biopharmaceutics classification system. https://doi.org/10.1016/j.ejpb.2004.03.001
12. Plosker GL, Croom KF (2005) Sulfasalazine: a review of its use in the management of rheumatoid arthritis. https://doi.org/10.2165/00003495-200565130-00008

13. Shadid M, Gurau G, Shamshina JL, Chuang BC, Hailu S, Guan E, Chowdhury SK, Wu JT, Rizvi SAA, Griffin RJ, Rogers RD (2015) Sulfasalazine in ionic liquid form with improved solubility and exposure. MedChemComm 6. https://doi.org/10.1039/c5md00290g
14. da Costa MAB, Villa ALV, Barros RDCDSA, Ricci-Júnior E, dos Santos EP (2015) Development, characterization and evaluation of the dissolution profile of sulfasalazine suspensions. Braz J Pharm Sci 51. https://doi.org/10.1590/S1984-82502015000200022
15. Tang P, Yang H, Tang B, Wu D, Du Q, Xu K, Li H (2017) Dimethyl-β-cyclodextrin/salazosulfapyridine inclusion complex-loaded chitosan nanoparticles for sustained release. Carbohydr Polym 156. https://doi.org/10.1016/j.carbpol.2016.09.038
16. Liu F, Antoniou J, Li Y, Majeed H, Liang R, Ma Y, Ma J, Zhong F (2016) Chitosan/sulfobutylether-β-cyclodextrin nanoparticles as a potential approach for tea polyphenol encapsulation. Food Hydrocolloids 57. https://doi.org/10.1016/j.foodhyd.2016.01.024
17. Mahmoud AA, El-Feky GS, Kamel R, Awad GEA (2011) Chitosan/sulfobutylether-β-cyclodextrin nanoparticles as a potential approach for ocular drug delivery. Int J Pharm 413. https://doi.org/10.1016/j.ijpharm.2011.04.031
18. Ji J, Hao S, Liu W, Zhang J, Wu D, Xu Y (2011) Preparation and evaluation of O-carboxymethyl chitosan/cyclodextrin nanoparticles as hydrophobic drug delivery carriers. Polym Bull 67. https://doi.org/10.1007/s00289-011-0449-4
19. Khalil SKH, El-Feky GS, El-Banna ST, Khalil WA (2012) Preparation and evaluation of warfarin-β-cyclodextrin loaded chitosan nanoparticles for transdermal delivery. Carbohydr Polym 90. https://doi.org/10.1016/j.carbpol.2012.06.056
20. Kawashima Y, Handa T, Kasai A, Takenaka H, Lin SY, Ando Y (1985) Novel method for the preparation of controlled-release theophylline granules coated with a polyelectrolyte complex of sodium polyphosphate-chitosan. J Pharm Sci 74. https://doi.org/10.1002/jps.2600740308
21. Nair AB, Attimarad M, Al-Dhubiab BE, Wadhwa J, Harsha S, Ahmed M (2014) Enhanced oral bioavailability of acyclovir by inclusion complex using hydroxypropyl-β-cyclodextrin. Drug Deliv 21. https://doi.org/10.3109/10717544.2013.853213
22. Kubbinga M, Nguyen MA, Staubach P, Teerenstra S, Langguth P (2015) The influence of chitosan on the oral bioavailability of acyclovir—a comparative bioavailability study in humans. Pharm Res 32. https://doi.org/10.1007/s11095-014-1613-y
23. Naik DR, Raval JP (2016) Amorphous polymeric binary blend pH-responsive nanoparticles for dissolution enhancement of antiviral drug. J Saudi Chem Soc 20. https://doi.org/10.1016/j.jscs.2012.09.020
24. Malik NS, Ahmad M, Minhas MU, Tulain R, Barkat K, Khalid I, Khalid Q (2020) Chitosan/xanthan gum based hydrogels as potential carrier for an antiviral drug: fabrication, characterization, and safety evaluation. Front Chem 8. https://doi.org/10.3389/fchem.2020.00050
25. van Vlierberghe S, Dubruel P, Schacht E (2011) Biopolymer-based hydrogels as scaffolds for tissue engineering applications: a review. https://doi.org/10.1021/bm200083n
26. Elgadir MA, Uddin MS, Ferdosh S, Adam A, Chowdhury AJK, Sarker MZI (2015) Impact of chitosan composites and chitosan nanoparticle composites on various drug delivery systems: a review. https://doi.org/10.1016/j.jfda.2014.10.008
27. Petri DFS (2015) Xanthan gum: a versatile biopolymer for biomedical and technological applications. J Appl Polym Sci 132. https://doi.org/10.1002/app.42035
28. Kumar A, Rao KM, Han SS (2018) Application of xanthan gum as polysaccharide in tissue engineering: a review. https://doi.org/10.1016/j.carbpol.2017.10.009
29. Shalviri A, Liu Q, Abdekhodaie MJ, Wu XY (2010) Novel modified starch-xanthan gum hydrogels for controlled drug delivery: synthesis and characterization. Carbohydr Polym 79. https://doi.org/10.1016/j.carbpol.2009.10.016
30. Kabiri K, Omidian H, Zohuriaan-Mehr MJ, Doroudiani S (2011) Superabsorbent hydrogel composites and nanocomposites: a review. https://doi.org/10.1002/pc.21046
31. Sohail M, Ahmad M, Minhas MU, Ali L, Khalid I, Rashid H (2015) Controlled delivery of valsartan by cross-linked polymeric matrices: synthesis, in vitro and in vivo evaluation. Int J Pharm 487. https://doi.org/10.1016/j.ijpharm.2015.04.013

32. Sethi A, Ahmad M, Huma T, Khalid I, Ahmad I (2021) Evaluation of low molecular weight cross linked chitosan nanoparticles, to enhance the bioavailability of 5-flourouracil. Dose-Response 19. https://doi.org/10.1177/15593258211025353
33. Xu X, Ho W, Zhang X, Bertrand N, Farokhzad O (2015) Cancer nanomedicine: from targeted delivery to combination therapy. https://doi.org/10.1016/j.molmed.2015.01.001
34. Mohammed MA, Syeda JTM, Wasan KM, Wasan EK (2017) An overview of chitosan nanoparticles and its application in non-parenteral drug delivery. https://doi.org/10.3390/pharmaceutics9040053
35. Kumari A, Yadav SK, Yadav SC (2010) Biodegradable polymeric nanoparticles based drug delivery systems. https://doi.org/10.1016/j.colsurfb.2009.09.001
36. Zheng T, Liang Y, Ye S, He Z (2009) Superabsorbent hydrogels as carriers for the controlled-release of urea: experiments and a mathematical model describing the release rate. Biosyst Eng 102. https://doi.org/10.1016/j.biosystemseng.2008.09.027
37. Qi L, Xu Z, Jiang X, Hu C, Zou X (2004) Preparation and antibacterial activity of chitosan nanoparticles. Carbohydr Res 339. https://doi.org/10.1016/j.carres.2004.09.007
38. Silva RM, Silva GA, Coutinho OP, Mano JF, Reis RL (2004) Preparation and characterisation in simulated body conditions of glutaraldehyde crosslinked chitosan membranes. J Mater Sci Mater Med 15. https://doi.org/10.1023/B:JMSM.0000046392.44911.46
39. Peppas NA, Bures P, Leobandung W, Ichikawa H (2000) Hydrogels in pharmaceutical formulations. https://doi.org/10.1016/S0939-6411(00)00090-4
40. Lin CC, Metters AT (2006) Hydrogels in controlled release formulations: network design and mathematical modeling. https://doi.org/10.1016/j.addr.2006.09.004
41. Kim JO, Kabanov AV, Bronich TK (2009) Polymer micelles with cross-linked polyanion core for delivery of a cationic drug doxorubicin. J Controlled Release 138. https://doi.org/10.1016/j.jconrel.2009.04.019
42. Geever LM, Cooney CC, Lyons JG, Kennedy JE, Nugent MJD, Devery S, Higginbotham CL (2008) Characterisation and controlled drug release from novel drug-loaded hydrogels. Eur J Pharm Biopharm 69. https://doi.org/10.1016/j.ejpb.2007.12.021
43. Al-Kassas R, Wen J, Cheng AEM, Kim AMJ, Liu SSM, Yu J (2016) Transdermal delivery of propranolol hydrochloride through chitosan nanoparticles dispersed in mucoadhesive gel. Carbohydr Polym 153. https://doi.org/10.1016/j.carbpol.2016.06.096
44. Kraisit P, Limmatvapirat S, Luangtana-Anan M, Sriamornsak P (2018) Buccal administration of mucoadhesive blend films saturated with propranolol loaded nanoparticles. Asian J Pharm Sci 13. https://doi.org/10.1016/j.ajps.2017.07.006
45. Hua S (2019) Advances in nanoparticulate drug delivery approaches for sublingual and buccal administration. https://doi.org/10.3389/fphar.2019.01328
46. Salamat-Miller N, Chittchang M, Johnston TP (2005) The use of mucoadhesive polymers in buccal drug delivery. https://doi.org/10.1016/j.addr.2005.07.003
47. Chen MC, Mi FL, Liao ZX, Hsiao CW, Sonaje K, Chung MF, Hsu LW, Sung HW (2013) Recent advances in chitosan-based nanoparticles for oral delivery of macromolecules. https://doi.org/10.1016/j.addr.2012.10.010
48. David KI, Jaidev LR, Sethuraman S, Krishnan UM (2015) Dual drug loaded chitosan nanoparticles-sugar-coated arsenal against pancreatic cancer. Colloids Surf B Biointerfaces 135. https://doi.org/10.1016/j.colsurfb.2015.08.038
49. Al-Nemrawi NK, Alsharif SSM, Dave RH (2018) Preparation of chitosan-TPP nanoparticles: the influence of chitosan polymeric properties and formulation variables. Int J Appl Pharm 10. https://doi.org/10.22159/ijap.2018v10i5.26375
50. Shu XZ, Zhu KJ (2002) The influence of multivalent phosphate structure on the properties of ionically cross-linked chitosan films for controlled drug release. Eur J Pharm Biopharm 54. https://doi.org/10.1016/S0939-6411(02)00052-8
51. Calvo P, Remuñán-López C, Vila-Jato JL, Alonso MJ (1997) Novel hydrophilic chitosan-polyethylene oxide nanoparticles as protein carriers. J Appl Polym Sci 63. https://doi.org/10.1002/(SICI)1097-4628(19970103)63:1<125::AID-APP13>3.0.CO;2-4

52. Kraisit P (2019) Impact of hydroxypropyl methylcellulose (HPMC) type and concentration on the swelling and release properties of propranolol hydrochloride matrix tablets usning a simplex centroid design. Int J Appl Pharm 11. https://doi.org/10.22159/ijap.2019v11i2.31127
53. Newton AMJ, Indana VL, Kumar J (2015) Chronotherapeutic drug delivery of tamarind gum, chitosan and okra gum controlled release colon targeted directly compressed propranolol HCl matrix tablets and in-vitro evaluation. Int J Biol Macromol 79. https://doi.org/10.1016/j.ijbiomac.2015.03.031
54. Duangjit S, Kraisit P, Luangtana-Anan M (2016) An investigation of propranolol-loaded chitosan nanoparticles for transmucosal delivery: physical characterization. Thai J Pharm Sci 40
55. Kittiwisut S, Kraisit P (2020) Physicochemical characterization of propranolol-loaded chitosan nanoparticles for a buccal drug delivery system. Int J Appl Pharm 12. https://doi.org/10.22159/ijap.2020v12i4.38009
56. Kircanski K, Joormann J, Gotlib IH (2012) Cognitive aspects of depression. Wiley Interdiscip Rev Cogn Sci 3. https://doi.org/10.1002/wcs.1177
57. Marcus M, Yasamy MT, van Ommeren M, Chisholm D. Depression, a global public health concern
58. Kilts CD (2003) Potential new drug delivery systems for antidepressants: an overview
59. Kumar M, Misra A, Mishra AK, Mishra PP, Pathak K (2008) Mucoadhesive nanoemulsion-based intranasal drug delivery system of olanzapine for brain targeting. J Drug Target 16. https://doi.org/10.1080/10611860802476504
60. Tong GF, Qin N, Sun LW (2017) Development and evaluation of desvenlafaxine loaded PLGA-chitosan nanoparticles for brain delivery. Saudi Pharm J 25. https://doi.org/10.1016/j.jsps.2016.12.003
61. Perry R, Cassagnol M (2009) Desvenlafaxine: a new serotonin-norepinephrine reuptake inhibitor for the treatment of adults with major depressive disorder. https://doi.org/10.1016/j.clinthera.2009.07.012
62. Seeman P (2002) Atypical antipsychotics: mechanism of action. https://doi.org/10.1177/070674370204700106
63. Soeiro-de-Souza MG, Dias VV, Missio G, Balanzá-Martinez V, Valiengo L, Carvalho AF, Moreno RA (2015) Role of quetiapine beyond its clinical efficacy in bipolar disorder: from neuroprotection to the treatment of psychiatric disorders (review). https://doi.org/10.3892/etm.2015.2213
64. Riedel M, Müller N, Stradding M, Spellmann I, Severus E, Möller HJ (2007) Quetiapine in the treatment of schizophrenia and related disordes. Neuropsychiatr Dis Treat 3. https://doi.org/10.2147/nedt.2007.3.2.219
65. Tajes M, Ramos-Fernández E, Weng-Jiang X, Bosch-Morató M, Guivernau B, Eraso-Pichot A, Salvador B, Fernàndez-Busquets X, Roquer J, Muñoz FJ (2014) The blood-brain barrier: structure, function and therapeutic approaches to cross it. https://doi.org/10.3109/09687688.2014.937468
66. Lu CT, Zhao YZ, Wong HL, Cai J, Peng L, Tian XQ (2014) Current approaches to enhance CNS delivery of drugs across the brain barriers. https://doi.org/10.2147/IJN.S61288
67. Casettari L, Illum L (2014) Chitosan in nasal delivery systems for therapeutic drugs. https://doi.org/10.1016/j.jconrel.2014.05.003
68. Touitou E, Illum L (2013) Nasal drug delivery. https://doi.org/10.1007/s13346-012-0111-1
69. Mittal D, Ali A, Md S, Baboota S, Sahni JK, Ali J (2014) Insights into direct nose to brain delivery: current status and future perspective. https://doi.org/10.3109/10717544.2013.838713
70. Alam MI, Beg S, Samad A, Baboota S, Kohli K, Ali J, Ahuja A, Akbar M (2010) Strategy for effective brain drug delivery. https://doi.org/10.1016/j.ejps.2010.05.003
71. Md S, Mustafa G, Baboota S, Ali J (2015) Nanoneurotherapeutics approach intended for direct nose to brain delivery. https://doi.org/10.3109/03639045.2015.1052081
72. Kreuter J (2012) Nanoparticulate systems for brain delivery of drugs. https://doi.org/10.1016/j.addr.2012.09.015

73. Costantino L, Boraschi D (2012) Is there a clinical future for polymeric nanoparticles as brain-targeting drug delivery agents? https://doi.org/10.1016/j.drudis.2011.10.028
74. Mittal D, Md S, Hasan Q, Fazil M, Ali A, Baboota S, Ali J (2016) Brain targeted nanoparticulate drug delivery system of rasagiline via intranasal route. Drug Deliv 23. https://doi.org/10.3109/10717544.2014.907372
75. Shadab MD, Khan RA, Mustafa G, Chuttani K, Baboota S, Sahni JK, Ali J (2013) Bromocriptine loaded chitosan nanoparticles intended for direct nose to brain delivery: pharmacodynamic, pharmacokinetic and scintigraphy study in mice model. Eur J Pharm Sci 48. https://doi.org/10.1016/j.ejps.2012.12.007
76. Bari NK, Fazil M, Hassan MQ, Haider MR, Gaba B, Narang JK, Baboota S, Ali J (2015) Brain delivery of buspirone hydrochloride chitosan nanoparticles for the treatment of general anxiety disorder. Int J Biol Macromol 81. https://doi.org/10.1016/j.ijbiomac.2015.07.041
77. Yeh TH, Hsu LW, Tseng MT, Lee PL, Sonjae K, Ho YC, Sung HW (2011) Mechanism and consequence of chitosan-mediated reversible epithelial tight junction opening. Biomaterials 32. https://doi.org/10.1016/j.biomaterials.2011.03.056
78. Shah B, Khunt D, Misra M, Padh H (2016) Application of Box–Behnken design for optimization and development of quetiapine fumarate loaded chitosan nanoparticles for brain delivery via intranasal route*. Int J Biol Macromol 89. https://doi.org/10.1016/j.ijbiomac.2016.04.076
79. Ahmad N, Umar S, Ashafaq M, Akhtar M, Iqbal Z, Samim M, Ahmad FJ (2013) A comparative study of PNIPAM nanoparticles of curcumin, demethoxycurcumin, and bisdemethoxycurcumin and their effects on oxidative stress markers in experimental stroke. Protoplasma 250. https://doi.org/10.1007/s00709-013-0516-9
80. Kim Y, So HS, Kim JK, Park C, Lee JH, Woo WH, Cho KH, Moon BS, Park R (2007) Anti-inflammatory effect of oyaksungisan in peripheral blood mononuclear cells from cerebral infarction patients. Biol Pharm Bull 30. https://doi.org/10.1248/bpb.30.1037
81. Ashafaq M, Raza SS, Khan MM, Ahmad A, Javed H, Ahmad ME, Tabassum R, Islam F, Siddiqui MS, Safhi MM, Islam F (2012) Catechin hydrate ameliorates redox imbalance and limits inflammatory response in focal cerebral ischemia. Neurochem Res 37. https://doi.org/10.1007/s11064-012-0786-1
82. Berner MD, Sura ME, Alves BN, Hunter KW (2005) IFN-γ primes macrophages for enhanced TNF-α expression in response to stimulatory and non-stimulatory amounts of microparticulate β-glucan. Immunol Lett 98. https://doi.org/10.1016/j.imlet.2004.10.020
83. Sun M, Zhao Y, Gu Y, Xu C (2009) Inhibition of nNOS reduces ischemic cell death through down-regulating calpain and caspase-3 after experimental stroke. Neurochem Int 54. https://doi.org/10.1016/j.neuint.2008.12.017
84. Wang X, Liu J, Zhu H, Tejima E, Tsuji K, Murata Y, Atochin DN, Huang PL, Zhang C, Lo EH (2008) Effects of neuroglobin overexpression on acute brain injury and long-term outcomes after focal cerebral ischemia. Stroke 39. https://doi.org/10.1161/STROKEAHA.107.506022
85. Li RC, Guo SZ, Lee SK, Gozal D (2010) Neuroglobin protects neurons against oxidative stress in global ischemia. J Cereb Blood Flow Metab 30. https://doi.org/10.1038/jcbfm.2010.90
86. Lao CJ, Lin JG, Kuo JS, Chao PDL, Cheng CY, Tang NY, Hsieh CL (2005) Microglia, apoptosis and interleukin-1β expression in the effect of *Sophora japonica* L. on cerebral infarct induced by ischemia-reperfusion in rats. Am J Chin Med 33. https://doi.org/10.1142/S0192415X0500303X
87. Liu Y, Lü L, Hettinger CL, Dong G, Zhang D, Rezvani K, Wang X, Wang H (2014) Ubiquilin-1 protects cells from oxidative stress and ischemic stroke caused tissue injury in mice. J Neurosci 34. https://doi.org/10.1523/JNEUROSCI.3541-13.2014
88. Badary OA, Taha RA, Gamal El-Din AM, Abdel-Wahab MH (2003) Thymoquinone is a potent superoxide anion scavenger. Drug Chem Toxicol 26. https://doi.org/10.1081/DCT-120020404
89. Annapurna A, Ansari MA, Manjunath PM (2013) Partial role of multiple pathways in infarct size limiting effect of quercetin and rutin against cerebral ischemia-reperfusion injury in rats. Eur Rev Med Pharmacol Sci 17

90. Ahmad N, Ahmad R, Naqvi AA, Alam MA, Ashafaq M, Samim M, Iqbal Z, Ahmad FJ (2016) Rutin-encapsulated chitosan nanoparticles targeted to the brain in the treatment of cerebral ischemia. Int J Biol Macromol 91. https://doi.org/10.1016/j.ijbiomac.2016.06.001
91. Witt KA, Davis TP (2006) CNS drug delivery: opioid peptides and the blood-brain barrier. https://doi.org/10.1208/aapsj080109
92. Lockman PR, Mumper RJ, Khan MA, Allen DD (2002) Nanoparticle technology for drug delivery across the blood-brain barrier. https://doi.org/10.1081/DDC-120001481
93. Silva GA (2007) Nanotechnology approaches for drug and small molecule delivery across the blood brain barrier. https://doi.org/10.1016/j.surneu.2006.08.033
94. Alam S, Khan ZI, Mustafa G, Kumar M, Islam F, Bhatnagar A, Ahmad FJ (2012) Development and evaluation of thymoquinone-encapsulated chitosan nanoparticles for nose-to-brain targeting: a pharmacoscintigraphic study. Int J Nanomed 7. https://doi.org/10.2147/IJN.S35329
95. Singh N, Pandey BR, Verma P (2011) An overview of phytotherapeutic approach in prevention and treatment of Alzheimer's syndrome & dementia
96. Pal A, Nayak S, Sahu PK, Swain T (2011) Piperine protects epilepsy associated depression: a study on role of monoamines
97. Zarai Z, Boujelbene E, Ben Salem N, Gargouri Y, Sayari A (2013) Antioxidant and antimicrobial activities of various solvent extracts, piperine and piperic acid from *Piper nigrum*. LWT Food Sci Technol 50. https://doi.org/10.1016/j.lwt.2012.07.036
98. Li S, Wang C, Wang M, Li W, Matsumoto K, Tang Y (2007) Antidepressant like effects of piperine in chronic mild stress treated mice and its possible mechanisms. Life Sci 80. https://doi.org/10.1016/j.lfs.2006.12.027
99. Lai LH, Fu QH, Liu Y, Jiang K, Guo QM, Chen QY, Yan B, Wang QQ, Shen JG (2012) Piperine suppresses tumor growth and metastasis in vitro and in vivo in a 4T1 murine breast cancer model. Acta Pharm Sin 33. https://doi.org/10.1038/aps.2011.209
100. Shrivastava P, Vaibhav K, Tabassum R, Khan A, Ishrat T, Khan MM, Ahmad A, Islam F, Safhi MM, Islam F (2013) Anti-apoptotic and anti-inflammatory effect of piperine on 6-OHDA induced Parkinson's rat model. J Nutr Biochem 24. https://doi.org/10.1016/j.jnutbio.2012.03.018
101. Chonpathompikunlert P, Wattanathorn J, Muchimapura S (2010) Piperine, the main alkaloid of Thai black pepper, protects against neurodegeneration and cognitive impairment in animal model of cognitive deficit like condition of Alzheimer's disease. Food Chem Toxicol 48. https://doi.org/10.1016/j.fct.2009.12.009
102. Vijayakumar RS, Surya D, Nalini N (2004) Antioxidant efficacy of black pepper (*Piper nigrum* L.) and piperine in rats with high fat diet induced oxidative stress. Redox Rep 9. https://doi.org/10.1179/135100004225004742
103. Mishra A, Punia JK, Bladen C, Zamponi GW, Goel RK (2015) Anticonvulsant mechanisms of piperine, a piperidine alkaloid. Channels 9. https://doi.org/10.1080/19336950.2015.1092836
104. Chen CY, Li W, Qu KP, Chen CR (2013) Piperine exerts anti-seizure effects via the TRPV1 receptor in mice. Eur J Pharmacol 714. https://doi.org/10.1016/j.ejphar.2013.07.041
105. Khameneh B, Iranshahy M, Ghandadi M, Ghoochi Atashbeyk D, Fazly Bazzaz BS, Iranshahi M (2015) Investigation of the antibacterial activity and efflux pump inhibitory effect of co-loaded piperine and gentamicin nanoliposomes in methicillin-resistant *Staphylococcus aureus*. Drug Dev Ind Pharm 41. https://doi.org/10.3109/03639045.2014.920025
106. Ratanavaraporn J, Kanokpanont S, Damrongsakkul S (2014) The development of injectable gelatin/silk fibroin microspheres for the dual delivery of curcumin and piperine. J Mater Sci Mater Med 25. https://doi.org/10.1007/s10856-013-5082-3
107. Wang JJ, Zeng ZW, Xiao RZ, Xie T, Zhou GL, Zhan XR, Wang SL (2011) Recent advances of chitosan nanoparticles as drug carriers. https://doi.org/10.2147/ijn.s17296
108. Rudzinski WE, Aminabhavi TM (2010) Chitosan as a carrier for targeted delivery of small interfering RNA. https://doi.org/10.1016/j.ijpharm.2010.08.022
109. Anissian D, Ghasemi-Kasman M, Khalili-Fomeshi M, Akbari A, Hashemian M, Kazemi S, Moghadamnia AA (2018) Piperine-loaded chitosan-STPP nanoparticles reduce neuronal loss

and astrocytes activation in chemical kindling model of epilepsy. Int J Biol Macromol 107. https://doi.org/10.1016/j.ijbiomac.2017.09.073
110. Jemal A, Siegel R, Xu J, Ward E (2010) Cancer statistics, 2010. CA Cancer J Clin 60. https://doi.org/10.3322/caac.20073
111. Siegel R, DeSantis C, Virgo K, Stein K, Mariotto A, Smith T, Cooper D, Gansler T, Lerro C, Fedewa S, Lin C, Leach C, Cannady RS, Cho H, Scoppa S, Hachey M, Kirch R, Jemal A, Ward E (2012) Cancer treatment and survivorship statistics, 2012. CA Cancer J Clin 62. https://doi.org/10.3322/caac.21149
112. Malaiya A, Yadav AK (2018) Hyaluronic acid anchored nanoparticulate carrier for delivery of methotrexate. Asian J Pharm Pharmacol 4. https://doi.org/10.31024/ajpp.2018.4.6.27
113. Toole BP (1990) Hyaluronan and its binding proteins, the hyaladherins. Curr Opin Cell Biol 2. https://doi.org/10.1016/0955-0674(90)90081-O
114. Bertolami CN, Berg S, Messadi DV (1992) Binding and internalization of hyaluronate by human cutaneous fibroblasts. Matrix 12. https://doi.org/10.1016/S0934-8832(11)80100-9
115. Misra R, Mohanty S (2014) Sustained release of methotrexate through liquid-crystalline folate nanoparticles. J Mater Sci Mater Med 25. https://doi.org/10.1007/s10856-014-5257-6
116. Sonneveld P, Schultz FW, Nooter K, Hählen K (1986) Pharmacokinetics of methotrexate and 7-hydroxy-methotrexate in plasma and bone marrow of children receiving low-dose oral methotrexate. Cancer Chemother Pharmacol 18. https://doi.org/10.1007/BF00262278
117. Lin J, Sun J, Wang Y, Ma Y, Chen W, Zhang Z, Gui G, Lin B (2015) Ocular pharmacokinetics of naringenin eye drops following topical administration to rabbits. J Ocul Pharmacol Ther 31. https://doi.org/10.1089/jop.2014.0047
118. Bloch SB, Larsen M, Munch IC (2012) Incidence of legal blindness from age-related macular degeneration in Denmark: year 2000 to 2010. Am J Ophthalmol 153. https://doi.org/10.1016/j.ajo.2011.10.016
119. Shen Y, Zhang WY, Chiou GCY (2010) Effect of naringenin on $NaIO_3$-induced retinal pigment epithelium degeneration and laser-induced choroidal neovascularization in rats. Int J Ophthalmol 3. https://doi.org/10.3980/j.issn.2222-3959.2010.01.02
120. Lin JL, Wang YD, Ma Y, Zhong CM, Zhu MR, Chen WP, Lin BQ (2014) Protective effects of naringenin eye drops on N-methyl-N-nitrosourea-induced photoreceptor cell death in rats. Int J Ophthalmol 7. https://doi.org/10.3980/j.issn.2222-3959.2014.03.01
121. Devasari N, Dora CP, Singh C, Paidi SR, Kumar V, Sobhia ME, Suresh S (2015) Inclusion complex of erlotinib with sulfobutyl ether-β-cyclodextrin: preparation, characterization, in silico, in vitro and in vivo evaluation. Carbohydr Polym 134. https://doi.org/10.1016/j.carbpol.2015.08.012
122. Trapani A, Lopedota A, Franco M, Cioffi N, Ieva E, Garcia-Fuentes M, Alonso MJ (2010) A comparative study of chitosan and chitosan/cyclodextrin nanoparticles as potential carriers for the oral delivery of small peptides. Eur J Pharm Biopharm 75. https://doi.org/10.1016/j.ejpb.2010.01.010
123. Zhang P, Liu X, Hu W, Bai Y, Zhang L (2016) Preparation and evaluation of naringenin-loaded sulfobutylether-β-cyclodextrin/chitosan nanoparticles for ocular drug delivery. Carbohydr Polym 149. https://doi.org/10.1016/j.carbpol.2016.04.115
124. Pittalà V, Salerno L, Romeo G, Siracusa MA, Modica MN, Romano GL, Salomone S, Drago F, Bucolo C (2015) Effects of novel hybrids of caffeic acid phenethyl ester and NSAIDs on experimental ocular inflammation. Eur J Pharmacol 752. https://doi.org/10.1016/j.ejphar.2015.02.012
125. Ferrari M (2005) Cancer nanotechnology: opportunities and challenges. https://doi.org/10.1038/nrc1566
126. Nagpal K, Singh SK, Mishra DN (2012) Nanoparticle mediated brain targeted delivery of gallic acid: in vivo behavioral and biochemical studies for improved antioxidant and antidepressant-like activity. Drug Deliv 19. https://doi.org/10.3109/10717544.2012.738437
127. Lai LF, Guo HX (2011) Preparation of new 5-fluorouracil-loaded zein nanoparticles for liver targeting. Int J Pharm 404. https://doi.org/10.1016/j.ijpharm.2010.11.025

128. Jahanshahi M, Williams S, Lyddiatt A, Shojaosadati SA (2004) Preparation and purification of synthetic protein nanoparticulates. IEE Proc Nanobiotechnol. https://doi.org/10.1049/ip-nbt:20041085
129. Langer K, Balthasar S, Vogel V, Dinauer N, von Briesen H, Schubert D (2003) Optimization of the preparation process for human serum albumin (HSA) nanoparticles. Int J Pharm 257. https://doi.org/10.1016/S0378-5173(03)00134-0
130. Rajera R, Nagpal K, Singh SK, Mishra DN (2013) Toxicological study of the Primaquine phosphate loaded chitosan nanoparticles in mice. Int J Biol Macromol 62. https://doi.org/10.1016/j.ijbiomac.2013.08.019
131. Nguyen HTT, Merlin D (2012) Homeostatic and innate immune responses: role of the transmembrane glycoprotein CD98. https://doi.org/10.1007/s00018-012-0963-z
132. Prager GW, Féral CC, Kim C, Han J, Ginsberg MH (2007) CD98hc (SLC3A2) interaction with the integrin β subunit cytoplasmic domain mediates adhesive signaling. J Biol Chem 282. https://doi.org/10.1074/jbc.M702877200
133. Xiao B, Laroui H, Viennois E, Ayyadurai S, Charania MA, Zhang Y, Zhang Z, Baker MT, Zhang B, Gewirtz AT, Merlin D (2014) Nanoparticles with surface antibody against CD98 and carrying CD98 small interfering RNA reduce colitis in mice. Gastroenterology 146. https://doi.org/10.1053/j.gastro.2014.01.056
134. Xiao B, Laroui H, Ayyadurai S, Viennois E, Charania MA, Zhang Y, Merlin D (2013) Mannosylated bioreducible nanoparticle-mediated macrophage-specific TNF-α RNA interference for IBD therapy. Biomaterials 34. https://doi.org/10.1016/j.biomaterials.2013.06.008
135. Xiong K, Zhang H, Du Y, Tian J, Ding S (2019) Identification of HDAC9 as a viable therapeutic target for the treatment of gastric cancer. Exp Mol Med 51. https://doi.org/10.1038/s12276-019-0301-8
136. Lu B, Xu XD, Zhang XZ, Cheng SX, Zhuo RX (2008) Low molecular weight polyethylenimine grafted N-maleated chitosan for gene delivery: properties and in vitro transfection studies. Biomacromolecules 9. https://doi.org/10.1021/bm8004676
137. Putnam D (2006) Polymers for gene delivery across length scales. https://doi.org/10.1038/nmat1645
138. Tseng YC, Mozumdar S, Huang L (2009) Lipid-based systemic delivery of siRNA. https://doi.org/10.1016/j.addr.2009.03.003
139. Choi B, Cui ZK, Kim S, Fan J, Wu BM, Lee M (2015) Glutamine-chitosan modified calcium phosphate nanoparticles for efficient siRNA delivery and osteogenic differentiation. J Mater Chem B 3. https://doi.org/10.1039/c5tb00843c
140. Xiao B, Merlin D (2012) Oral colon-specific therapeutic approaches toward treatment of inflammatory bowel disease. Expert Opin Drug Deliv 9. https://doi.org/10.1517/17425247.2012.730517
141. Xiao B, Ma P, Viennois E, Merlin D (2016) Urocanic acid-modified chitosan nanoparticles can confer anti-inflammatory effect by delivering CD98 siRNA to macrophages. Colloids Surf B Biointerfaces 143. https://doi.org/10.1016/j.colsurfb.2016.03.035
142. Fiejdasz S, Szczubiałka K, Lewandowska-Łańcucka J, Osyczka AM, Nowakowska M (2013) Biopolymer-based hydrogels as injectable materials for tissue repair scaffolds. Biomed Mater (Bristol) 8. https://doi.org/10.1088/1748-6041/8/3/035013
143. Gutiérrez MC, García-Carvajal ZY, Jobbágy M, Rubio F, Yuste L, Rojo F, Ferrer ML, del Monte F (2007) Poly(vinyl alcohol) scaffolds with tailored morphologies for drug delivery and controlled release. Adv Funct Mater 17. https://doi.org/10.1002/adfm.200700093
144. Loh QL, Choong C (2013) Three-dimensional scaffolds for tissue engineering applications: role of porosity and pore size. https://doi.org/10.1089/ten.teb.2012.0437
145. Griffith LG (2002) Emerging design principles in biomaterials and scaffolds for tissue engineering. Ann N Y Acad Sci. https://doi.org/10.1111/j.1749-6632.2002.tb03056.x
146. Barnes CP, Sell SA, Boland ED, Simpson DG, Bowlin GL (2007) Nanofiber technology: designing the next generation of tissue engineering scaffolds. https://doi.org/10.1016/j.addr.2007.04.022

147. Ramakrishna S, Fujihara K, Teo WE, Yong T, Ma Z, Ramaseshan R (2006) Electrospun nanofibers: solving global issues. Mater Today 9. https://doi.org/10.1016/S1369-7021(06)71389-X
148. Persano L, Camposeo A, Tekmen C, Pisignano D (2013) Industrial upscaling of electrospinning and applications of polymer nanofibers: a review. https://doi.org/10.1002/mame.201200290
149. Rogina A (2014) Electrospinning process: versatile preparation method for biodegradable and natural polymers and biocomposite systems applied in tissue engineering and drug delivery. Appl Surf Sci 296. https://doi.org/10.1016/j.apsusc.2014.01.098
150. Zhou W, Feng Y, Yang J, Fan J, Lv J, Zhang L, Guo J, Ren X, Zhang W (2015) Electrospun scaffolds of silk fibroin and poly(lactide-co-glycolide) for endothelial cell growth. J Mater Sci Mater Med 26. https://doi.org/10.1007/s10856-015-5386-6
151. Lin T, Fang J, Wang H, Cheng T, Wang X (2006) Using chitosan as a thickener for electrospinning dilute PVA solutions to improve fibre uniformity. Nanotechnology 17. https://doi.org/10.1088/0957-4484/17/15/017
152. Yang J, Tian F, Wang Z, Wang Q, Zeng YJ, Chen SQ (2008) Effect of chitosan molecular weight and deacetylation degree on hemostasis. J Biomed Mater Res Part B Appl Biomater 84. https://doi.org/10.1002/jbm.b.30853
153. Tang C, Saquing CD, Harding JR, Khan SA (2010) In situ cross-linking of electrospun poly(vinyl alcohol) nanofibers. Macromolecules 43. https://doi.org/10.1021/ma902269p
154. Alavarse AC, de Oliveira Silva FW, Colque JT, da Silva VM, Prieto T, Venancio EC, Bonvent JJ (2017) Tetracycline hydrochloride-loaded electrospun nanofibers mats based on PVA and chitosan for wound dressing. Mater Sci Eng C 77. https://doi.org/10.1016/j.msec.2017.03.199
155. Cerchiara T, Chidichimo G, Gallucci MC, Vuono D (2010) Effects of extraction methods on the morphology and physico-chemical properties of Spanish Broom (*Spartium junceum* L.) fibres. Fibres Text East Eur 80
156. Boateng JS, Pawar HV, Tetteh J (2013) Polyox and carrageenan based composite film dressing containing anti-microbial and anti-inflammatory drugs for effective wound healing. Int J Pharm 441. https://doi.org/10.1016/j.ijpharm.2012.11.045
157. Ozcan AV, Demir M, Onem G, Goksin I, Baltalarli A, Topkara VK, Kaleli I (2006) Topical versus systemic vancomycin for deep sternal wound infection caused by methicillin-resistant *Staphylococcus aureus* in a rodent experimental model. Tex Heart Inst J 33
158. Yan H, He J, Chen S, Yu S, Fan C (2014) Intrawound application of vancomycin reduces wound infection after open release of post-traumatic stiff elbows: a retrospective comparative study. J Shoulder Elbow Surg 23. https://doi.org/10.1016/j.jse.2014.01.049
159. Aramwit P, Yamdech R, Ampawong S (2016) Controlled release of chitosan and sericin from the microspheres-embedded wound dressing for the prolonged anti-microbial and wound healing efficacy. AAPS J 18. https://doi.org/10.1208/s12248-016-9897-y
160. Cerchiara T, Abruzzo A, Ñahui Palomino RA, Vitali B, de Rose R, Chidichimo G, Ceseracciu L, Athanassiou A, Saladini B, Dalena F, Bigucci F, Luppi B (2017) Spanish Broom (*Spartium junceum* L.) fibers impregnated with vancomycin-loaded chitosan nanoparticles as new antibacterial wound dressing: preparation, characterization and antibacterial activity. Eur J Pharm Sci 99. https://doi.org/10.1016/j.ejps.2016.11.028
161. Kirsner RS, Orsted H, Wright JB (2001) The role of silver in wound healing-part 3: matrix metalloproteinases in normal and impaired wound healing: a potential role of nanocrystalline silver. Wounds 13
162. Salas Campos L, Fernándes Mansilla M, Martínez de la Chica AM (2005) Topical chemotherapy for the treatment of burns. Rev Enferm (Barcelona, Spain) 28
163. Adhya A, Bain J, Dutta G, Hazra A, Majumdar B, Ray O, Ray S, Adhikari S (2015) Healing of burn wounds by topical treatment: a randomized controlled comparison between silver sulfadiazine and nano-crystalline silver. J Basic Clin Pharm 6. https://doi.org/10.4103/0976-0105.145776
164. Rosen J, Landriscina A, Kutner A, Adler BL, Krausz AE, Nosanchuk JD, Friedman AJ (2015) Silver sulfadiazine retards wound healing in mice via alterations in cytokine expression. https://doi.org/10.1038/jid.2015.21

165. Dobaria NB, Badhan AC, Mashru RC (2009) A novel itraconazole bioadhesive film for vaginal delivery: design, optimization, and physicodynamic characterization. AAPS PharmSciTech 10. https://doi.org/10.1208/s12249-009-9288-0
166. Boateng JS, Matthews KH, Stevens HNE, Eccleston GM (2008) Wound healing dressings and drug delivery systems: a review. https://doi.org/10.1002/jps.21210
167. Agnihotri S, Bajaj G, Mukherji S, Mukherji S (2015) Arginine-assisted immobilization of silver nanoparticles on ZnO nanorods: an enhanced and reusable antibacterial substrate without human cell cytotoxicity. Nanoscale 7. https://doi.org/10.1039/c4nr06913g
168. Jayakumar R, Menon D, Manzoor K, Nair SV, Tamura H (2010) Biomedical applications of chitin and chitosan based nanomaterials—a short review. https://doi.org/10.1016/j.carbpol.2010.04.074
169. El-Feky GS, Sharaf SS, El Shafei A, Hegazy AA (2017) Using chitosan nanoparticles as drug carriers for the development of a silver sulfadiazine wound dressing. Carbohydr Polym 158. https://doi.org/10.1016/j.carbpol.2016.11.054

Chapter 8
Applications of Modified Chitosan Nanoparticles for Chemotherapy

8.1 Applications of Modified Chitosan Nanoparticles (CSNPs) for General Chemotherapy

Chitosan nanoparticles (CSNPs) exhibit exceptional properties including pharmacokinetic and pharmacodynamic modulation, transport of hydrophobic and hydrophilic drugs, controlled and sustained release, rapid penetration through biological membranes, enhanced accumulation at diseased sites, low immunogenicity, sustainability, and surface modification capability (via amine and hydroxyl groups). The key benefits involve targeted drug delivery, improved bioavailability, reduced systemic toxicity, enhanced therapeutic efficacy, and customizable surface chemistry. The advantages in drug delivery system included encapsulation of various drug types (e.g., proteins, peptides, nucleic acids), protection from degradation and metabolism, pH-dependent release and mucoadhesive properties applicable for cancer therapy, infectious diseases, gene therapy, vaccine delivery, and tissue engineering. The CSNPs' unique combination of properties makes them an attractive platform for developing advanced drug delivery systems.

8.1.1 Doxorubicin-Loaded **Auricularia auricula** *Polysaccharide CSNPs*

Doxorubicin hydrochloride (DOX-HCl), a kind of anthracycline, is connected. It functions as a broad anticancer agent and is an antibacterial. However, because of the mechanism that generates free radicals and causes cardio-toxic effects, its applications are restricted [1]. To counteract the harmful effects of doxorubicin, nanodrug delivery methods such nanoparticles, micelles, and liposomes have been created [2]. However, triethylamine must neutralize DOX during the creation of these drug delivery systems in order to eliminate HCl and make DOX hydrophobic in organic

K. M. Zia, *Chitosan*, Smart Nanomaterials Technology,
https://doi.org/10.1007/978-981-96-2629-8_8

solvents. The anticancer effects of DOX-HCl are diminished by this therapy [3]. The majority of materials utilized in drug delivery systems, such PLGA or block copolymers, include both hydrophobic and hydrophilic sections, which is the basic justification for this treatment [4]. As a result, these materials reduce the ability to load medicines with hydrophilic binders. Scientists employed hydrophilic substances like CS and alginate to boost the drug-loading capacity and activity of DOX-HCl in order to solve this issue. For the delivery of DOX-HCl, Xiong et al. synthesized *Auricularia auricula* polysaccharide (AAP) chitosan nanoparticles (CSNPs) [5]. The monosaccharides glucose, mannose, galactose, xylose, rhamnose, and arabinose make up the biological molecule known as AAP [6]. The AAP is ecologically benign and effective for entrapping chemotherapy medications because of its biocompatibility, biodegradability, bioavailability, low toxicity, cheap cost of manufacture, high safety, and water solubility [5]. The modified scheme [5–9] for the preparation of DOX AAP-CS-NPs and DOX TPP-CS-NPs has been presented (Fig. 8.1).

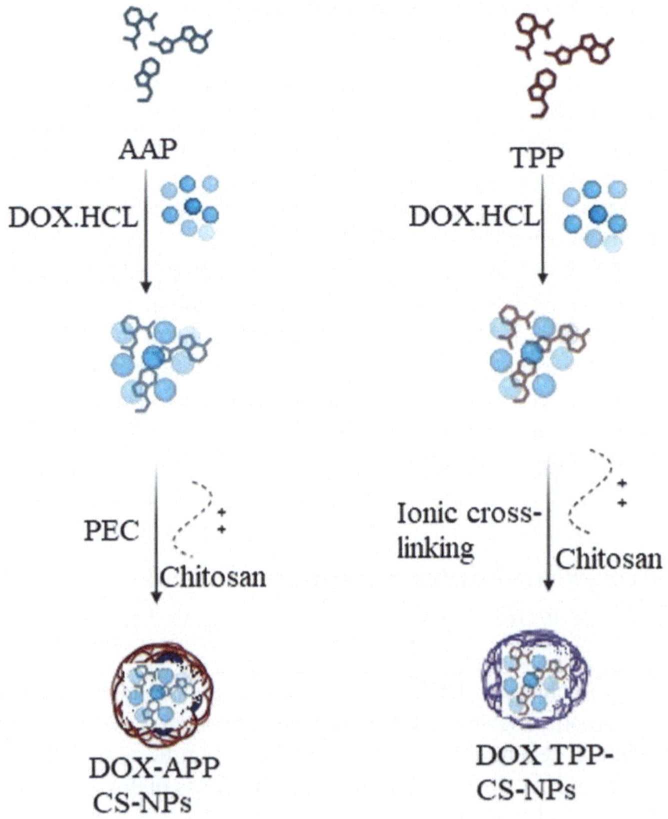

Fig. 8.1 Scheme for the preparation of DOX AAP-CS-NPs and DOX TPP-CS-NPs [5–9]

8.1.2 Collage/Chitosan Film Containing Biofinylated Glycol CSNPs

In most cases, localized chemotherapy is utilized to boost the immune system and stop the growth of secondary tumors. It may lessen adverse effects and deliver medication at a physical location for an extended amount of time [9]. Polymers including collagen, chitosan, poly(lactide-co-glycoside), and poly(lactic acid) (PLA) are employed for localized drug administration because these are biocompatible and biodegradable [10]. Because of its extended drug release, greater targeted effect, less systemic toxicity, and higher treatment effectiveness, collagen-based films containing drug-loaded nanoparticles (NPs) have drawn a lot of attention as a promising drug delivery technology for localized chemotherapy. It is standard practice to get collagen for medicinal uses from pig and cow skin and bones, which is connected to ethical dilemmas and to bovine spongiform encephalopathy [11].

Fish is an alternative collagen source [12]. There has been a lot of effort done to make polymeric nanoparticles using natural polysaccharides such pullulan, curdlan, and chitosan [13]. A new drug delivery system uses glycol chitosan, a water-soluble chitosan derivative that is safe, biodegradable, and biocompatible. Amphiphilic in nature, hydrophobically modified glycol chitosan (HGC) derivatives (cholesterol, deoxycholic acid, and 5-cholanic acid-containing glycol) produce self-assembled nanoparticles in aqueous conditions and act as carriers for hydrophobic medicines and genes [14]. Because of this, an HGC/fish collagen film combination is a suitable delivery method for hydrophobic medicines. By using the freeze-drying process, Chen et al. created an effective drug delivery system based on collage/chitosan film containing biotinylated glycol CSNPs (Fig. 8.2). The conjugation can be enhanced by the addition of biotin to improve medication delivery that targets tumors. The amphiphilic feature of biotinylated cholesterol-modified glycol chitosan (bio-CHGC) allows it to self-assemble nanoparticles via the hydrophobic interactions of the cholesterol moieties [15].

8.1.3 Doxorubicin-Loaded Polyethylene Glycoylated and Poloxemer-Modified CSNPs

Polyethylene glycol (PEG) is a hydrophilic polymer that has the capacity to lengthen the period that nanocarriers circulate in blood. The PEG therefore modifies chitosan nanoparticles (CSNPs). As a result, the nanoparticles (NPs) may accumulate in tumor areas with improved permeability and retention, increasing the tumor's exposure to medication [6, 16, 17]. For pH-triggered DOX release, Scheeren et al. synthesized PEGylated and poloxemer-modified CSNPs [18].

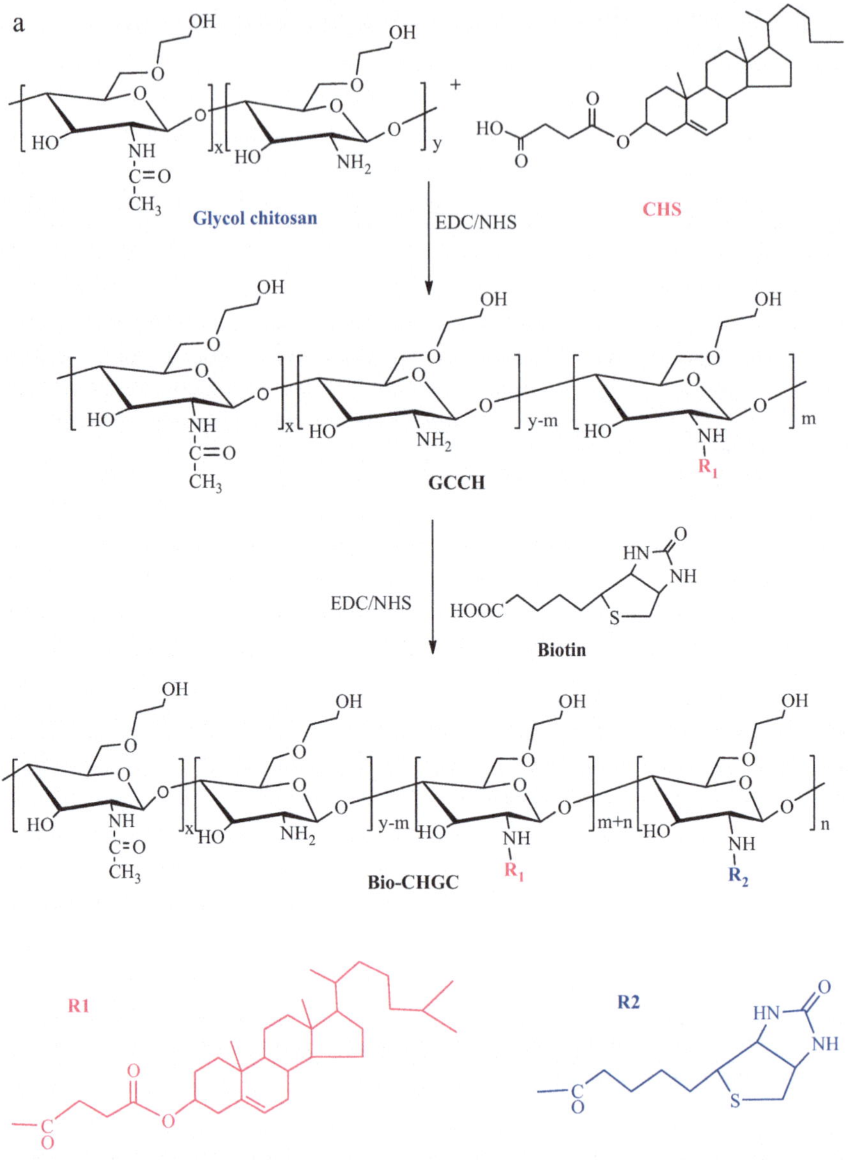

Fig. 8.2 **a** Synthetic route of biotinylated cholesterol-modified glycol chitosan (bio-CHGC). **b** Schematic illustration of the assembly of DOX-loaded nanoparticles, and the preparation and release of drug film [15]

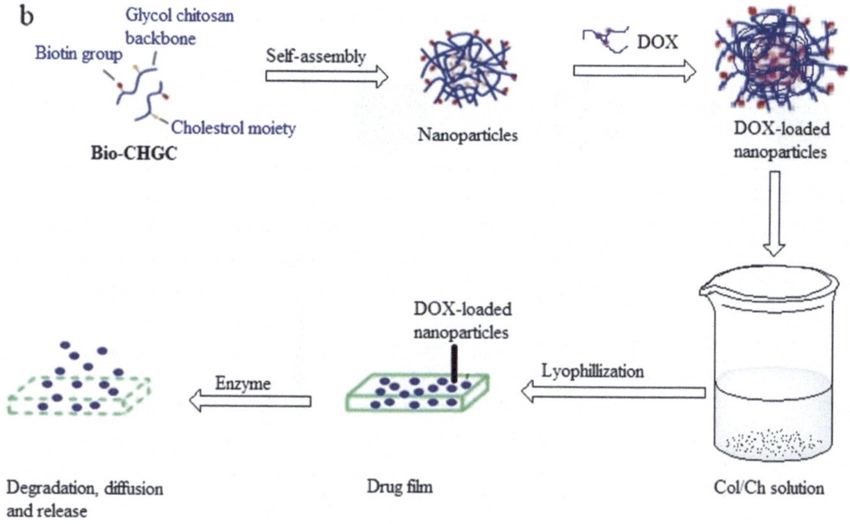

Fig. 8.2 (continued)

8.1.4 Doxorubicin-Loaded Magnetic CSNPs

Since doxorubicin-loaded magnetic CSNPs may specifically target tumor cells while under a magnetic field, magnetic nanoparticles (MNPs) have become more significant in recent years in chemotherapy [19]. MNPs, which may be loaded with medicinal medicines and coated with a variety of natural polymers, include metallic, bimetallic, and superparamagnetic iron oxide NPs [20]. To address the issues of solubility in aqueous medium, toxicity, quick clearance, and lack of selectivity [21], the researchers [21, 22] synthesized DOX-loaded magnetic CSNPs (Fig. 8.3). As the CSNPs respond to pH, so releasing drugs at a pH of 5.5 begin which is produced by the anaerobic breakdown of glucose in cancer cells [22]. A recent study [23] demonstrated the potential of chitosan-coated magnetic nanoparticles (MNP-CS) as carriers for the anticancer drug Telmisartan (TEL). The key findings include high drug-loading capacity (~ 50%) due to large BET surface area (134 m^2/g) of mesoporous MNP-CS, amide bond formation between chitosan's amino groups and TEL's carboxylic groups enables efficient drug loading, pH-responsive controlled release characteristics of MNP-CS-TEL, dose-dependent cytotoxicity toward PC-3 human prostate cancer cells, and significant antitumor activity of the designed nanoformulation. The advantages include targeted drug delivery, improved bioavailability, enhanced therapeutic efficacy, and reduced systemic toxicity. The mechanism involves chitosan's amino groups forming amide bonds with TEL's carboxylic groups. Mesoporous MNP-CS provides a high surface area for drug loading, while its pH-responsive release enables controlled TEL delivery. These findings suggest an effective treatment for prostate cancer, with potential applications in other cancer

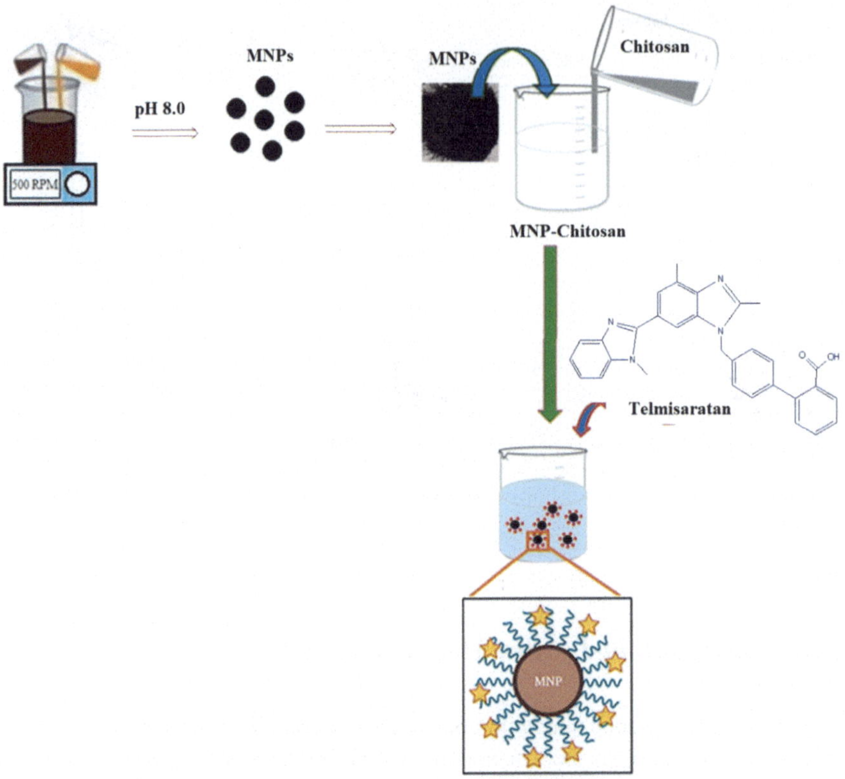

Fig. 8.3 Schematic for chitosan-coated magnetic nanoparticles as carriers of anticancer drug Telmisartan [23]

types. Further research is needed to optimize nanoformulation design and evaluate its in vivo efficacy [23].

8.1.5 Paclitaxel (PTX)-Loaded CSNPs

A chemotherapic medication, paclitaxel belongs to the alkaloid drug family. However, owing to its short circulation half-life, subpar physiochemical characteristics, low stability, and toxicity, its usefulness is constrained. In order to get around these issues and maintain PTX's therapeutic effectiveness, Nag et al. and other researchers [24–26] created functionalized chitosan-PEG nanoparticles for the transfer of anticancer drugs (Fig. 8.4). As the cancer cells overexpress the transferrin receptor, so proven as desirable and efficient target for the delivery of anticancer drugs to tumor sites [25].

Fig. 8.4 Schematic representation for method: **a** preparation of chitosan nanoparticles (CSNPs). **b** Transferrin-conjugated PEGylated chitosan nanoparticles [24, 26]

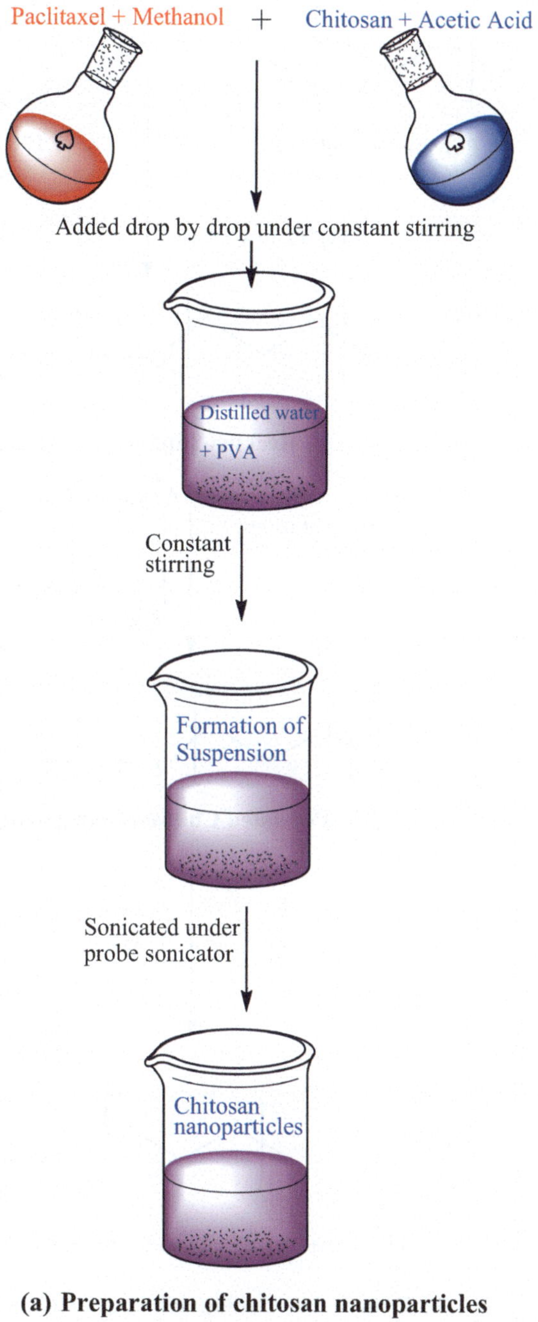

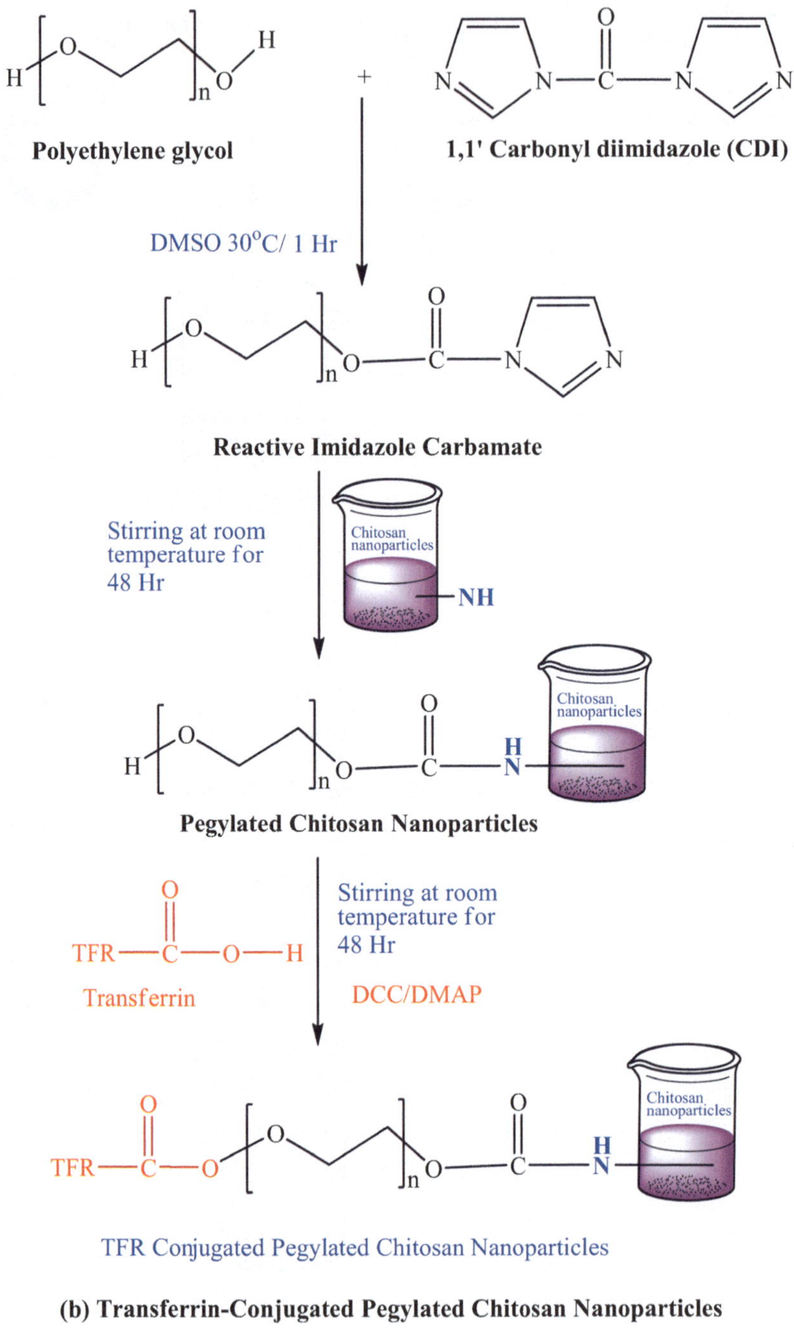

Fig. 8.4 (continued)

8.1.6 Vincristine-Loaded CSNPs

Vincristine, an anticancer alkaloid, treats various cancers, including testicular cancer, Hodgkin's disease, breast cancer, and Hodgkin's lymphoma [27]. The mechanism being followed involved binds to tubulin protein and preventing mitosis with established limitation including resistance in cancer cells, including; SGC 7901 (human gastric carcinoma), KB cell VJ-300 (human cancer), and PC-9 subline (human lung cancer) [28]. The solution may include chitosan nanoparticles for hydrophobic drug delivery, stable, biocompatible, and non-toxic [29], and target-specific ligands, such as folic acid, can be conjugated. The breakthrough in the system was reported by Salar and Kumar whom developed folic acid-CSNPs for vincristine delivery [16] and exploits folate receptors on cancer cell surfaces for targeted delivery with potential enhanced solubility and bioavailability of vincristine and effective against resistant cancer cells with modifications. The future directions listed are optimization of folic acid-CSNPs for vincristine delivery, in vitro and in vivo studies for efficacy and safety evaluation and exploration of other targeting ligands and nanoparticle designs.

8.1.7 Methotrexate Polyamido Amine-Loaded CSNPs

The chemotherapeutic medication methotrexate has been used to treat a variety of cancers, including lung cancer, breast cancer, choriocarcinoma, leukemia, and tumors connected to trophoblastic cells [30]. However, its drawbacks include drug resistance and acute and chronic damage to normal cells (hepatotoxicity) [31]. Polyamido amine (PAMAM) is an artificial macromolecule with a wide range of uses, including gene transfection, tissue engineering, and drug delivery [32]. The CSNPs with PAMAM enhanced methotrexate's anticancer effectiveness. For the administration of methotrexate, Leng et al. created polyamido dendrimer conjugated CSNPs [33].

8.1.8 Docetaxel-Loaded CSNPs

Docetaxel (DTX) is an effective therapy for lung, head, neck, breast, and ovarian cancer. DTX inhibits the formation of microtubules and inhibits cell growth since it is a microtubule stabilizing agent [34]. Due to the lipophilic nature of DTX, its use in medicine is not entirely risk-free [35] to combat the negative effects brought on by DTX usage. Jain et al. developed DTX-loaded CSNPs (Fig. 8.5) by water-in-oil emulsion method [36].

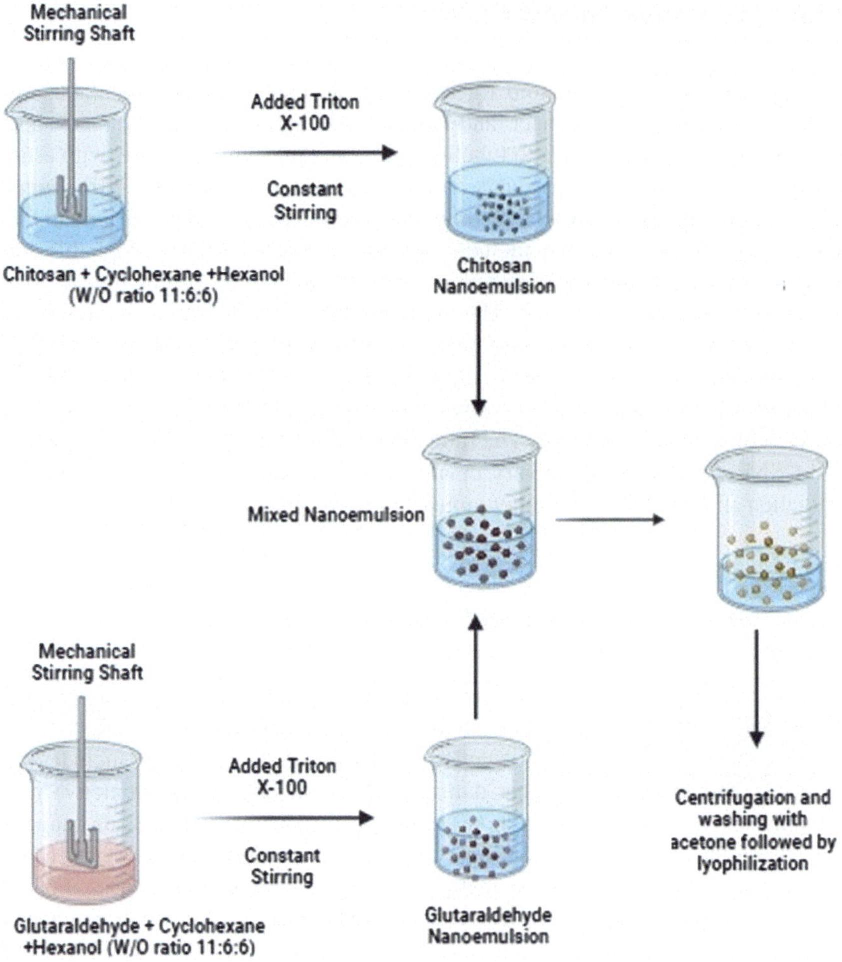

Fig. 8.5 Preparation of docetaxel-loaded chitosan nanoparticles by water-in-oil nanoemulsion method [36]

8.1.9 PEGylated CS-Poly(Ethylene Imine) NPs

For the delivery of siRNA to tumor cells, Ragelle et al. produced PEGylated CS-Poly(ethylene imine) NPs [37]. To boost the therapeutic effectiveness of ligands, NPs are coupled with arginine-glycine aspartate (RGD) peptide or an RGD peptidomimetic (RGDP) [38].

8.1.10 Small Interfering RNA Glycol CSNPs

Multidrug resistance (MDR), which reduces the effectiveness of anticancer medications, has been previously suggested as the primary reason why cancer treatments fail. Genetic changes that affect DNA repair ability or drug-targeted enzyme levels are among the reasons and mechanisms that underlie MDR, as include enhanced drug abrupt withdrawal, reduced cellular uptake, and increased drug efflux [39]. P-glycoprotein (Pgp)-mediated drug efflux is, out of all of them, the most crucial MDR mechanism. The Pgp is an MDR-1 protein that releases the majority of medicines and reduces drug buildup in cancer cells [40]. The chemo-sensitivity of cancer cells to anticancer drugs can be reduced by Pgp overexpression.

The Pgp inhibition has previously been attempted using small compounds such as verapamil and cyclosporine A [41]. These compounds have the ability to regulate Pgp function and expression, but it also has long-term toxicity and a number of systemic adverse effects, such as immunosuppression and bradycardia [42]. Today, sequence-specific gene silencing has been investigated as a way to stop the production of the MDR-1 gene in cancer cells that are resistant to treatment [43]. The sequence-specific gene is essential for managing Pgp without causing any negative side effects. The siRNA molecules naturally exhibit poor cellular absorption and decreased stability near the nucleus. Therefore, efficient carriers are necessary for the transport of siRNA. Liposomes and cationic polymers are typically utilized as siRNA carriers. These are highly toxic and show less accumulation [44]. The modification of siRNA improves its physiochemical properties.

Yhee et al. created glycol CSNPs for siRNA delivery (Fig. 8.6). Sulfhydryl groups can be used to modify the chitosan and glycol polymers, and the resultant polymers and psi-Pgp generated stable and condensed nanoparticles (psi-Pgp-tGl NPs) serve as an efficient treatment tool to manage the drawbacks of chemotherapy for MDR cancer [45].

8.1.11 Hydrotropic Oligomer-Conjugated Glycol CSNPs

The most effective hydrotropic agent is N,N-diethylnicotinamide (DENA) derivatives, and DENA (HO) is used with glycol CSNPs to increase drug-loading amounts [44]. To increase the solubility of paclitaxel (PTX), Koo et al. created hydrotropic oligomer-conjugated glycol CSNPs as shown in the accompanying Fig. 8.7 [46]. Following the incorporation of PTX, HO-based glycol CSNPs have notable anticancer effects [46].

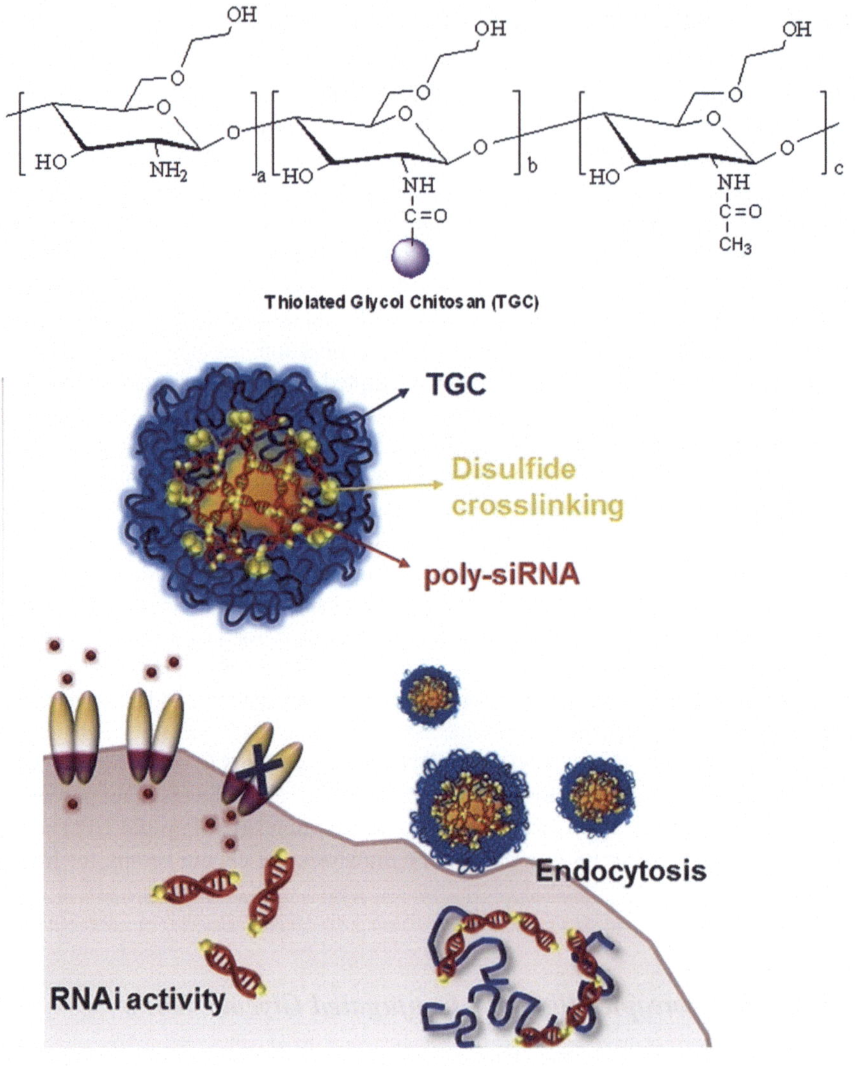

Fig. 8.6 Formation of poly-siRNA-thiolated glycol chitosan (tGC) NPs and their intracellular delivery for target gene silencing [45]

8.2 Applications of Modified CSNPs for Localized Chemotherapy

Chitosan's exceptional properties make it an ideal material for cancer treatment. CSNPs offer numerous benefits like easy manufacturing, biocompatibility, biodegradability, mucoadhesiveness, enhanced permeation, economic viability, and

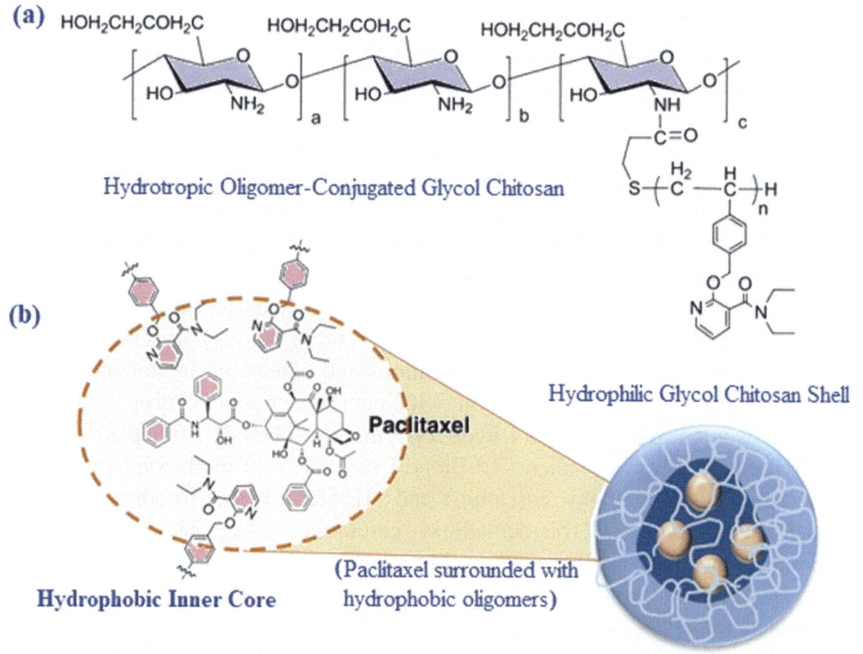

Fig. 8.7 Hydrotropic oligomer-conjugated glycol chitosan nanoparticles (HO-CNPs). **a** Chemical structure of HO-CNPs. **b** Amphiphilic self-assembly of PTX-encapsulated HO-CNPs (PTX-HO-CNPs) [46]

eco-friendliness. The key features involved cationic functional groups and electrostatic interaction with chemotherapeutic agents. The benefits in cancer therapy involve efficient drug delivery, targeted tumor accumulation, improved therapeutic efficacy, reduced systemic toxicity, and enhanced patient compliance. The CSNPs' role encompasses encapsulation of chemotherapeutic agents, controlled release and sustained delivery, tumor-specific targeting, and minimizing side effects. By leveraging CSNPs' unique characteristics, researchers can develop innovative cancer treatments with improved outcomes.

8.2.1 Cervical Cancer Treatment

Cervical cancer is the second most prevalent malignancy in women worldwide, primarily caused by Human Papillomavirus (HPV) [47]. Chemotherapy's limitations, including multidrug resistance and toxicity, necessitate innovative solutions [48]. Chitosan Nanoparticles (CSNPs) are biodegradable, linear polymer composed of N-acetyl-D-glucosamine and D-glucosamine [49]. Ideal carrier system due to non-toxicity, biocompatibility, mucoadhesiveness, and degradability enhances intracellular transport of hydrophilic macromolecules [50]. Curcumin is low molecular weight, poorly water-soluble drug. Three active molecules: curcumin, demethoxy curcumin, and bis-demethoxy curcumin [51] exhibit antioxidant, anti-inflammatory, antiproliferative, antiangiogenic, and anticancer properties [52]. Archana et al. developed chitosan-curcumin nanoparticles (CS-CUR-NPs) to enhance stability and enable sustained drug release [53]. CSNPs hinder rapid drug clearance from inflammation sites [54]. The formulation using electrostatic interaction with sodium tripolyphosphate (TPP) [55] is applicable for loaded with various drugs and vaccines involving curcumin [56], meningococcal C conjugate [51], and diphtheria [57] and tetanus toxoid [58]. Chitosan nanoparticles offer a promising platform for curcumin delivery in cervical cancer therapy, addressing chemotherapy's limitations and enhancing treatment efficacy. Curcumin is a low molecular weight with poor water-soluble drug with three active molecules like: 1,7-bis(4-hydroxy-3-methoxyphenyl)-1,6-heptadiene-3,5-dione (curcumin); 1,6-Heptadiene-3,5-dione,1-(4-hydroxy-3-methoxyphenyl)-7-(4-hydroxyphenyl) (demethoxy curcumin) and (1E,6E)-1,7-bis (4-hydroxy phenyl) hepta-1,6-diene-3,5-dione (bis-demethoxy curcumin) [51]. Curcuminoids have antioxidant, anti-inflammatory, antiproliferative, antiangiogenic [52] antitumorigenic, anticoagulant, and antibacterial, anticancer, and anti-ischemic and wound healing activities [59].

To achieve this, Khan et al. concentrated their curcumin research on the creation of possible delivery systems to increase the drug's aqueous solubility, stability, and bioavailability as well as to administer the medication in a regulated manner close to cancer cells and tissues [60]. While Shaikh et al. [59] reported a ninefold increase in curcumin bioavailability with the same polymeric nanoparticles, Maiti et al. [61] observed a 2.5-fold increase in curcumin bioavailability while delivering with phospholipid complex as opposed to curcumin alone, Anand et al. recently achieved a twofold increase in curcumin bioavailability with PLGA nanoparticles [62]. By lengthening the residence period, Khan et al. are able to boost the drug's bioavailability and consequently facilitate drug absorption through adhesion with cellular surfaces [60].

Chitosan nanoparticles (CsNPs) are the finest option for encasing curcumin to increase its bioavailability. Khan et al. devised a chitosan-based nanoparticle delivery system [60], in order to solubilize curcumin in aqueous conditions, safeguard it from hydrolytic degradation, and administer curcumin in a regulated manner; as a consequence, administered curcumin will have a higher bioavailability, maximizing its therapeutic effectiveness for tumors while minimizing pharmacokinetic problems. The potentiality of the formulated curcumin-loaded chitosan nanoparticles can be assessed in this regard by analyzing their smaller size to assess in vitro release kinetics, stability, cellular uptake, cytotoxicity, and apoptosis inducing activity in cervical cancer cells [60].

8.2.2 Curcumin-Loaded CSNPs

Cervical cancer is currently the second most prevalent malignancy in women and the leading cause of cancer death in developing nations [63]. The viral oncogenes E6 and E7 are expressed in the majority of cervical malignancies with papillomaviruses. Important cellular regulatory proteins that regulate apoptosis, survival, cell proliferation, differentiation, and genome stability are suppressed by the interaction between HPV E6 and E7 [64]. Cervical cancer is treated with curcumin, an anticancer medication. However, because it is less water soluble and drug resistant, its uses are constrained [65]. Curcumin is packed with biodegradable and biocompatible NPs, including chitosan, to get around these drawbacks [65]. In order to improve the inflammatory action against cervical cancer, Khan et al. synthesized curcumin-loaded CSNPs [60].

8.2.3 Fore Brain Cancer Treatment

Due to their versatility and high cargo entrapment effectiveness, the CS derivatives make for effective and adaptable nanodelivery systems. Through solid-state deacetylation or enzymatic hydrolysis of chitin deacetylases, chitin monomers are linearized in an alkaline environment [66]. The bipolyaminosaccharide structure consists of a carbohydrate backbone, numerous –OH and $-NH_2$ functional groups, which serve as easily accessible moieties for functional plasmid or cargo release efficiency mediated by endosome osmotic pressure (caused by hydrogen protons influx), and lastly, a low toxicity index because of its biocompatibility and biodegradability across various biological applications [67]. Either complexed, contained within a CS particle, or scattered within a CS matrix, is how cargo is often described. Cargo is frequently characterized as complexed, enclosed within a CS particle, or dispersed within a CS matrix. Through the processes of cross-linking, cationic salt solvation, emulsification, ionic complexation, or gelation, one may create these particulate systems by engaging with different functional groups on proteins, antibodies, drugs, DNA/RNA, or other pH-sensitive moieties [67].

Microparticles/microspheres and NPs/nanospheres, also known as "nanocapsules," are vesicular particulate systems with a hollow sphere consisting of an oil or water core (that may include active cargo) and a polymeric shell. Particles are classified according to their spherical diameter and spatial composition. This structure serves as a link between the adsorption of cargo on the particle surface and cargo trapping in the core. While matrixial structures are particulate systems in which the active molecule is integrated into the polymer network, these are also known as nanospheres [68]. Additionally, "solid-lipid" NPs are systems that emulsify lipids with a surfactant for structural stability after using lipids in the solid phase [69]. When conveying material that is just marginally water soluble, this structure is advantageous. Two immiscible liquids combined to create nanoemulsions, which are lipid-like structures stabilized by a surfactant. Micelles are particles without an aqueous center that are created from a single lipid layer [70].

Chitosan (CS) nanosystems are tailored for optimal administration based on cargo polarity, solubility, molecular weight, and route of administration. The advantages of CS nanocarriers include efficient CNS delivery, enhanced membrane adhesion, high transfection efficiency, and targeted tumor accumulation. The CS nanoparticle size and composition critically impact the tumor targeting, biodistribution, ideal applications, hypervascularized disease models (e.g., brain tumors), nanosized carriers leverage improved permeability and retention effect, passive diffusion in intratumoral region and reduced off-target toxicity. The key findings involve CS coatings/formulations enhance particle systems' efficiency [71], size, and composition optimize tumor targeting and biodistribution [71] and improve permeability and retention effect benefits brain tumor treatment [72]. Recent developments in non-viral gene delivery strategies utilizing chitosan (CS) nanoparticles have shown promise in brain cancer treatment. Additionally, CS-based encapsulation approaches for drugs and chemotherapeutics have demonstrated potential.

To circumvent CNS drug delivery restrictions, CS nanoparticles (CSNPs) modified by poly(lactic-co-glycolic acid) (PLGA) are coupled with Arg-Gly-Asp RGD-linked peptide and loaded with clinically authorized paclitaxel (PTX) chemotherapeutic medication for glioblastomas (GBM) treatment. The PTX-PLGACSNP-RGD particle showed superior tumor targeting and uptake by integrin receptor-mediated endocytosis, caused cell-cycle arrest at G2/M, and accelerated lung tumor cell killing via emulsion-solvent evaporation [73]. The in vivo brain pharmacokinetics of common GBM treatment, such temozolomide (TMZ), are greatly enhanced when polyamidoamine (PAMAM) dendrimer is coated with chitosan. Chitosan-coated PAMAM coupled to TMZ enhanced GBM tumor targeting in U-251 and T-98G cell lines at lower TMZ doses [74]. In another study, the CS nanospheres can be produced via ionotropic gelation by complexing pDNA with CS tripolyphosphate (TPP) and hyaluronic acid (HA).

The production of the sol–gel transition is facilitated by ionic gelation, and TPP reduces particle size while stabilizing the complex in biological fluids [75].

To assess the affinity and transport of RA to the U87 human GBM line, a trimethylated solid-lipid CS formulation is created [76]. In vivo tumor development is inhibited by the combination of CS micelles with myricetin when compared to the free medication and controls [77]. Based on overexpressed receptors on the cell surface, glioma cells can be selectively targeted by integrating conjugated CS into micelles. The transferrin receptor on glioma cells may be targeted by combining CS with d-u-tocopheryl glycol succinate 1000 (TPGS) and adding it to micelles that are loaded with docetaxel [78]. This strategy of glioma cell targeting for chemotherapy displays higher cell uptake and stability in vivo over time and is more than 200 times more effective on C6 rat glioma cell survival than free Docel [78]. This may represent a possible future use for DNA technology in CNS cancer, according to the findings of the study using CS-containing micelles in chemotherapy administration [79].

8.2.4 Docetaxel-Loaded D-α-Tocopherol Polyethylene Glycol 1000 Succinate-Loaded CSNPs

A plant alkaloid called docetaxel is used to treat brain cancer. Many individuals worldwide are affected with brain cancer, which is extremely difficult to cure [80]. The most prevalent and severe kind of brain cancer is glioma, which has a twelve to fifteen-month median survival time and survival rates of fewer than 5% over a five-year period [81]. Chemotherapy effectiveness is decreased because of the blood–brain barrier (BBB), which prevents drugs from entering the brain [82]. Another obstacle that lessens medication accumulation in brain tumors is the blood tumor barrier [83]. Multidrug resistance, which develops through a number of pathways, including one in which p-glycoprotein functions as an efflux pump for the majority of hydrophobic anticancer medicines, is the main cause of therapy failure [84].

Bioadhesive nanocarrier systems are made using chitosan derivatives. Its positive charge reacts electrostatically with the cell membrane's negative charge, causing the cancer cell to become bioadhesive [85]. Chitosan breaks down into glutamic acid, which may be used in regular cellular metabolism. Additionally, D-tocopherol polyethylene glycol 1000 succinate (TPGS) and chitosan, both of which are pH responsive, can cause glioma cells to apoptose [86]. Here, chitosan and TPGS are conjugated to improve drug solubility in an acidic environment. Additionally, it offers stability and a longer retention duration. It is further altered with transferring, which serves as a receptor for the administration of a certain medication. Docetaxel-loaded D-tocopherol polyethylene glycol 1000 succinate linked chitosan (TPGS) nanoparticles were created by Agrawal et al. [78, 80] for the treatment of cancer (Fig. 8.8).

8.2.5 Temozolomide (TMZ) Poly(Ɛ-Caprolactone Diol)-Based Polyurethane-Loaded CSNPs

Since temozolomide (TMZ) is an alkylating agent, it can penetrate the blood–brain barrier. Therefore, it is a viable option for the treatment of glioblastoma, a kind of brain cancer [87]. Its ineffectiveness is hampered by its brief 1.8 h plasma half-life. Polymeric nanoparticles have been used to encapsulate TMZ in order to achieve regulated release [88]. For the prolonged release of TMZ, Irani et al. developed TMZ-embedded CSNPs loaded into PCL-Diol-b-PU [89].

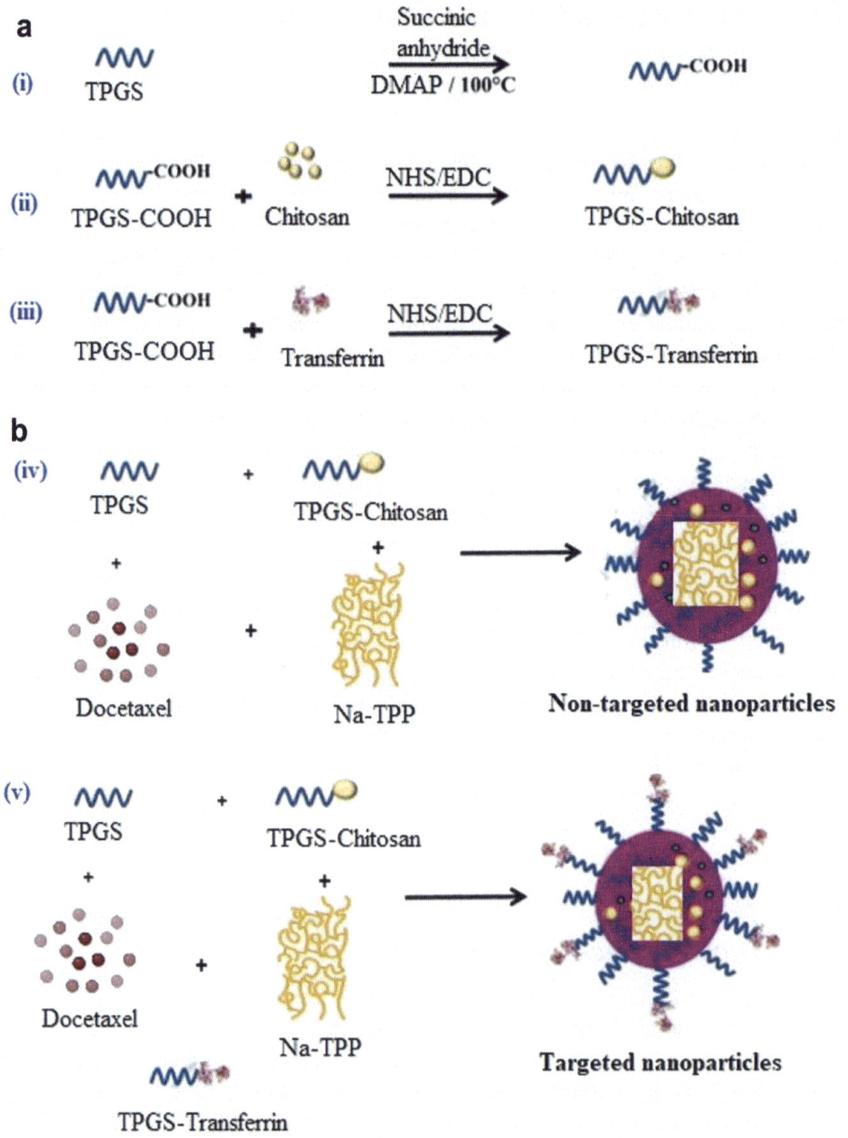

Fig. 8.8 Schematic diagram for **a** synthesis of **i** TPGS-COOH, **ii** TPGS-chitosan, **iii** TPGS-transferrin and **b** preparation of **iv** non-targeted nanoparticles (DTX-TCS-NP) and **v** targeted nanoparticles (DTX-TCS-Tf-NP)

8.2.6 The SiRNA-Loaded Galectin/CSNPs

For the purpose of treating glioblastoma, van Woensel et al. produced siRNA-loaded CSNPs with galectin [90]. Natural protein called galectin exhibits excellent accumulation in glioblastoma (GBM) and a positive immune response [91]. Gal-1 causes CD8+ T lymphocytes to undergo apoptosis and inhibits the release of pro-inflammatory cytokines [92].

8.2.7 For Breast Cancer Treatment

For the treatment of different malignancies, several hydrophilic [93] and hydrophobic [94] medicines have been encapsulated in CSNPs and evaluated in vitro and in vivo. Doxorubicin (DOX), a strong anthracycline medication that is hydrophilic and authorized for use against a variety of tumors, can be conjugated to CS using a succinic anhydride spacer [93]. Due to their affinity for HER^{2+} cancer cells, trastuzumab conjugated CS-DOX NPs ought to be more readily absorbed than CS-DOX and free medication. In a different investigation, DOX can be contained in chitosan-pluronic micelles. The NPs can be created with a high drug-loading capacity and a particle size of 50 nm. The CS-pluronic micelle containing DOX (DOX-NP) displayed 77.33% better in vitro therapeutic effectiveness than free DOX in MCF7 breast cancer cell lines [95].

For medication delivery to specific tumors, paclitaxel-loaded CS and PEG-coated poly-d,L-lactide-co-glycolide (PLGA) NPs can be created. In vitro cytotoxicity and uptake effectiveness of the produced NPs are examined in retinoblastoma, breast cancer, and pancreatic cancer cell lines. Regarding the antiproliferative impact and cell-cycle inhibition, PLGA-CS NPs and PLGA-Cs PEG NPs beat PLGA NPs and natural paclitaxel. In light of the findings reported [96], the use of produced NPs as carriers for the delivery of tumor drugs is recommended after going through series of experiments.

8.2.8 Lapatinib-Loaded Dextran Sulfate CSNPs

Lapatinib is a tyrosine kinase inhibitor that targets HER-1 and HER-2, two members of the EGFR family (epidermal growth factor receptors). In various breast tumors, HER-1 and HER-2 overexpression has been studied. However, all chemical medications are hazardous to patients and have several negative effects, according to medical experts [97]. So, biopolymers NPs, particularly carbohydrates, are used to encapsulate these medications. Dextran sulfate and chitosan are two carbohydrates that are regarded as effective nanocarriers for the delivery of drugs [98]. The toxicity produced by the positive charge of chitosan is lessened by the negative charge of dextran, and DS-CSNPs have the potential to load hydrophilic medicines [99, 100]. For the administration of the hydrophobic medication, Mobasseri et al. produced hydrophobic lapatinib encapsulated dextran-chitosan nanoparticles using a toxic solvent-free method [101]. Research on cancer treatment highlights key advancements, including the identification of multidrug resistance in vincristine-resistant cells, the role of HPV proteins in cervical cancer progression, and the development of an HPLC-UV method for precise lapatinib monitoring [102–104].

8.3 Conclusion

Cancer treatment is complex and often requires a combination of parallel and sequential therapies, tailored to the type and stage of the disease. Common multimodal approaches include; surgery cum chemotherapy, surgery cum radiotherapy, and chemotherapy cum radiotherapy. Despite its widespread use, chemotherapy has limitations, including systemic toxicity, drug resistance, and side effects. Beyond chemotherapy, alternative approaches include immunotherapy, targeted therapy, hormone therapy, gene therapy, radiation therapy, surgical oncology, and integrative therapies (e.g., nutrition, mind–body therapies). Effective cancer treatment requires a personalized approach, considering: tumor biology, patient genetics, lifestyle factors, and disease stage. Advances in cancer research continue to uncover new treatment strategies, including: nanotechnology, stem cell therapy, CAR-T cell therapy, and precision medicine. This chapter provides an overview of modified chitosan nanoparticles (CSNPs) for chemotherapy applications. Doxorubicin hydrochloride (Dox-HCl), a widely used anticancer antibiotic, has been extensively studied. To address localized chemotherapy challenges, researchers have developed collagen/chitosan films containing biofunctionalized glycol CSNPs. Localized chemotherapy aims to boost the immune system, inhibit secondary tumor growth, minimize side effects, and provide sustained medication delivery. Doxorubicin-loaded polyethylene glycolated and poloxamer-modified CSNPs have shown increased drug exposure. Magnetic nanoparticles (MNPs) have gained popularity in chemotherapy due to their ability to target tumor cells under magnetic guidance. Several anticancer medications have been explored with chitosan-loaded nanoparticles, including paclitaxel (alkaloid), vincristine (anticancer alkaloid), and methotrexate (chemotherapeutic agent for lung, breast, and other cancers). Modified CSNPs have shown promise in cervical cancer treatment in curcumin-loaded CSNPs exhibit anticervical cancer inflammatory activity, brain cancer treatment, and breast cancer treatment. This chapter highlights the potential of modified chitosan nanoparticles in enhancing chemotherapy efficacy and reducing side effects.

References

1. Jin H, Xu CX, Kim HW, Chung YS, Shin JY, Chang SH, Park SJ, Lee ES, Hwang SK, Kwon JT, Minai-Tehrani A, Woo M, Noh MS, Youn HJ, Kim DY, Yoon BI, Lee KH, Kim TH, Cho CS, Cho MH (2008) Urocanic acid-modified chitosan-mediated PTEN delivery via aerosol suppressed lung tumorigenesis in K-rasLA1 mice. Cancer Gene Ther 15. https://doi.org/10.1038/sj.cgt.7701116
2. Xu W, Siddiqui IA, Nihal M, Pilla S, Rosenthal K, Mukhtar H, Gong S (2013) Aptamer-conjugated and doxorubicin-loaded unimolecular micelles for targeted therapy of prostate cancer. Biomaterials 34. https://doi.org/10.1016/j.biomaterials.2013.03.006
3. Yahuafai J, Asai T, Nakamura G, Fukuta T, Siripong P, Hyodo K, Ishihara H, Kikuchi H, Oku N (2014) Suppression in mice of immunosurveillance against PEGylated liposomes by encapsulated doxorubicin. J Control Release 192. https://doi.org/10.1016/j.jconrel.2014.07.010
4. Park J, Fong PM, Lu J, Russell KS, Booth CJ, Saltzman WM, Fahmy TM (2009) PEGylated PLGA nanoparticles for the improved delivery of doxorubicin. Nanomedicine 5. https://doi.org/10.1016/j.nano.2009.02.002
5. Xiong W, Li L, Wang Y, Yu Y, Wang S, Gao Y, Liang Y, Zhang G, Pan W, Yang X (2016) Design and evaluation of a novel potential carrier for a hydrophilic antitumor drug: *Auricularia auricula* polysaccharide-chitosan nanoparticles as a delivery system for doxorubicin hydrochloride. Int J Pharm 511. https://doi.org/10.1016/j.ijpharm.2016.07.026
6. Guo H, Yang C, Hu Z, Wang W, Wu Y, Lai Q, Yuan Z (2013) Ethylene glycol oligomer modified-sodium alginate for efficiently improving the drug loading and the tumor therapeutic effect. J Mater Chem B 1. https://doi.org/10.1039/c3tb20968g
7. Zhang C, Wang W, Liu T, Wu Y, Guo H, Wang P, Tian Q, Wang Y, Yuan Z (2012) Doxorubicin-loaded glycyrrhetinic acid-modified alginate nanoparticles for liver tumor chemotherapy. Biomaterials 33. https://doi.org/10.1016/j.biomaterials.2011.11.045
8. Ma Z, Wang J, Zhang L, Zhang Y, Ding K (2010) Evaluation of water soluble β-d-glucan from *Auricularia auricula*-judae as potential anti-tumor agent. Carbohydr Polym 80. https://doi.org/10.1016/j.carbpol.2010.01.015
9. Khaskheli SG, Zheng W, Sheikh SA, Khaskheli AA, Liu Y, Soomro AH, Feng X, Sauer MB, Wang YF, Huang W (2015) Characterization of *Auricularia auricula* polysaccharides and its antioxidant properties in fresh and pickled product. Int J Biol Macromol 81. https://doi.org/10.1016/j.ijbiomac.2015.08.020
10. Park YJ, Ku Y, Chung CP, Lee SJ (1998) Controlled release of platelet-derived growth factor from porous poly(L-lactide) membranes for guided tissue regeneration. J Control Release 51. https://doi.org/10.1016/S0168-3659(97)00169-7
11. Fernández-Díaz MD, Montero P, Gómez-Guillén MC, (2001) Gel properties of collagens from skins of cod (*Gadus morhua*) and hake (*Merluccius merluccius*) and their modification by the coenhancers magnesium sulphate, glycerol and transglutaminase. Food Chem 7. https://doi.org/10.1016/S0308-8146(01)00110-8
12. Huang YR, Shiau CY, Chen HH, Huang BC (2011) Isolation and characterization of acid and pepsin-solubilized collagens from the skin of balloon fish (*Diodon holocanthus*). Food Hydrocoll 25. https://doi.org/10.1016/j.foodhyd.2011.02.011
13. Wang Y, Liu Y, Liu Y, Wang Y, Wu J, Li R, Yang J, Zhang N (2014) pH-sensitive pullulan-based nanoparticles for intracellular drug delivery. Polym Chem 5. https://doi.org/10.1039/c3py00817g
14. Li L, Gao FP, Tang HB, Bai YG, Li RF, Li XM, Liu LR, Wang YS, Zhang QQ (2010) Self-assembled nanoparticles of cholesterol-conjugated carboxymethyl curdlan as a novel carrier of epirubicin. Nanotechnology 21. https://doi.org/10.1088/0957-4484/21/26/265601
15. Chen M-M, Huang Y-Q, Cao H, Liu Y, Guo H, Chen LS, Wang J-H, Qi-Qing Zhang Q-Q (2015) Collagen/chitosan film containing biotinylated glycol chitosan nanoparticles for localized drug delivery. Colloids Surf B Biointerfaces 128:339–346. https://doi.org/10.1016/j.colsurfb.2015.02.024

16. Salar RK, Kumar N (2016) Synthesis and characterization of vincristine loaded folic acid–chitosan conjugated nanoparticles. Resour-Efficient Technol 2. https://doi.org/10.1016/j.reffit.2016.10.006
17. Yu J, Xie X, Zheng M, Yu L, Zhang L, Zhao J, Jiang D, Che X (2012) Fabrication and characterization of nuclear localization signal-conjugated glycol chitosan micelles for improving the nuclear delivery of doxorubicin. Int J Nanomed 7. https://doi.org/10.2147/IJN.S36150
18. Scheeren LE, Nogueira DR, Macedo LB, Vinardell MP, Mitjans M, Infante MR, Rolim CMB (2016) PEGylated and poloxamer-modified chitosan nanoparticles incorporating a lysine-based surfactant for pH-triggered doxorubicin release. Colloids Surf B Biointerfaces 138. https://doi.org/10.1016/j.colsurfb.2015.11.049
19. Jeong YI, Kim DG, Jang MK, Nah JW (2008) Preparation and spectroscopic characterization of methoxy poly(ethylene glycol)-grafted water-soluble chitosan. Carbohydr Res 343. https://doi.org/10.1016/j.carres.2007.10.025
20. Maeda H (2001) The enhanced permeability and retention (EPR) effect in tumor vasculature: the key role of tumor-selective macromolecular drug targeting. Adv Enzyme Regul 41. https://doi.org/10.1016/S0065-2571(00)00013-3
21. Sun C, Lee JSH, Zhang M (2008) Magnetic nanoparticles in MR imaging and drug delivery. https://doi.org/10.1016/j.addr.2008.03.018
22. Unsoy G, Khodadust R, Yalcin S, Mutlu P, Gunduz U (2014) Synthesis of doxorubicin loaded magnetic chitosan nanoparticles for pH responsive targeted drug delivery. Eur J Pharm Sci 62. https://doi.org/10.1016/j.ejps.2014.05.021
23. Dhavale RP, Dhavale RP, Sahoo SC, Kollu P, Jadhav SU, Patil PS, Dongale TD, Chougale AD, Patil PB (2021) Chitosan coated magnetic nanoparticles as carriers of anticancer drug Telmisartan: pH-responsive controlled drug release and cytotoxicity studies. J Phys Chem Solids 148:109749. https://doi.org/10.1016/j.jpcs.2020.109749
24. Nag M, Gajbhiye V, Kesharwani P, Jain NK (2016) Transferrin functionalized chitosan-PEG nanoparticles for targeted delivery of paclitaxel to cancer cells. Colloids Surf B Biointerfaces 148. https://doi.org/10.1016/j.colsurfb.2016.08.059
25. Luong D, Kesharwani P, Deshmukh R, Mohd Amin MCI, Gupta U, Greish K, Iyer AK (2016) PEGylated PAMAM dendrimers: enhancing efficacy and mitigating toxicity for effective anticancer drug and gene delivery. https://doi.org/10.1016/j.actbio.2016.07.015
26. Kursa M, Walker GF, Roessler V, Ogris M, Roedl W, Kircheis R, Wagner E (2003) Novel shielded transferrin-polyethylene glycol-polyethylenimine/DNA complexes for systemic tumor-targeted gene transfer. Bioconjug Chem 14. https://doi.org/10.1021/bc0256087
27. Towle MJ, Salvato KA, Wels BF, Aalfs KK, Zheng W, Seletsky BM, Zhu X, Lewis BM, Kishi Y, Yu MJ, Littlefield BA (2011) Eribulin induces irreversible mitotic blockade: implications of cell-based pharmacodynamics for in vivo efficacy under intermittent dosing conditions. Cancer Res 71. https://doi.org/10.1158/0008-5472.CAN-10-1874
28. Yang YX, Xiao ZQ, Chen ZC, Zhang GY, Yi H, Zhang PF, Ii JL, Zhu G (2006) Proteome analysis of multidrug resistance in vincristine-resistant human gastric cancer cell line SGC7901/VCR. Proteomics 6. https://doi.org/10.1002/pmic.200402031
29. Hu B, Pan C, Sun Y, Hou Z, Ye H, Hu B, Zeng X (2008) Optimization of fabrication parameters to produce chitosan-tripolyphosphate nanoparticles for delivery of tea catechins. J Agric Food Chem 56. https://doi.org/10.1021/jf801111c
30. Salmaso S, Bersani S, Semenzato A, Caliceti P (2007) New cyclodextrin bioconjugates for active tumour targeting. J Drug Target 15. https://doi.org/10.1080/10611860701349752
31. Natale RB, Yagoda A, Watson RC, Whitmore WF, Blumenreich M, Braun DW (1981) Methotrexate: an active drug in bladder cancer. Cancer 47. https://doi.org/10.1002/1097-0142(19810315)47:6<1246::AID-CNCR2820470603>3.0.CO;2-G
32. Cheng Y, Zhao L, Li Y, Xu T (2011) Design of biocompatible dendrimers for cancer diagnosis and therapy: current status and future perspectives. Chem Soc Rev 40. https://doi.org/10.1039/c0cs00097c
33. Leng ZH, Zhuang QF, Li YC, He Z, Chen Z, Huang SP, Jia HY, Zhou JW, Liu Y, Du LB (2013) Polyamidoamine dendrimer conjugated chitosan nanoparticles for the delivery of methotrexate. Carbohydr Polym 98. https://doi.org/10.1016/j.carbpol.2013.07.021

34. Svenson S (2009) Dendrimers as versatile platform in drug delivery applications. https://doi.org/10.1016/j.ejpb.2008.09.023
35. Band Horwitz S (1992) Mechanism of action of taxol. Trends Pharmacol Sci 13. https://doi.org/10.1016/0165-6147(92)90048-B
36. Jain A, Thakur K, Kush P, Jain UK (2014) Docetaxel loaded chitosan nanoparticles: formulation, characterization and cytotoxicity studies. Int J Biol Macromol 69. https://doi.org/10.1016/j.ijbiomac.2014.06.029
37. Ragelle H, Colombo S, Pourcelle V, Vanvarenberg K, Vandermeulen G, Bouzin C, Marchand-Brynaert J, Feron O, Foged C, Préat V (2015) Intracellular siRNA delivery dynamics of integrin-targeted, PEGylated chitosan-poly(ethylene imine) hybrid nanoparticles: a mechanistic insight. J Control Release 211. https://doi.org/10.1016/j.jconrel.2015.05.274
38. Tucker GC (2006) Integrins: molecular targets in cancer therapy. https://doi.org/10.1007/s11912-006-0043-3
39. Sakurai Y, Hatakeyama H, Sato Y, Hyodo M, Akita H, Ohga N, Hida K, Harashima H (2014) RNAi-mediated gene knockdown and anti-angiogenic therapy of RCCs using a cyclic RGD-modified liposomal-siRNA system. J Control Release 173. https://doi.org/10.1016/j.jconrel.2013.10.003
40. Wu CP, Hsieh CH, Wu YS (2011) The emergence of drug transporter-mediated multidrug resistance to cancer chemotherapy. Mol Pharm 8. https://doi.org/10.1021/mp200261n
41. Bradley G, Ling V (1994) P-glycoprotein, multidrug resistance and tumor progression. Cancer Metast Rev 13. https://doi.org/10.1007/BF00689638
42. Kim RB (2002) Drugs as P-glycoprotein substrates, inhibitors, and inducers. https://doi.org/10.1081/DMR-120001389
43. Yadav S, van Vlerken LE, Little SR, Amiji MM (2009) Evaluations of combination MDR-1 gene silencing and paclitaxel administration in biodegradable polymeric nanoparticle formulations to overcome multidrug resistance in cancer cells. Cancer Chemother Pharmacol 63. https://doi.org/10.1007/s00280-008-0790-y
44. Li YT, Chua MJ, Kunnath AP, Chowdhury EH (2012) Reversing multidrug resistance in breast cancer cells by silencing ABC transporter genes with nanoparticle-facilitated delivery of target siRNAs. Int J Nanomed 7. https://doi.org/10.2147/IJN.S30500
45. Yhee JY, Song S, Lee SJ, Park SG, Kim KS, Kim MG, Son S, Koo H, Kwon IC, Jeong JH, Jeong SY, Kim SH, Kim K (2015) Cancer-targeted MDR-1 siRNA delivery using self-cross-linked glycol chitosan nanoparticles to overcome drug resistance. J Control Release 198. https://doi.org/10.1016/j.jconrel.2014.11.019
46. Koo H, Min KH, Lee SC, Park JH, Park K, Jeong SY, Choi K, Kwon IC, Kim K (2013) Enhanced drug-loading and therapeutic efficacy of hydrotropic oligomer-conjugated glycol chitosan nanoparticles for tumor-targeted paclitaxel delivery. J Control Release 172. https://doi.org/10.1016/j.jconrel.2013.08.297
47. Parkin DM (2001) Global cancer statistics in the year 2000. https://doi.org/10.1016/S1470-2045(01)00486-7
48. Das M, Mohanty C, Sahoo SK (2009) Ligand-based targeted therapy for cancer tissue. https://doi.org/10.1517/17425240902780166
49. Anitha A, Divya Rani VV, Krishna R, Sreeja V, Selvamurugan N, Nair SV, Tamura H, Jayakumar R (2009) Synthesis, characterization, cytotoxicity and antibacterial studies of chitosan, O-carboxymethyl and N,O-carboxymethyl chitosan nanoparticles. Carbohydr Polym 78. https://doi.org/10.1016/j.carbpol.2009.05.028
50. Dodane V, Amin Khan M, Merwin JR (1999) Effect of chitosan on epithelial permeability and structure. Int J Pharm 182. https://doi.org/10.1016/S0378-5173(99)00030-7
51. Baudner BC, Balland O, Giuliani MM, von Hoegen P, Rappuoli R, Betbeder D, del Giudice G (2002) Enhancement of protective efficacy following intranasal immunization with vaccine plus a nontoxic LTK63 mutant delivered with nanoparticles. Infect Immun 70. https://doi.org/10.1128/IAI.70.9.4785-4790.2002
52. Aggarwal BB, Sung B (2009) Pharmacological basis for the role of curcumin in chronic diseases: an age-old spice with modern targets. https://doi.org/10.1016/j.tips.2008.11.002

53. Archana D, Dutta PK, Dutta J (2015) Chitosan: a potential therapeutic dressing material for wound healing. In: Chitin and chitosan for regenerative medicine. https://doi.org/10.1007/978-81-322-2511-9_8
54. Karri VVSR, Kuppusamy G, Talluri SV, Mannemala SS, Kollipara R, Wadhwani AD, Mulukutla S, Raju KRS, Malayandi R (2016) Curcumin loaded chitosan nanoparticles impregnated into collagen-alginate scaffolds for diabetic wound healing. Int J Biol Macromol 93. https://doi.org/10.1016/j.ijbiomac.2016.05.038
55. Calvo P, Remuñán-López C, Vila-Jato JL, Alonso MJ (1997) Novel hydrophilic chitosan-polyethylene oxide nanoparticles as protein carriers. J Appl Polym Sci 63. https://doi.org/10.1002/(SICI)1097-4628(19970103)63:1<125::AID-APP13>3.0.CO;2-4
56. Akhtar F, Rizvi MMA, Kar SK (2012) Oral delivery of curcumin bound to chitosan nanoparticles cured *Plasmodium yoelii* infected mice. Biotechnol Adv 30. https://doi.org/10.1016/j.biotechadv.2011.05.009
57. van der Lubben IM, Kersten G, Fretz MM, Beuvery C, Verhoef JC, Junginger HE (2003) Chitosan microparticles for mucosal vaccination against diphtheria: oral and nasal efficacy studies in mice. Vaccine 21. https://doi.org/10.1016/S0264-410X(02)00686-2
58. Jaganathan KS, Rao YUB, Singh P, Prabakaran D, Gupta S, Jain A, Vyas SP (2005) Development of a single dose tetanus toxoid formulation based on polymeric microspheres: a comparative study of poly(D,L-lactic-co-glycolic acid) versus chitosan microspheres. Int J Pharm 294. https://doi.org/10.1016/j.ijpharm.2004.12.026
59. Shaikh J, Ankola DD, Beniwal V, Singh D, Kumar MNVR (2009) Nanoparticle encapsulation improves oral bioavailability of curcumin by at least 9-fold when compared to curcumin administered with piperine as absorption enhancer. Eur J Pharm Sci 37. https://doi.org/10.1016/j.ejps.2009.02.019
60. Khan MA, Zafaryab M, Mehdi SH, Ahmad I, Rizvi MMA (2016) Characterization and anti-proliferative activity of curcumin loaded chitosan nanoparticles in cervical cancer. Int J Biol Macromol 93. https://doi.org/10.1016/j.ijbiomac.2016.08.050
61. Maiti K, Mukherjee K, Gantait A, Saha BP, Mukherjee PK (2007) Curcumin-phospholipid complex: preparation, therapeutic evaluation and pharmacokinetic study in rats. Int J Pharm 330. https://doi.org/10.1016/j.ijpharm.2006.09.025
62. Anand P, Nair HB, Sung B, Kunnumakkara AB, Yadav VR, Tekmal RR, Aggarwal BB (2010) Design of curcumin-loaded PLGA nanoparticles formulation with enhanced cellular uptake, and increased bioactivity in vitro and superior bioavailability in vivo. Biochem Pharmacol 79. https://doi.org/10.1016/j.bcp.2009.09.003
63. Saravanakumar G, Min KH, Min DS, Kim AY, Lee CM, Cho YW, Lee SC, Kim K, Jeong SY, Park K, Park JH, Kwon IC (2009) Hydrotropic oligomer-conjugated glycol chitosan as a carrier of paclitaxel: synthesis, characterization, and in vivo biodistribution. J Control Release 140. https://doi.org/10.1016/j.jconrel.2009.06.015
64. zur Hausen H (2002) Papillomaviruses and cancer: from basic studies to clinical application. https://doi.org/10.1038/nrc798
65. DiMaio D, Liao JB (2006) Human papillomaviruses and cervical cancer. Adv Virus Res 66:125–159. https://doi.org/10.1016/S0065-3527(06)66003-X
66. Ahmed TA, Aljaeid BM (2016) Preparation, characterization, and potential application of chitosan, chitosan derivatives, and chitosan metal nanoparticles in pharmaceutical drug delivery. https://doi.org/10.2147/DDDT.S99651
67. Agnihotri SA, Mallikarjuna NN, Aminabhavi TM (2004) Recent advances on chitosan-based micro- and nanoparticles in drug delivery. https://doi.org/10.1016/j.jconrel.2004.08.010
68. Zorzi GK, Carvalho ELS, von Poser GL, Teixeira HF (2015) On the use of nanotechnology-based strategies for association of complex matrices from plant extracts. https://doi.org/10.1016/j.bjp.2015.07.015
69. Bayón-Cordero L, Alkorta I, Arana L (2019) Application of solid lipid nanoparticles to improve the efficiency of anticancer drugs. Nanomaterials 9. https://doi.org/10.3390/nano9030474

70. Torchilin VP (2005) Recent advances with liposomes as pharmaceutical carriers. https://doi.org/10.1038/nrd1632
71. Tan JKY, Sellers DL, Pham B, Pun SH, Horner PJ (2016) Non-viral nucleic acid delivery strategies to the central nervous system. https://doi.org/10.3389/fnmol.2016.00108
72. Rosenblum D, Joshi N, Tao W, Karp JM, Peer D (2018) Progress and challenges towards targeted delivery of cancer therapeutics. https://doi.org/10.1038/s41467-018-03705-y
73. Babu A, Amreddy N, Muralidharan R, Pathuri G, Gali H, Chen A, Zhao YD, Munshi A, Ramesh R (2017) Chemodrug delivery using integrin-targeted PLGA-chitosan nanoparticle for lung cancer therapy. Sci Rep 7. https://doi.org/10.1038/s41598-017-15012-5
74. Sharma AK, Gupta L, Sahu H, Qayum A, Singh SK, Nakhate KT, Ajazuddin, Gupta U (2018) Chitosan engineered PAMAM dendrimers as nanoconstructs for the enhanced anti-cancer potential and improved in vivo brain pharmacokinetics of temozolomide. Pharm Res 35. https://doi.org/10.1007/s11095-017-2324-y
75. Raja MAG, Katas H, Wen TJ (2015) Stability, intracellular delivery, and release of siRNA from chitosan nanoparticles using different cross-linkers. PLoS ONE 10. https://doi.org/10.1371/journal.pone.0128963
76. Ying M, Wang S, Sang Y, Sun P, Lal B, Goodwin CR, Guerrero-Cazares H, Quinones-Hinojosa A, Laterra J, Xia S (2011) Regulation of glioblastoma stem cells by retinoic acid: role for notch pathway inhibition. Oncogene 30. https://doi.org/10.1038/onc.2011.58
77. Wang G, Wang JJ, Tang XJ, Du L, Li F (2016) In vitro and in vivo evaluation of functionalized chitosan-pluronic micelles loaded with myricetin on glioblastoma cancer. Nanomedicine 12. https://doi.org/10.1016/j.nano.2016.02.004
78. Agrawal P, Sonali, Singh RP, Sharma G, Mehata AK, Singh S, Rajesh CV, Pandey BL, Koch B, Muthu MS (2017) Bioadhesive micelles of d-α-tocopherol polyethylene glycol succinate 1000: synergism of chitosan and transferrin in targeted drug delivery. Colloids Surf B Biointerfaces 152. https://doi.org/10.1016/j.colsurfb.2017.01.021
79. Lara-Velazquez M, Alkharboosh R, Norton ES, Ramirez-Loera C, Freeman WD, Guerrero-Cazares H, Forte AJ, Quiñones-Hinojosa A, Sarabia-Estrada R (2020) Chitosan-based non-viral gene and drug delivery systems for brain cancer. https://doi.org/10.3389/fneur.2020.00740
80. Agrawal P, Singh RP, Sonali, Kumari L, Sharma G, Koch B, Rajesh CV, Mehata AK, Singh S, Pandey BL, Muthu MS (2017) TPGS-chitosan cross-linked targeted nanoparticles for effective brain cancer therapy. Mater Sci Eng C 74. https://doi.org/10.1016/j.msec.2017.02.008
81. American Cancer Society (2014) Cancer fact and figure 2014. American Cancer Society
82. Sonali, Singh RP, Sharma G, Kumari L, Koch B, Singh S, Bharti S, Rajinikanth PS, Pandey BL, Muthu MS (2016) RGD-TPGS decorated theranostic liposomes for brain targeted delivery. Colloids Surf B Biointerfaces 147. https://doi.org/10.1016/j.colsurfb.2016.07.058
83. Saenz del Burgo L, Hernández RM, Orive G, Pedraz JL (2014) Nanotherapeutic approaches for brain cancer management. https://doi.org/10.1016/j.nano.2013.10.001
84. Ren WH, Chang J, Yan CH, Qian XM, Long LX, He B, Yuan XB, Kang CS, Betbeder D, Sheng J, Pu PY (2010) Development of transferrin functionalized poly(ethylene glycol)/poly(lactic acid) amphiphilic block copolymeric micelles as a potential delivery system targeting brain glioma. J Mater Sci Mater Med 21. https://doi.org/10.1007/s10856-010-4106-5
85. Guo Y, Chu M, Tan S, Zhao S, Liu H, Otieno BO, Yang X, Xu C, Zhang Z (2014) Chitosan-g-TPGS nanoparticles for anticancer drug delivery and overcoming multidrug resistance. Mol Pharm 11. https://doi.org/10.1021/mp400514t
86. Zhang Z, Tan S, Feng SS (2012) Vitamin E TPGS as a molecular biomaterial for drug delivery. https://doi.org/10.1016/j.biomaterials.2012.03.046
87. Kutty RV, Tay CY, Lim CS, Feng SS, Leong DT (2015) Anti-migratory and increased cytotoxic effects of novel dual drug-loaded complex hybrid micelles in triple negative breast cancer cells. Nano Res 8. https://doi.org/10.1007/s12274-015-0760-8
88. Liu Y, Wang W, Yang J, Zhou C, Sun J (2013) pH-sensitive polymeric micelles triggered drug release for extracellular and intracellular drug targeting delivery. https://doi.org/10.1016/j.ajps.2013.07.021

89. Irani M, Mir Mohamad Sadeghi G, Haririan I (2017) A novel biocompatible drug delivery system of chitosan/temozolomide nanoparticles loaded PCL-PU nanofibers for sustained delivery of temozolomide. Int J Biol Macromol 97. https://doi.org/10.1016/j.ijbiomac.2017.01.073
90. van Woensel M, Wauthoz N, Rosière R, Mathieu V, Kiss R, Lefranc F, Steelant B, Dilissen E, van Gool SW, Mathivet T, Gerhardt H, Amighi K, de Vleeschouwer S (2016) Development of siRNA-loaded chitosan nanoparticles targeting galectin-1 for the treatment of glioblastoma multiforme via intranasal administration. J Control Release 227. https://doi.org/10.1016/j.jconrel.2016.02.032
91. Revia RA, Zhang M (2016) Magnetite nanoparticles for cancer diagnosis, treatment, and treatment monitoring: recent advances. https://doi.org/10.1016/j.mattod.2015.08.022
92. Toussaint LG, Nilson AE, Goble JM, Ballman KV, James CD, Lefranc F, Kiss R, Uhm JH (2012) Galectin-1, a gene preferentially expressed at the tumor margin, promotes glioblastoma cell invasion. Mol Cancer 11. https://doi.org/10.1186/1476-4598-11-32
93. Yousefpour P, Atyabi F, Vasheghani-Farahani E, Movahedi AA, Dinarvand R (2011) Targeted delivery of doxorubicin-utilizing chitosan nanoparticles surface-functionalized with anti-Her2 trastuzumab. Int J Nanomed 6. https://doi.org/10.2147/ijn.s21523
94. Li F, Li J, Wen X, Zhou S, Tong X, Su P, Li H, Shi D (2009) Anti-tumor activity of paclitaxel-loaded chitosan nanoparticles: an in vitro study. Mater Sci Eng C 29. https://doi.org/10.1016/j.msec.2009.07.001
95. Naruphontjirakul P, Viravaidya-Pasuwat K (2011) Development of doxorubicin—core shell chitosan nanoparticles to treat cancer. Eng Technol 11
96. Parveen S, Sahoo SK (2010) Evaluation of cytotoxicity and mechanism of apoptosis of doxorubicin using folate-decorated chitosan nanoparticles for targeted delivery to retinoblastoma. Cancer Nanotechnol 1. https://doi.org/10.1007/s12645-010-0006-0
97. Nelson MH, Dolder CR (2007) A review of lapatinib ditosylate in the treatment of refractory or advanced breast cancer. https://doi.org/10.2147/tcrm.s1216044
98. Liang XJ, Chen C, Zhao Y, Wang PC (2010) Circumventing tumor resistance to chemotherapy by nanotechnology. Methods Mol Biol 596. https://doi.org/10.1007/978-1-60761-416-6_21
99. Li F, Danquah M, Singh S, Wu H, Mahato RI (2011) Paclitaxel- and lapatinib-loaded lipopolymer micelles overcome multidrug resistance in prostate cancer. Drug Deliv Transl Res 1. https://doi.org/10.1007/s13346-011-0042-2
100. Leung SYL, Jackson J, Miyake H, Burt H, Gleave ME (2000) Polymeric micellar paclitaxel phosphorylates Bcl-2 and induces apoptotic regression of androgen-independent LNCaP prostate tumors. Prostate 44. https://doi.org/10.1002/1097-0045(20000701)44:2<156::AID-PROS8>3.0.CO;2-8
101. Mobasseri R, Karimi M, Tian L, Naderi-Manesh H, Ramakrishna S (2017) Hydrophobic lapatinib encapsulated dextran-chitosan nanoparticles using a toxic solvent free method: fabrication, release property & in vitro anti-cancer activity. Mater Sci Eng C 74. https://doi.org/10.1016/j.msec.2016.12.027
102. Kohno K, Kikuchi J, Sato SI, Takano H, Saburi Y, Asoh KI, Kuwano M (1988) Vincristine-resistant human cancer KB cell line and increased expression of multidrug-resistance gene. Jpn J Cancer Res 79. https://doi.org/10.1111/j.1349-7006.1988.tb01550.x
103. Lagunas-Martínez A, Madrid-Marina V, Gariglio P (2010) Modulation of apoptosis by early human papillomavirus proteins in cervical cancer. https://doi.org/10.1016/j.bbcan.2009.03.005
104. Escudero-Ortiz V, Pérez-Ruixo JJ, Valenzuela B (2013) Development and validation of a high-performance liquid chromatography ultraviolet method for lapatinib quantification in human plasma. Ther Drug Monit 35. https://doi.org/10.1097/FTD.0b013e3182959080

Chapter 9
Applications of Modified Chitosan Nanoparticles in Water, Food, and Agriculture

9.1 Applications of Modified Chitosan Nanoparticles (CSNPs) in Water Treatment

Chitosan exhibits exceptional properties, making it an effective adsorbent material for wastewater pollutant removal because of key properties like hydrophobicity, biocompatibility, biodegradability, non-toxicity, and reactive functional groups ($-NH_2$ and $-OH$). The use of chitosan in wastewater treatment is advantageous because high adsorption capacity, selective removal of pollutants, environmental sustainability, cost-effectiveness and reusability. The chitosan has potential for pollutant removal applications heavy metals (e.g., Pb, Cu, Cd), organic dyes, pesticides, pharmaceuticals, and industrial effluents. The chitosan's amino and hydroxyl groups interact with pollutants through the mechanism of electrostatic attraction, hydrogen bonding and complexation.

9.1.1 Polyacrylamide-Grafted (PG) CSNPs for the Removal of Cu(II)

The polymers with the greatest proportion of polyacrylamide-grafted (PG) are chosen in order to investigate their capacity to extract Cu(II) ions from water [1]. Using a batch equilibration approach, the rate of interaction between Cu(II) ions and polymers is examined as a consequence of time spent in contact at pH = 7. For all polymers, a typical relationship on metal ion absorption on contact time. The findings show that a rapid rate of equilibration—roughly 31–42% of metal ion uptake—was reached after 6 h, and that a stable state was reached after 12–18 h. Some polymers have the greatest ability to absorb Cu(II) ions, whereas others have the lowest capacity. This might rely on the proportion of acrylamide grafting onto chitosan. Functional groups

K. M. Zia, *Chitosan*, Smart Nanomaterials Technology,
https://doi.org/10.1007/978-981-96-2629-8_9

that are in the chain of framework structure rise with an increase in the grafting % of chitosan-g-PAA, which in turn increases the Cu(II) ion ability to be absorbed [1].

9.1.2 Chitosan-Alginate Nanoparticles (CANPs) for the Removal of Hg

Mercury (Hg) is a highly toxic transition element commonly found in industrial wastewater, posing significant environmental and health risks [2]. Chitosan and alginate, two biocompatible polymers, exhibits cationic and anionic nature, enabling adsorption of both anionic and cationic toxic metal ions and electrostatic interactions with metal ions [3]. Dubey et al. [4] developed CSANPs as effective adsorbents for Hg(II) ions removal from aqueous solutions. Mechanism of Hg(II) adsorption involve electrostatic attraction between negatively charged alginate chains and positively charged Hg(II) ions and coordination of Hg(II) ions with oxygen and nitrogen atoms of chitosan's amine group. The key interactions are electrostatic holding of chitosan and alginate chains with tripolyphosphate (TPP) and calcium ions and complexation of Hg(II) ions with alginate and chitosan. Possible interaction of Hg(II) with CSANPs is shown in Fig. 9.1. This study demonstrates the potential of CSANPs as efficient adsorbents for mercury removal from industrial wastewater.

The method also shows that because of their respective positive and negative charges, the alginate and chitosan are likewise kept together by electrostatic attraction forces [4]. Comparison of adsorption capacity of CANPs with other adsorbents is summarized in Table 9.1.

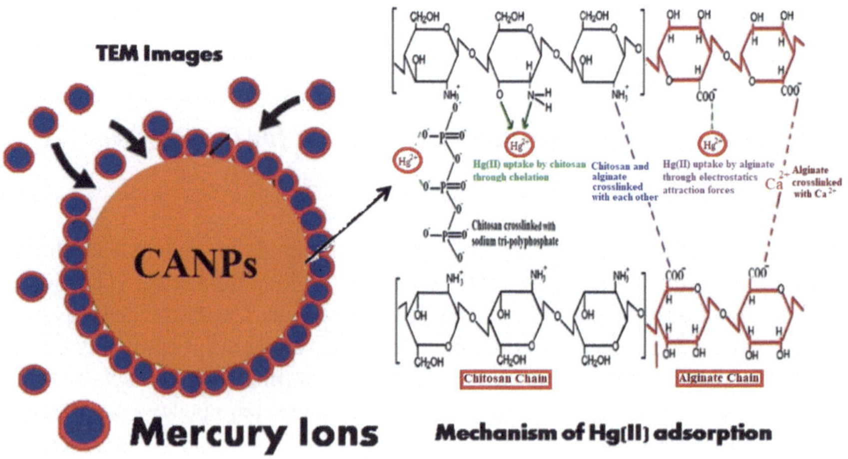

Fig. 9.1 Preparation of CSANPs and possible interaction with Hg(II) [4]

Table 9.1 Comparison of adsorption capacity of CANPs with other adsorbents

Adsorbent	Q_0 (mg/g)	References
Chitosan-coated magnetite nanoparticles	9.34	[5]
Chitosan/cotton fibers (via C–N single bond)	96.28	[6]
Chitosan–thioglyceraldehyde Schiff's base	98	[7]
CSTU	135	[8]
Hg-PET-TSC	137.13	[9]
CANPs	217.39	[4]

9.1.3 Biomass of Cu and Co-Doped Alginate-Coated CSNPs for Removal of Heavy Metals

Sodium alginate is a natural polysaccharide used as an absorbent for water purification because of its low production cost, biocompatibility and low toxicity [10]. Different studies show that both CS and alginate are used to remove heavy metals [11]. Esmaeili et al. prepared Cu and Co-doped alginate-coated CSNPs because alginate-coated CSNPs can remove efficiently heavy metals from wastewater than either CS or alginate alone. The pseudo-second order-kinetic model is employed to describe the mechanisms of the bio-sorption process [12].

9.1.4 Applications of Modified Chitosan Nanoparticles CSNPs in Food Industry

The CSNPs combine the benefits of chitosan (CS) and nanoparticles (NPs), exhibiting key characteristics like surface and interface effects, small size, and quantum size effects. The expanding applications involve food preservation (antimicrobial and antioxidant properties), active packaging (controlled release and barrier properties); pharmaceutical delivery (targeted drug release and bioavailability enhancement) and bioactive Encapsulation (protection and controlled release of nutrients and phytochemicals).

9.1.5 Anthocyanins Loaded CSNPs for Food Products

Anthocyanins are pigments which show solubility in water and have antioxidant and biological properties that oppose cancer, Angiocardiopathy, diabetes, neurodegenerative diseases, inflammation, and senium [13, 14]. These are used as natural colorants in food products [15]. For the purpose of encapsulating the anthocyanins in the two varieties of red-fleshed apples, CSNPs are produced using nanoprecipitation. The impact of varying anthocyanin percentages on CSNP encapsulation is investigated. A range of values, from 20 to 100 mg/L, are investigated. The anthocyanin content of RL (only 141.1 mg/kg) led to the selection of the maximal level of 100 mg/L. The greatest anthocyanin content is found to yield the largest loading capacity in both XJ4 and RL. The XJ4 exhibited a 1.77-fold increase in anthocyanin loading ability (81.73 ug/mg) in comparison to RL (46.23 ug/mg). Consequently, the concentration of 100 mg/L was selected for additional study [16].

9.1.6 Novel Nano-fibrillated Cellulose/Chitosan Nanoparticles for Food Packaging

Featuring distinctive physical features including low density, openness, chemically altered ability, biodegradability, biocompatibility, high tensile strength, film-forming ability, and oxygen barrier qualities, nano-fibrillated cellulose (NFC) is a promising nanomaterial [17, 18]. That's why it is used in many fields such as paper-coating, tissue engineering and drug delivery system [19]. But some microorganisms can attack on cellulose due to the absence of antimicrobial functional groups and polysaccharide nature. On the other side CSNPs have antimicrobial properties due to their cationic nature that destroy the membrane of bacteria and kill them [20]. Hassan et al. prepared novel nano-fibrillated cellulose/chitosan nanoparticles nanocomposites films. The CSNPs are coated on the surface of oxidized cellulose to develop antimicrobial cellulosic material. Coating of paper sheets with glycerol-plasticized NFC/CSNPs can improve the tensile strength, increase grease proof properties and antimicrobial properties, and have potential applications in food packaging [21].

9.1.7 Fish Gelatin/CSNPs for Food Packaging

Using naturally occurring antimicrobial compounds in food packaging materials to prevent unintended alterations in the nutritional value of food is the foundation of antimicrobial food packaging, a potentially effective type of active packaging [22]. Some researcher synthesized fish gelatin/CSNPs composite films for food packaging [23]. Gelatin is a protein obtained from the partial hydrolysis of collagen and is used for the formation of flexible and transparent films due to its good film-forming ability. It is a hydrophilic polymer due to which its films show poor water vapor barrier and mechanical properties [24]. To solve this problem and to improve the performance of gelatin, bio-nanocomposites (biopolymer matrix reinforced with NPs) are used [25]. In addition, organ essential oils and their derivatives are also added due to their strong antioxidant and antimicrobial activities against *Staphylococcus aureus* and *listeria* monocytogenes (gram positive) and *Escherichia coli* and *Salmonella* (gram negative) microorganisms [23].

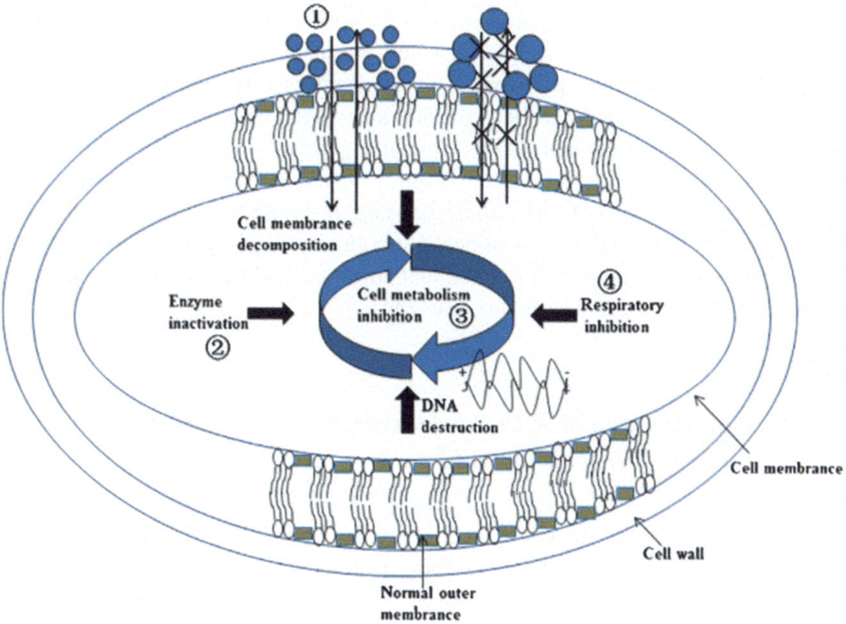

Fig. 9.2 Suggested antibacterial mechanism of nanoparticles [27]

9.1.8 Catechins and Catechins-Zn Complex Loaded β-CSNPs for Food Products

Catechins (CAT) are a group of polyphenolic compounds, broadly applied as nutraceuticals for enhancing antioxidant, antibacterial, and anti-inflammatory actions [26]. But CAT have less residence time, low solubility and less stability in gut and intestinal track due to low pH. Zhang et al. synthesized catechins-Zn complex loaded β-CSNPs to understand the antibacterial mechanism of NPs for food and biological applications as shown in Fig. 9.2 [27].

*9.1.9 CSNPs with Summer Savory (***Satureja hortensis** *L.) Essential Oil*

Satureja hortensis L. is named as summer savory, found in the south European region [28] and is also extensively distributed in Eastern Anatolia [29]. Summer savory essential oil has antimicrobial, anti-inflammatory, and antioxidant properties [29–32]. During the preparation of food packaging stuff, essential oil can easily decomposed under pressure and temperature [33]. Feyzioglu et al. [34] synthesized CSNPs with summer savory (*Satureja hortensis* L.) essential oil. While EOs are poorly soluble in water, their application in the food business is restricted. Therefore, by making EOs more soluble in water, encapsulation is a useful technique to expand their application potential. This work effectively created water-soluble chitosan nanoparticles (NPs) loaded with summer savory essential oil (EO) utilizing the ionic gelation process and STPP, achieving a notably high encapsulation efficiency. Depending on the EO concentration, EO-loaded NPs demonstrated potent in vitro antibacterial and DPPH radical scavenging properties. The chitosan NPs' ability to retain water and withstand heat are both improved by the addition of EO. The results of this investigation showed that chitosan might be used to encapsulate summer savory to expand its culinary uses [34].

9.1.10 CSNPs Loaded with 2,5-Dihydroxybenzoic Acid and Protocatechuic Acid

Madureira et al. prepared CSNPs loaded with 2, 5-dihydroxybenzoic acid and protocatechuic acid [35]. Protocatechuic acid found in green tea is the important metabolite of antioxidant phenolic compounds [36]. It has many bioactivities like anticancer, anti-ulcerogenic, antimicrobial, and anti-aging [37] and 2, 5-DHBA has antimicrobial, antimutagenic and anti-inflammatory properties. That's why these are extensively used in food industry [38]. The CSNPs increase the stability of these phenolic acids in gastrointestinal track, and have enough time to stick on the intestinal epithelium where phenolic acid release and adsorbed in blood. Phenolic acid loaded CSNPs increase the nutritional content of food and provide health benefits [35].

9.1.11 Tea Polyphenole (TP) Zn Complex-Loaded β-Chitosan Nanoparticles

Zinc (Zn) is an essential micronutrient with various biological functions, including treating allotriophagy and dermatitis, enhancing growth rate and improving appetite [28]. The combination of tea polyphenol (TP)-Zn complex activates aromatic hydroxyl groups and increases antioxidant activity and bioactivity. The limitations and solution involve low oral bioavailability of TP and TP-Zn complex due to gastrointestinal digestion and oral mucosa. Encapsulation in biodegradable polymers, such as β-chitosan (β-CS), overcomes this limitation. Zhang et al. [28] prepared tea polyphenol Zn complex loaded β-chitosan (β-CS) nanoparticles (Fig. 9.3) as potential antioxidant substances in various food products. The β-chitosan nanoparticles (CSNPs) derived from squid pens, exhibit higher reactivity, affinity, solubility and weaker intermolecular hydrogen bonding than α-CS. The TP-Zn complex loaded CS-NPs demonstrate encapsulation and antioxidant activity with enhanced antioxidant activity (DPPH radical scavenging, ORAC value, reducing power), improved

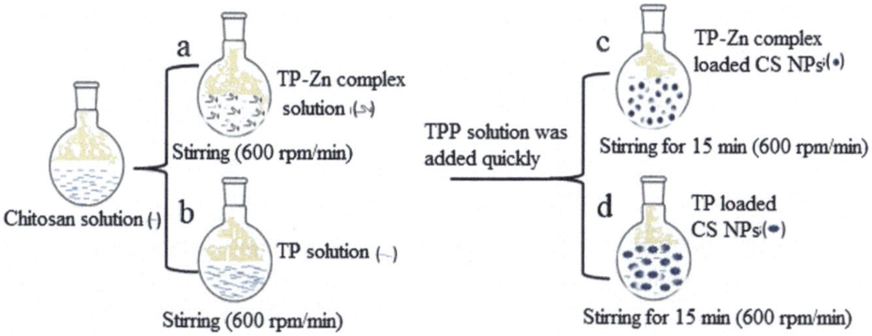

Fig. 9.3 The synthetic procedure the **tea polyphenole (TP)**-Zn complex and TP-loaded CSNPs. The TP-Zn complex and TP solution added to the churning chitosan solution are represented by letters **a** and **b**, respectively; the TP-Zn complex and TP loaded CS NPs suspension are represented by letters **c** and **d**. *NPs* nanoparticles; *CS* chitosan [28]

bioavailability and delayed degradation rate. The TP-Zn complex encapsulated in β-CS nanoparticles is a promising antioxidant substance for food products, offering enhanced bioactivity, and bioavailability [28].

9.2 Applications of Modified Chitosan Nanoparticles (CSNPs) in the Field of Agriculture

Typical applications of CSNPs offer a promising solution for precision agriculture, enabling targeted and controlled delivery of agrochemicals to improve crop health and productivity [39–41].

9.2.1 Chitosan-Based Polymers and Nanoparticles for Fruits and Seeds Coating

Banana fruit quality and shelf life are improved by the application of CSNPs [42], while fresh blueberry quality is preserved and microbiological development is suppressed by the CS and procyanidins coating [43]. Through antibacterial action, coating with *Ficus hirta* Vahl. Fruits extract decreased postharvest loss and enhanced Xinyu tangerine storability during cold storage [44]. By reducing microbiological contamination and prolonging shelf life without compromising the appearance or quality of the item, strawberries are coated with a photo-activated chlorophyll in-CS conjugate [45]. By boosting the expression of antioxidant genes, a CS/nano-silica/sodium alginate composite film is applied to winter jujube fruits to extend their shelf life [46]. Additionally, CS/nano-TiO_2 and chitosan/nano-SiO_2 coated films improved the activities of scavenging antioxidant enzymes, preserving the quality of Ginco biloba seeds [47].

9.2.2 Growth Promotion and Improved Disease Resistance

Due to their buildup in the environment and in living things, chemicals that are widely used in agriculture to increase plant output and resilience to diseases have the potential to permanently harm the ecology. Additionally, it may lead to plant disease resistance, calling for the addition of new agrochemicals or an increase in the dose of already existing ones [48]. The expenses and dangers are clearly increased. In the hunt for fresh solutions to this conundrum, the use of nanotechnology seems to be very hopeful. Chitosan (CS), one of the most valuable polymers in nanotechnology, is harmless to people, biocompatible, and degradable. Furthermore, CS may be readily manipulated without impairing its fundamental properties, in contrast to other biopolymers as chitin, starch, gelatin, cellulose, and glucans [49]. As a consequence of altering its molecular and biological characteristics, CS has been extensively utilized for a variety of uses. Recent research has revealed that CS-based compounds have a variety of antibacterial and regulating actions in plants. For instance, CS nanoparticles (CSNPs) with CS characteristics and nanoparticle traits like surface and interface effect, small size, and quantum size effects are able to act as a germination elicitor of *Oryza sativa* L. [50] and to positively affect wheat seed germination and seedling growth (*Triticum aestivum* L.) [51]. Furthermore, these nanoparticles are able to induce the expression of pathogenesis-related proteins, thereby increasing tomato resistance to *F. andiyazi* [52] and to inhibit *Fusarium oxysporum* growth in *Zingiber officinale* [53]. It is common practice to employ CSNPs in combination with other compounds to promote plant development and create resistance to phytopathogens. The CSNPs' ability to entrap nitrogen, phosphate, and potassium enhanced the absorption of nutrients, photosynthesis, and development of the coffee plants [54] and induced starch and major proteins accumulation in *Pisum sativum* plants [55]. The ability of CSNPs to promote development and to induce resilience is examined when it is complexed with different metals. The amount of gluten in durum wheat increased, and maize crop output increased, after foliar and seed treatment with zinc-loaded CSNPs. While Cu- and Zn-CSNPs inhibited *Rhizoctonia solani*-caused cotton seedling illness, zinc-loaded CSNPs enhanced the amount of gluten in durum wheat and boosted maize crop production [56, 57]. The Cu-CSNPs boosted plant growth and nutritional content, improved onion bulb output, and induced resistance to *Pyricularia grisea* blast disease in finger millet plants [58]. Fusarium wilt, a soil- and seed-borne disease of chickpeas brought on by *F. oxysporum*, is stopped by the application of several metal-CSNPs (CuO-CSNPs, ZnO-CSNPs, and Ag-CSNPs) [59]. While *Phytophthora capsici* development on pepper plants is considerably influenced by Ag-CSNPs treatment [60]. The Ag-CSNPs application successfully prevented leaf fall disease caused by *Corynespora cassiicola* on rubber trees [61], and Lanthanum-CSNPs efficiently stimulated growth and enhanced disease resistance of rice [62]. Investigation on CSNPs complexes with elicitors, secondary compounds, and phytoregulators. When tomato plants are treated with harpin-CSNPs, the expression of numerous genes involved in the defense response is changed, and *R. solani* infection is reduced [63]. At the right dosage, tripolyphosphate-CSNPs promoted micro-propagation plant development on *Capsicum annuum*, however hazardous concentrations markedly hindered plant growth [64]. Thiamine-CSNPs enthused development and persuaded fighting against *F. oxysporum* in chickpea seedlings [65], whereas salicylic acid-CSNPs positively modulated plant evolution and enhanced resistance responses on lettuce seedlings [66], indorsed growth and defense in maize [67]. *Cymbopogon martinii* essential oil-encapsulated CSNPs demonstrated antifungal and antimycotoxin effects against *F. graminearum* in maize grains in a lab setting [68]. Growth of *A. solani*, *F. oxysporum*, and *F. moniliforme* are impacted by coumarin-functionalized chitosan derivatives [39]. In wheat, hybrids of chitosan, vanillin, and salicylic acid successfully generated resistance to *Puccinia recondita* and *Cochliobolus sativus* [69]. A bentonite-chitosan nanoclay exhibited elicitor activity on tomato leaflets and antibacterial activity against *P. syringe* and *F. solani* [40]. Derivatives of CSNPs also have a favorable effect on how the body reacts to abiotic stressors. Cu-CSNPs, for instance, improved the development and proliferation of defense genes in tomato plants that are subjected to salt stress [70], and S-nitrosoglutathione-CSNPs enhanced sugarcane plants' ability to withstand droughts [41].

9.2.3 Crop Productivity and Conservation of Agronomic Commodities

Chitosan-based films have shown great potential in extending shelf life and controlling senescence and decay including peach (*P. persica* L.), strawberry, lychee (*Litchi sinensis* Sonn.), and grapes [71–73], as well as for minimally processed products such as sliced mango (*Mangifera indica* L.), apple (*Malus domestica* Borkh.), and papaya (*Carica papaya* L.) to retain color, original taste, and texture [74, 75]. Later studies blended chitosan with other biopolymers or included natural antibacterial agents in chitosan compositions for consumable films. Some of these have stronger antioxidant or antibacterial qualities. *E. coli*, *S. aureus*, *S. typhimurium*, *L. monocytogenes*, and *B. cereus*, all of which are food pathogenic bacteria, are examined for antimicrobial activity [76]. Formulations based on biopolymer-chitosan mixtures are created to produce a matrix for the regulated release of nutrients, medicines, tastes, and antioxidants, helping to control respiration rate and lessen enzymatic browning in fruits [77–80]. A novel approach to extending shelf life and enhancing microbial control in fresh-cut fruits applicable for fresh-cut cantaloupe (*Cucumis melo* L.) and pineapple (*Ananas comosus* L.). The innovative films are established as chitosan/methyl cellulose film and chitosan/methyl cellulose film with vanillin (natural antimicrobial agent). The key benefits involved improved microbial control, enhanced fruit quality, and extended storage life at 10 °C [81]. These biodegradable and non-toxic films offer a promising solution for reducing food waste and maintaining freshness in the fruit industry. New physiological and biological properties are currently being added to this copolymer, which has great potential in a variety of industries, including pharmacology and medicine (fibers, fabrics, drugs, membranes, artificial organs), cosmetology (lotions, hair additives, facial and body creams), food (coating, preservative, antioxidant, antimicrobial), biotechnology (chelator, emulsifier, flocculant), and agricultural (soil modifier, films, fungic coating) [82–91]. With regard to the agricultural industry, the increased solubility of chitosan makes it simple to employ in aqueous solutions at different concentrations [92]. The functionalized chitosans developed for controlled medication release can also be used as delivery systems for nutrients, pesticides, herbicides, nematicides, and insects to be delivered in soil under regulated conditions. Inducing disease-resistance genes in plants and preventing microbial infections may both be accomplished by covering seeds and leaves with chitosan derivative films [93].

9.2.4 Recent Advances of Chitosan Nanoparticles Applications in Plants

The hunt for different biological methods to meet this demand has recently been driven by the rising global food demand, continuing climate change, the hazardous consuming of farmlands, and consumers' increased interest in high-quality, secure, and ecologically friendly food products. One of the options being investigated to prevent the use of toxic products to manage plant illnesses and improve agricultural yield in biopolymer-based materials. In several instances, these substances showed sufficient action against microbes with negligible harm to animals and negligible environmental effect. These polymers also have the potential to reduce the use of possibly dangerous farming methods and large quantities of toxic fertilizers while increasing the output of many agricultural plants. The evaluated polymers based on chitosan, a biopolymer, delivered the greatest outcomes. Chitin, a copolymer of N-acetyl-D-glucosamine and D-glucosamine is the primary component of insect exoskeletons, is the source of CS, a straight unbranched polymer of 1,4-D-glucosamine. Diatoms, fungus, mollusks, freshwater and marine sponges, as well as diatoms, all contain chitin [94]. As a byproduct of the seafood industry's processing of prawn and crab shells, this natural polymer is practical and readily accessible. Chitin is actually the second-largest sustainable carbon source in the world after cellulose, with annual output of over 1011 tones. As a consequence, it can be utilized in the manufacturing of CS for profit. Over 2000 tons of CS are produced annually by industrial processing globally [94]. For this straightforward process, crystalline chitin is heated to 120–150 °C and treated with 40–50% (w/v) NaOH. The majority of the acetyl groups in N-acetyl-D-glucosamine are removed during this process, turning it into 1,4-D-glucosamine. Chitin cannot be directly used since it is insoluble in the primary solvents, whereas CS is readily soluble in weak organic acids like acetic acid and may have its restricted solubility in water increased by chemical modification such as carboxy-methylation [95, 96]. The word "chitosan," however, appears to apply to a collection of polymers with different degrees of mixed de-acetylation and polymerization, viscosity, molecular mass, acidity, and dissociation constant. The biological characteristics of chitosan might be significantly impacted by this variability; it should be underlined [97]. Chitosan has a number of benefits over other biopolymers (cellulose, starch, galactomannans, etc.): it is secure, affordable, and its molecular structure makes it simple to incorporate particular molecules into polymers to create those with particular uses. These qualities provide chitosan a key role for a wide variety of prospective customers, including the medical and biotechnology sectors as well as agricultural applications [98, 99]. More experts have been looking into how CS-based chemicals affect vegetation in recent years. The most current studies on the effects of chitosan spraying on plant production, disease resistance, and the lengthening of the economic life of detached fruits are evaluated by Malerba et al. [100].

Numerous studies looked into the use of CS-based materials in agriculture, and in a number of situations, these compounds are able to raise plant production (Table 9.2).

Table 9.2 Chitosan effects on plant growth

Plant species	Chitosan formulation and administration	Chitosan effect	References
Strawberry (*Fragaria* × *ananassa* Duch.)	0.025–0.05% (plant spraying)	Higher fruit yield	[101]
Potato (*Solanum tuberosum* L.)	200, 325 and 558 mg/ha (foliar spraying)	Enhancement of tuber size	[102]
Bell pepper (*Capsicum annuum*)	0.3–0.5% (leaves and fruits spraying)	Increase in fruit weight, fruit diameter, and yield	[103]
Basil (*Ocimum ciliatum* and *Ocimum basilicum*)	0.02–0.04% (foliar spraying)	Increase in plant growth and total phenol content	[104]
Rice (*Oryza sativa* L.)	0.05% (plants soaking and spraying)	Higher photosynthesis rate	[105]
Rape (*Brassica rapa* L.)	0.05–0.1% (foliar spraying)	Increase in plant growth and leaf chlorophyll content	[106]
Barley (*Hordeum vulgare* L.)	0.1–1% (germinating seeds)	Higher phenolic content	[107]
Maize (*Zea mays* L.)	0.04–0.16% (seeds soaking and plant spraying)	Promotion of plant growth and grain weight	[108]

9.2.5 Developing Nano-delivery Systems for Agriculture

It's possible that conventional farming methods can no longer produce enough food to feed a growing populace without having a detrimental effect on the ecosystem. In order to address some of the pressing challenges of toxic pesticides, pests with resistance, and increasing soil pollutants, sustainable agriculture practices must be applied as soon as feasible in all sectors [109]. Green nanotechnology and renewable nanomaterials are being used in modern agriculture to execute precision farming, which attempts to boost production sustainably with little resources [110]. Nanotechnology can benefit agribusiness in two key ways: by raising agricultural outputs and by improving resource utilization effectiveness, mainly by decreasing wastage. Only 45% of P and 30–40% of N from the applied fertilizers are absorbed, which is less than 50% of the usage efficiency of the fertilizers now in use [111]. For a variety of uses, including as nano-herbicides, nano-detectors, and nano-fertilizers, nature-derived polymeric nanoparticles can be used to solve common farming problems like environmental pollution and safety issues. These designed nano-products contribute to an increase in food output and nutritious value by enhancing the quality of herbicides, fertilizers, and growth factors. For instance, to accomplish "precision farming," which focuses only on crop production while leaving the water and soil alone, nano-carriers are utilized to carry and distribute pesticides in a more regulated and gradual release profile [112]. In order to improve nutrient exchange ratios and decrease mineral wastage, fertilizer packaging is the main objective. Agriculture still makes extensive use of polymer-coated controlled-release fertilizers that contain N, P, K, and other minerals. Pesticides are encapsulated using nano-encapsulation more frequently than fertilizers. When compared to encapsulating these nutrients into nano-particulate systems in fertilizer applications, inorganic nanoparticles of zinc (Zn), copper (Cu), iron (Fe), cerium (Ce), and titanium (Ti) have been widely researched in laboratory and field settings [113]. Chitosan is a material found in nature with numerous uses in horticulture. Chitosan's innate ability to promote development and antibacterial qualities make it suitable for use in agriculture as a standalone substance or as a framework for nutrition packaging. It is better to use chitosan nanoparticles in nano form rather than macro form due to their higher dispersibility. Chitosan nanoparticles loaded with nutrients N, P, and K have been shown by Heba et al. to accelerate plant development, cutting the life cycle of wheat plants tested from 170 to 130 days [114]. Chitosan nanoparticles alone have been shown in field testing on barley to increase grain production when applied to drought-stressed barley plants [115]. By solubilizing them and enabling a regulated release of these chemicals at the target location, pesticide nano-encapsulation significantly lowers the quantity of pesticides used. To facilitate simpler agricultural application, the cyclo-dextrins are then used to functionalize chitosan before creating chitosan nanoparticles. Natural chemicals are very volatile, hence nano-complexation extended their exposure period to the environment [116]. In another study, zein, a protein generated from nature, is used to create nanoencapsulates of insecticides originating from plants, including citronella oil, geraniol, eugenol, and cinnamaldehyde [117]. The use of such plant preparations also helps to avoid the development of pesticide tolerance in insects. It has been demonstrated that encapsulation can extend compound release, reduce IC50, and aid avoid compound breakdown. To provide a prolonged and regulated release of encapsulated carbendazim with excellent bio-efficacy and suppression against fungus like *Fusarium oxysporum* and *Aspergillus parasiticus*, Sandhya et al. created a biopolymeric nanocarrier made of natural polymers, chitosan, and pectin. Additionally, according to phyto-toxicity studies, carbendazim-loaded polymeric nano-formulation enhances seed germination and root development in *Zea mays* and *Cucumis sativa* seeds [118]. Imazapic and imazapyr, two herbicides used to control weeds in maize and peanut crops, may be made less toxic and more effective by encapsulation in chitosan alginate systems [119]. Similar research has previously shown that the herbicide paraquat may have its action maintained while having its toxicity reduced by four times by nanoencapsulation in chitosan nanoparticles [120]. The polyelectrolyte complexation of chitosan and alginate is used to create nanocapsules that are used to delay the release of the water-soluble pesticide acetamiprid. The carrier's ability to slow down the chemical's release made it possible for the chemical to be released in a regulated manner for 36 h. The polyelectrolyte complexation of chitosan and alginate is used to create nanocapsules that are used to delay the release of the water-soluble pesticide acetamiprid [121]. Additionally, it has been demonstrated that chitosan nanoparticles increase plants' natural defenses, resulting in disease resilience. By boosting antioxidant and defense enzyme activity, maize plants treated with Cu-chitosan nanoparticles are able to fend off Curvularia leaf spot disease. Similar to this, chitosan nanoparticles are found in a different research to provide finger millet plants evaluated antifungal characteristics by causing reactive oxygen species and activating peroxidases in the plant [122].

9.3 Conclusion

This chapter explores the diverse applications of chitosan nanoparticles (CSNPs) in three key sectors: water treatment, food industry, and agriculture. The CSNPs have been utilized for effective removal of heavy metals from water and industrial wastewater such as polyacrylamide-grafted CSNPs for Cu removal, chitosan-alginate nanoparticles (CANPs) for Hg removal and Cu and Co-doped alginate-coated CSNPs for heavy metal removal. The CSNPs have been employed in food packaging due to their antimicrobial properties, biodegradability and biocompatibility. Notable applications include nano-fibrillated cellulose/CSNPs for food packaging, fish gelatin/CSNPs composite films and catechins and catechins-Zn complex loaded β-CSNPs. Chitosan-based polymers and nanoparticles have been used for fruits and seeds coating, developing films and coatings to extend shelf life and reducing toxic fertilizer use and increasing crop yields. Nano-delivery systems for agriculture have also been explored, leveraging green nanotechnology and sustainable nanoparticles for precision farming. The CSNPs demonstrate potential in water treatment, food packaging, and agriculture. Chitosan-based films and coatings show promise in extending shelf life and reducing toxic fertilizer use. Nano-delivery systems and green nanotechnology enable precision farming and sustainable agriculture practices.

References

1. Al-Karawi AJM, Al-Qaisi ZHJ, Abdullah HI, Al-Mokaram AMA, Al-Heetimi DTA (2011) Synthesis, characterization of acrylamide grafted chitosan and its use in removal of copper(II) ions from water. Carbohydr Polym 83. https://doi.org/10.1016/j.carbpol.2010.08.017
2. Wang LJ, Wang JP, Yuan SJ, Zhang SJ, Tang Y, Yu HQ (2009) Gamma radiation-induced dispersion polymerization in aqueous salts solution for manufacturing a cationic flocculant. Chem Eng J 149. https://doi.org/10.1016/j.cej.2008.10.010
3. Pillai CKS, Paul W, Sharma CP (2009) Chitin and chitosan polymers: chemistry, solubility and fiber formation. https://doi.org/10.1016/j.progpolymsci.2009.04.001
4. Dubey R, Bajpai J, Bajpai AK (2016) Chitosan-alginate nanoparticles (CANPs) as potential nanosorbent for removal of Hg(II) ions. Environ Nanotechnol Monit Manag 6. https://doi.org/10.1016/j.enmm.2016.06.008
5. Rahbar N, Jahangiri A, Boumi S, Khodayar MJ (2014) Mercury removal from aqueous solutions with chitosan-coated magnetite nanoparticles optimized using the Box–Behnken design. Jundishapur J Nat Pharm Prod 9. https://doi.org/10.17795/jjnpp-15913
6. Qu R, Sun C, Ma F, Zhang Y, Ji C, Xu Q, Wang C, Chen H (2009) Removal and recovery of Hg(II) from aqueous solution using chitosan-coated cotton fibers. J Hazard Mater 167. https://doi.org/10.1016/j.jhazmat.2009.01.043
7. Monier M (2012) Adsorption of Hg^{2+}, Cu^{2+} and Zn^{2+} ions from aqueous solution using formaldehyde cross-linked modified chitosan-thioglyceraldehyde Schiff's base. Int J Biol Macromol 50. https://doi.org/10.1016/j.ijbiomac.2011.11.026
8. Monier M, Abdel-Latif DA (2012) Preparation of cross-linked magnetic chitosan-phenylthiourea resin for adsorption of Hg(II), Cd(II) and Zn(II) ions from aqueous solutions. J Hazard Mater 209–210. https://doi.org/10.1016/j.jhazmat.2012.01.015
9. Monier M, Abdel-Latif DA (2013) Synthesis and characterization of ion-imprinted chelating fibers based on PET for selective removal of Hg^{2+}. Chem Eng J 221. https://doi.org/10.1016/j.cej.2013.02.003
10. Zhou L, Wang Y, Liu Z, Huang Q (2009) Characteristics of equilibrium, kinetics studies for adsorption of Hg(II), Cu(II), and Ni(II) ions by thiourea-modified magnetic chitosan microspheres. J Hazard Mater 161. https://doi.org/10.1016/j.jhazmat.2008.04.078
11. Lakouraj MM, Mojerlou F, Zare EN (2014) Nanogel and superparamagnetic nanocomposite based on sodium alginate for sorption of heavy metal ions. Carbohydr Polym 106. https://doi.org/10.1016/j.carbpol.2014.01.092
12. Esmaeili A, Khoshnevisan N (2016) Optimization of process parameters for removal of heavy metals by biomass of Cu and Co-doped alginate-coated chitosan nanoparticles. Bioresour Technol 218. https://doi.org/10.1016/j.biortech.2016.07.005
13. Li H, Deng Z, Zhu H, Hu C, Liu R, Young JC, Tsao R (2012) Highly pigmented vegetables: anthocyanin compositions and their role in antioxidant activities. Food Res Int 46. https://doi.org/10.1016/j.foodres.2011.12.014
14. Bagchi D, Sen CK, Bagchi M, Atalay M (2004) Anti-angiogenic, antioxidant, and anti-carcinogenic properties of a novel anthocyanin-rich berry extract formula. Biochemistry (Moscow). https://doi.org/10.1023/B:BIRY.0000016355.19999.93
15. Nichenametla SN, Taruscio TG, Barney DL, Exon JH (2006) A review of the effects and mechanisms of polyphenolics in cancer. https://doi.org/10.1080/10408390591000541
16. Xu J, Li X, Liu S, Zhao P, Huo H, Zhang Y (2019) Effect of nanocrystallization of anthocyanins extracted from two types of red-fleshed apple varieties on its stability and antioxidant activity. Molecules 24. https://doi.org/10.3390/molecules24183366

17. Lee PS, Yim SG, Choi Y, van Anh Ha T, Ko S (2012) Physiochemical properties and prolonged release behaviours of chitosan-denatured β-lactoglobulin microcapsules for potential food applications. Food Chem 134. https://doi.org/10.1016/j.foodchem.2012.03.006
18. Abdul Khalil HPS, Davoudpour Y, Islam MN, Mustapha A, Sudesh K, Dungani R, Jawaid M (2014) Production and modification of nanofibrillated cellulose using various mechanical processes: a review. https://doi.org/10.1016/j.carbpol.2013.08.069
19. Jorfi M, Foster EJ (2015) Recent advances in nanocellulose for biomedical applications. https://doi.org/10.1002/app.41719
20. Avadi MR, Sadeghi AMM, Tahzibi A, Bayati K, Pouladzadeh M, Zohuriaan-Mehr MJ, Rafiee-Tehrani M (2004) Diethylmethyl chitosan as an antimicrobial agent: synthesis, characterization and antibacterial effects. Eur Polym J 40. https://doi.org/10.1016/j.eurpolymj.2004.02.015
21. Hassan EA, Hassan ML, Abou-Zeid RE, El-Wakil NA (2016) Novel nanofibrillated cellulose/chitosan nanoparticles nanocomposites films and their use for paper coating. Ind Crops Prod 93. https://doi.org/10.1016/j.indcrop.2015.12.006
22. Sung SY, Sin LT, Tee TT, Bee ST, Rahmat AR, Rahman WAWA, Tan AC, Vikhraman M (2013) Antimicrobial agents for food packaging applications. https://doi.org/10.1016/j.tifs.2013.08.001
23. Hosseini SF, Rezaei M, Zandi M, Farahmandghavi F (2016) Development of bioactive fish gelatin/chitosan nanoparticles composite films with antimicrobial properties. Food Chem 194. https://doi.org/10.1016/j.foodchem.2015.09.004
24. Rojas-Graü MA, Salvia-Trujillo L, Soliva-Fortuny R, Martín-Belloso O (2012) Edible films and coatings. In: Decontamination of fresh and minimally processed produce. https://doi.org/10.1002/9781118229187.ch14
25. Fakhreddin Hosseini S, Rezaei M, Zandi M, Ghavi FF (2013) Preparation and functional properties of fish gelatin-chitosan blend edible films. Food Chem 136. https://doi.org/10.1016/j.foodchem.2012.09.081
26. Solórzano-Santos F, Miranda-Novales MG (2012) Essential oils from aromatic herbs as antimicrobial agents. https://doi.org/10.1016/j.copbio.2011.08.005
27. Zhang H, Jung J, Zhao Y (2016) Preparation, characterization and evaluation of antibacterial activity of catechins and catechins-Zn complex loaded β-chitosan nanoparticles of different particle sizes. Carbohydr Polym 137. https://doi.org/10.1016/j.carbpol.2015.10.036
28. Zhang H, Zhao Y (2015) Preparation, characterization and evaluation of tea polyphenol-Zn complex loaded β-chitosan nanoparticles. Food Hydrocoll 48. https://doi.org/10.1016/j.foodhyd.2015.02.015
29. Sefidkon F, Abbasi K, Khaniki GB (2006) Influence of drying and extraction methods on yield and chemical composition of the essential oil of *Satureja hortensis*. Food Chem 99. https://doi.org/10.1016/j.foodchem.2005.07.026
30. Güllüce M, Sökmen M, Daferera D, Ağar G, Özkan H, Kartal N, Polissiou M, Sökmen A, Şahin F (2003) In vitro antibacterial, antifungal, and antioxidant activities of the essential oil and methanol extracts of herbal parts and callus cultures of *Satureja hortensis* L. J Agric Food Chem 51. https://doi.org/10.1021/jf0340308
31. Boyraz N, Özcan M (2006) Inhibition of phytopathogenic fungi by essential oil, hydrosol, ground material and extract of summer savory (*Satureja hortensis* L.) growing wild in Turkey. Int J Food Microbiol 107. https://doi.org/10.1016/j.ijfoodmicro.2005.10.002
32. Dikbas N, Kotan R, Dadasoglu F, Sahin F (2008) Control of *Aspergillus flavus* with essential oil and methanol extract of *Satureja hortensis*. Int J Food Microbiol 124. https://doi.org/10.1016/j.ijfoodmicro.2008.03.034
33. Şahin F, Karaman I, Güllüce M, Öğütçü H, Şengül M, Adigüzel A, Öztürk S, Kotan R (2003) Evaluation of antimicrobial activities of *Satureja hortensis* L. J Ethnopharmacol 87. https://doi.org/10.1016/S0378-8741(03)00110-7
34. Feyzioglu GC, Tornuk F (2016) Development of chitosan nanoparticles loaded with summer savory (*Satureja hortensis* L.) essential oil for antimicrobial and antioxidant delivery applications. LWT Food Sci Technol 70. https://doi.org/10.1016/j.lwt.2016.02.037

35. Madureira AR, Pereira A, Pintado M (2016) Chitosan nanoparticles loaded with 2,5-dihydroxybenzoic acid and protocatechuic acid: properties and digestion. J Food Eng 174. https://doi.org/10.1016/j.jfoodeng.2015.11.007
36. Chen F, Shi Z, Neoh KG, Kang ET (2009) Antioxidant and antibacterial activities of eugenol and carvacrol-grafted chitosan nanoparticles. Biotechnol Bioeng 104. https://doi.org/10.1002/bit.22363
37. Kakkar S, Bais S (2014) A review on protocatechuic acid and its pharmacological potential. ISRN Pharmacol 2014. https://doi.org/10.1155/2014/952943
38. Boaventura BCB, Murakami ANN, Prudêncio ES, Maraschin M, Murakami FS, Amante ER, Amboni RDDMC (2013) Enhancement of bioactive compounds content and antioxidant activity of aqueous extract of mate (*Ilex paraguariensis* A. St. Hil.) through freeze concentration technology. Food Res Int 53. https://doi.org/10.1016/j.foodres.2012.07.042
39. Yang G, Jin Q, Xu C, Fan S, Wang C, Xie P (2018) Synthesis, characterization and antifungal activity of coumarin-functionalized chitosan derivatives. Int J Biol Macromol 106. https://doi.org/10.1016/j.ijbiomac.2017.08.009
40. Merino D, Mansilla AY, Casalongué CA, Alvarez VA (2018) Preparation, characterization, and in vitro testing of nanoclay antimicrobial activities and elicitor capacity. J Agric Food Chem 66. https://doi.org/10.1021/acs.jafc.8b00049
41. Silveira NM, Seabra AB, Marcos FCC, Pelegrino MT, Machado EC, Ribeiro RV (2019) Encapsulation of S-nitrosoglutathione into chitosan nanoparticles improves drought tolerance of sugarcane plants. Nitric Oxide 84. https://doi.org/10.1016/j.niox.2019.01.004
42. Lustriane C, Dwivany FM, Suendo V, Reza M (2018) Effect of chitosan and chitosan-nanoparticles on post harvest quality of banana fruits. J Plant Biotechnol 45. https://doi.org/10.5010/JPB.2018.45.1.036
43. Mannozzi C, Tylewicz U, Chinnici F, Siroli L, Rocculi P, Dalla Rosa M, Romani S (2018) Effects of chitosan based coatings enriched with procyanidin by-product on quality of fresh blueberries during storage. Food Chem 251. https://doi.org/10.1016/j.foodchem.2018.01.015
44. Chen C, Nie Z, Wan C, Chen J (2019) Preservation of Xinyu tangerines with an edible coating using *Ficus hirta* Vahl. Fruits extract-incorporated chitosan. Biomolecules 9. https://doi.org/10.3390/biom9020046
45. Luksiene Z, Buchovec I (2019) Impact of chlorophyllin-chitosan coating and visible light on the microbial contamination, shelf life, nutritional and visual quality of strawberries. Innov Food Sci Emerg Technol 52. https://doi.org/10.1016/j.ifset.2019.02.003
46. Kou X, He Y, Li Y, Chen X, Feng Y, Xue Z (2019) Effect of abscisic acid (ABA) and chitosan/nano-silica/sodium alginate composite film on the color development and quality of postharvest Chinese winter jujube (*Zizyphus jujuba* Mill. cv. Dongzao). Food Chem 270. https://doi.org/10.1016/j.foodchem.2018.06.151
47. Tian F, Chen W, Wu CE, Kou X, Fan G, Li T, Wu Z (2019) Preservation of *Ginkgo biloba* seeds by coating with chitosan/nano-TiO_2 and chitosan/nano-SiO_2 films. Int J Biol Macromol 126. https://doi.org/10.1016/j.ijbiomac.2018.12.177
48. Kumaraswamy RV, Kumari S, Choudhary RC, Pal A, Raliya R, Biswas P, Saharan V (2018) Engineered chitosan based nanomaterials: bioactivities, mechanisms and perspectives in plant protection and growth. https://doi.org/10.1016/j.ijbiomac.2018.02.130
49. Divya K, Jisha MS (2018) Chitosan nanoparticles preparation and applications. https://doi.org/10.1007/s10311-017-0670-y
50. Divya K, Vijayan S, Nair SJ, Jisha MS (2019) Optimization of chitosan nanoparticle synthesis and its potential application as germination elicitor of *Oryza sativa* L. Int J Biol Macromol 124. https://doi.org/10.1016/j.ijbiomac.2018.11.185
51. Li R, He J, Xie H, Wang W, Bose SK, Sun Y, Hu J, Yin H (2019) Effects of chitosan nanoparticles on seed germination and seedling growth of wheat (*Triticum aestivum* L.). Int J Biol Macromol 126. https://doi.org/10.1016/j.ijbiomac.2018.12.118
52. Chun SC, Chandrasekaran M (2019) Chitosan and chitosan nanoparticles induced expression of pathogenesis-related proteins genes enhances biotic stress tolerance in tomato. Int J Biol Macromol 125. https://doi.org/10.1016/j.ijbiomac.2018.12.167

53. Athawale V, Paralikar P, Ingle AP, Rai M (2018) Biogenically engineered nanoparticles inhibit *Fusarium oxysporum* causing soft-rot of ginger. IET Nanobiotechnol 12. https://doi.org/10.1049/iet-nbt.2018.5086
54. Ha NMC, Nguyen TH, Wang SL, Nguyen AD (2019) Preparation of NPK nanofertilizer based on chitosan nanoparticles and its effect on biophysical characteristics and growth of coffee in green house. Res Chem Intermed 45. https://doi.org/10.1007/s11164-018-3630-7
55. Khalifa NS, Hasaneen MN (2018) The effect of chitosan–PMAA–NPK nanofertilizer on *Pisum sativum* plants. 3 Biotech 8. https://doi.org/10.1007/s13205-018-1221-3
56. Choudhary RC, Kumaraswamy RV, Kumari S, Sharma SS, Pal A, Raliya R, Biswas P, Saharan V (2019) Zinc encapsulated chitosan nanoparticle to promote maize crop yield. Int J Biol Macromol 127. https://doi.org/10.1016/j.ijbiomac.2018.12.274
57. Abd-Elsalam KA, Vasil'kov AY, Said-Galiev EE, Rubina MS, Khokhlov AR, Naumkin AV, Shtykova EV, Alghuthaymi MA (2018) Bimetallic blends and chitosan nanocomposites: novel antifungal agents against cotton seedling damping-off. Eur J Plant Pathol 151. https://doi.org/10.1007/s10658-017-1349-8
58. Sathiyabama M, Manikandan A (2018) Application of copper-chitosan nanoparticles stimulate growth and induce resistance in finger millet (*Eleusine coracana* Gaertn.) plants against blast disease. J Agric Food Chem 66. https://doi.org/10.1021/acs.jafc.7b05921
59. Kaur P, Duhan JS, Thakur R (2018) Comparative pot studies of chitosan and chitosan-metal nanocomposites as nano-agrochemicals against fusarium wilt of chickpea (*Cicer arietinum* L.). Biocatal Agric Biotechnol 14. https://doi.org/10.1016/j.bcab.2018.04.014
60. Nhien LTA, Luong ND, Tien LTT, Luan LQ (2018) Radiation synthesis of silver nanoparticles/chitosan for controlling leaf fall disease on rubber trees causing by *Corynespora cassiicola*. J Nanomater 2018. https://doi.org/10.1155/2018/7121549
61. Luan LQ, Xo DH (2018) In vitro and in vivo fungicidal effects of γ-irradiation synthesized silver nanoparticles against *Phytophthora capsici* causing the foot rot disease on pepper plant. J Plant Pathol 100. https://doi.org/10.1007/s42161-018-0064-4
62. Liang W, Yu A, Wang G, Zheng F, Hu P, Jia J, Xu H (2018) A novel water-based chitosan-La pesticide nanocarrier enhancing defense responses in rice (*Oryza sativa* L.) growth. Carbohydr Polym 199. https://doi.org/10.1016/j.carbpol.2018.07.042
63. Nadendla SR, Rani TS, Vaikuntapu PR, Maddu RR, Podile AR (2018) HarpinPss encapsulation in chitosan nanoparticles for improved bioavailability and disease resistance in tomato. Carbohydr Polym 199. https://doi.org/10.1016/j.carbpol.2018.06.094
64. Asgari-Targhi G, Iranbakhsh A, Ardebili ZO (2018) Potential benefits and phytotoxicity of bulk and nano-chitosan on the growth, morphogenesis, physiology, and micropropagation of *Capsicum annuum*. Plant Physiol Biochem 127. https://doi.org/10.1016/j.plaphy.2018.04.013
65. Muthukrishnan S, Murugan I, Selvaraj M (2019) Chitosan nanoparticles loaded with thiamine stimulate growth and enhances protection against wilt disease in Chickpea. Carbohydr Polym 212. https://doi.org/10.1016/j.carbpol.2019.02.037
66. Martin-Saldaña S, Chevalier MT, Iglesias MJ, Colman SL, Casalongué CA, Álvarez VA, Chevalier AA (2018) Salicylic acid loaded chitosan microparticles applied to lettuce seedlings: recycling shrimp fishing industry waste. Carbohydr Polym 200. https://doi.org/10.1016/j.carbpol.2018.08.019
67. Kumaraswamy RV, Kumari S, Choudhary RC, Sharma SS, Pal A, Raliya R, Biswas P, Saharan V (2019) Salicylic acid functionalized chitosan nanoparticle: a sustainable biostimulant for plant. Int J Biol Macromol 123. https://doi.org/10.1016/j.ijbiomac.2018.10.202
68. Kalagatur NK, Nirmal Ghosh OS, Sundararaj N, Mudili V (2018) Antifungal activity of chitosan nanoparticles encapsulated with *Cymbopogon martinii* essential oil on plant pathogenic fungi *Fusarium graminearum*. Front Pharmacol 9. https://doi.org/10.3389/fphar.2018.00610
69. Popova EV, Domnina NS, Kovalenko NM, Sokornova SV, Tyuterev SL (2018) Influence of chitosan hybrid derivatives on induced wheat resistance to pathogens with different nutrition strategies. Appl Biochem Microbiol 54. https://doi.org/10.1134/S0003683818050150

70. Hernández-Hernández H, Juárez-Maldonado A, Benavides-Mendoza A, Ortega-Ortiz H, Cadenas-Pliego G, Sánchez-Aspeytia D, González-Morales S (2018) Chitosan-PVA and copper nanoparticles improve growth and overexpress the SOD and JA genes in tomato plants under salt stress. Agronomy 8. https://doi.org/10.3390/agronomy8090175
71. Romanazzi G (2010) Chitosan treatment for the control of postharvest decay of table grapes, strawberries and sweet cherries. Fresh Prod 4
72. Du J, Gemma H, Iwahori S (1997) Effects of chitosan coating on the storage of peach, Japanese pear, and kiwifruit. J Jpn Soc Horticult Sci 66. https://doi.org/10.2503/jjshs.66.15
73. Zhang D, Quantick PC (1998) Antifungal effects of chitosan coating on fresh strawberries and raspberries during storage. J Horticult Sci Biotechnol 73. https://doi.org/10.1080/14620316.1998.11511045
74. Liao FH, Shieh MJ, Chang NC, Chien YW (2007) Chitosan supplementation lowers serum lipids and maintains normal calcium, magnesium, and iron status in hyperlipidemic patients. Nutr Res 27. https://doi.org/10.1016/j.nutres.2007.01.009
75. Erazo Solórzano C, Salazar Daza D, Vera Chang J, Tuárez García D (2020) Aplicación de bacterias ácido-lácticas provenientes del mucilago de cacao como agente de conservación de la papaya. Univ Cienc Tecnol 24. https://doi.org/10.47460/uct.v24i107.412
76. Dutta PK, Tripathi S, Mehrotra GK, Dutta J (2009) Perspectives for chitosan based antimicrobial films in food applications. https://doi.org/10.1016/j.foodchem.2008.11.047
77. Chen T, Embree HD, Brown EM, Taylor MM, Payne GF (2003) Enzyme-catalyzed gel formation of gelatin and chitosan: potential for in situ applications. Biomaterials 24. https://doi.org/10.1016/S0142-9612(03)00096-6
78. Badawy MEI, Rabea EI (2011) A biopolymer chitosan and its derivatives as promising antimicrobial agents against plant pathogens and their applications in crop protection. Int J Carbohydr Chem 2011. https://doi.org/10.1155/2011/460381
79. Park HJ (1995) Edible coatings and films to improve food quality. Trends Food Sci Technol 6. https://doi.org/10.1016/s0924-2244(00)88935-9
80. Casariego A, Souza BWS, Cerqueira MA, Teixeira JA, Cruz L, Díaz R, Vicente AA (2009) Chitosan/clay films' properties as affected by biopolymer and clay micro/nanoparticles' concentrations. Food Hydrocoll 23. https://doi.org/10.1016/j.foodhyd.2009.02.007
81. Sangsuwan J, Rattanapanone N, Rachtanapun P (2008) Effect of chitosan/methyl cellulose films on microbial and quality characteristics of fresh-cut cantaloupe and pineapple. Postharvest Biol Technol 49. https://doi.org/10.1016/j.postharvbio.2008.02.014
82. Roller S, Covill N (1999) The antifungal properties of chitosan in laboratory media and apple juice. Int J Food Microbiol 47. https://doi.org/10.1016/S0168-1605(99)00006-9
83. Sapers GM (1992) Chitosan enhances control of enzymatic browning in apple and pear juice by filtration. J Food Sci 57. https://doi.org/10.1111/j.1365-2621.1992.tb11296.x
84. Ochoa E, Saucedo-Pompa S, Rojas-Molina R, de la Garza H, Charles-Rodríguez AV, Aguilar CN (2011) Evaluation of a candelilla wax-based edible coating to prolong the shelf-life quality and safety of apples. Am J Agric Biol Sci 6. https://doi.org/10.3844/ajabssp.2011.92.98
85. Bhatnagar A, Sillanpää M (2009) Applications of chitin- and chitosan-derivatives for the detoxification of water and wastewater—a short review. https://doi.org/10.1016/j.cis.2009.09.003
86. Calero N, Muñoz J, Ramírez P, Guerrero A (2010) Flow behaviour, linear viscoelasticity and surface properties of chitosan aqueous solutions. Food Hydrocoll 24. https://doi.org/10.1016/j.foodhyd.2010.03.009
87. Sashiwa H, Aiba SI (2004) Chemically modified chitin and chitosan as biomaterials. https://doi.org/10.1016/j.progpolymsci.2004.04.001
88. Siripatrawan U, Harte BR (2010) Physical properties and antioxidant activity of an active film from chitosan incorporated with green tea extract. Food Hydrocoll 24. https://doi.org/10.1016/j.foodhyd.2010.04.003
89. Liu J, Tian S, Meng X, Xu Y (2007) Effects of chitosan on control of postharvest diseases and physiological responses of tomato fruit. Postharvest Biol Technol 44. https://doi.org/10.1016/j.postharvbio.2006.12.019

90. Thanh Hoai T, Kim Nga N (2019) Synthesis and investigation into apatite-forming ability of hydroxyapatite/chitosan-based scaffold. VNU J Sci Nat Sci Technol 35. https://doi.org/10.25073/2588-1140/vnunst.4896
91. Pati MK, Nayak P (2011) Grafting vinyl monomers onto chitosan: IV: graft copolymerized of acrylic acid onto chitosan using ceric ammonium nitrate as the initiator—characterization and antimicrobial activities. Mater Sci Appl 02. https://doi.org/10.4236/msa.2011.212232
92. Hemantaranjan A (2014) A future perspective in crop protection: chitosan and its oligosaccharides. Adv Plants Agric Res 1. https://doi.org/10.15406/apar.2014.01.00006
93. Plackett D (2011) Biopolymers—new materials for sustainable films and coatings. https://doi.org/10.1002/9781119994312
94. Kurita K (2006) Chitin and chitosan: functional biopolymers from marine crustaceans. https://doi.org/10.1007/s10126-005-0097-5
95. Rinaudo M (2006) Chitin and chitosan: properties and applications. https://doi.org/10.1016/j.progpolymsci.2006.06.001
96. van den Broek LAM, Boeriu CG (2019) Chitin and chitosan: properties and applications. https://doi.org/10.1002/9781119450467
97. Choi C, Nam JP, Nah JW (2016) Application of chitosan and chitosan derivatives as biomaterials. https://doi.org/10.1016/j.jiec.2015.10.028
98. Anitha A, Sowmya S, Kumar PTS, Deepthi S, Chennazhi KP, Ehrlich H, Tsurkan M, Jayakumar R (2014) Chitin and chitosan in selected biomedical applications. https://doi.org/10.1016/j.progpolymsci.2014.02.008
99. Malerba M, Cerana R (2016) Chitosan effects on plant systems. https://doi.org/10.3390/ijms17070996
100. Malerba M, Cerana R (2018) Recent advances of chitosan applications in plants. https://doi.org/10.3390/polym10020118
101. Akter Mukta J, Rahman M, As Sabir A, Gupta DR, Surovy MZ, Rahman M, Islam MT (2017) Chitosan and plant probiotics application enhance growth and yield of strawberry. Biocatal Agric Biotechnol 11. https://doi.org/10.1016/j.bcab.2017.05.005
102. Falcón-Rodríguez AB, Costales D, Gónzalez-Peña D, Morales D, Mederos Y, Jerez E, Cabrera JC (2017) Chitosans of different molecular weight enhance potato (*Solanum tuberosum* L.) yield in a field trial. Span J Agric Res 15. https://doi.org/10.5424/sjar/2017151-9288
103. Mahmood N, Abbasi NA, Hafiz IA, Ali I, Zakia S (2017) Effect of biostimulants on growth, yield and quality of bell pepper cv. Yolo wonder. Pak J Agric Sci 54. https://doi.org/10.21162/PAKJAS/17.5653
104. Ghasemi Pirbalouti A, Malekpoor F, Salimi A, Golparvar A (2017) Exogenous application of chitosan on biochemical and physiological characteristics, phenolic content and antioxidant activity of two species of basil (*Ocimum ciliatum* and *Ocimum basilicum*) under reduced irrigation. Sci Hortic 217. https://doi.org/10.1016/j.scienta.2017.01.031
105. Phothi R, Theerakarunwong CD (2017) Effect of chitosan on physiology, photosynthesis and biomass of rice (*Oryza sativa* L.) under elevated ozone. Aust J Crop Sci 11. https://doi.org/10.21475/ajcs.17.11.05.p578
106. Zong H, Liu S, Xing R, Chen X, Li P (2017) Protective effect of chitosan on photosynthesis and antioxidative defense system in edible rape (*Brassica rapa* L.) in the presence of cadmium. Ecotoxicol Environ Saf 138. https://doi.org/10.1016/j.ecoenv.2017.01.009
107. Ramakrishna R, Sarkar D, Manduri A, Iyer SG, Shetty K (2017) Improving phenolic bioactive-linked anti-hyperglycemic functions of dark germinated barley sprouts (*Hordeum vulgare* L.) using seed elicitation strategy. J Food Sci Technol 54. https://doi.org/10.1007/s13197-017-2828-9
108. Choudhary RC, Kumaraswamy RV, Kumari S, Sharma SS, Pal A, Raliya R, Biswas P, Saharan V (2017) Cu-chitosan nanoparticle boost defense responses and plant growth in maize (*Zea mays* L.). Sci Rep 7. https://doi.org/10.1038/s41598-017-08571-0
109. Lowry GV, Avellan A, Gilbertson LM (2019) Opportunities and challenges for nanotechnology in the agri-tech revolution. Nat Nanotechnol 14. https://doi.org/10.1038/s41565-019-0461-7

110. Fraceto LF, Grillo R, de Medeiros GA, Scognamiglio V, Rea G, Bartolucci C (2016) Nanotechnology in agriculture: which innovation potential does it have? Front Environ Sci 4. https://doi.org/10.3389/fenvs.2016.00020
111. Su Y, Ashworth V, Kim C, Adeleye AS, Rolshausen P, Roper C, White J, Jassby D (2019) Delivery, uptake, fate, and transport of engineered nanoparticles in plants: a critical review and data analysis. https://doi.org/10.1039/c9en00461k
112. Duhan JS, Kumar R, Kumar N, Kaur P, Nehra K, Duhan S (2017) Nanotechnology: the new perspective in precision agriculture. https://doi.org/10.1016/j.btre.2017.03.002
113. Raliya R, Saharan V, Dimkpa C, Biswas P (2018) Nanofertilizer for precision and sustainable agriculture: current state and future perspectives. https://doi.org/10.1021/acs.jafc.7b02178
114. Abdel-Aziz HMM, Hasaneen MNA, Omer AM (2016) Nano chitosan-NPK fertilizer enhances the growth and productivity of wheat plants grown in sandy soil. Span J Agric Res 14. https://doi.org/10.5424/sjar/2016141-8205
115. Behboudi F, Sarvestani TT, Kassaee ZZ, Sanavi MM, Sorooshzadeh S, Ahmadi AB (2018) Evaluation of chitosan nanoparticles effects on yield and yield components of barley (*Hordeum vulgare* L.) under late season drought stress. J Water Environ Nanotechnol 3
116. Campos EVR, Proença PLF, Oliveira JL, Melville CC, Vechia JFD, de Andrade DJ, Fraceto LF (2018) Chitosan nanoparticles functionalized with β-cyclodextrin: a promising carrier for botanical pesticides. Sci Rep 8. https://doi.org/10.1038/s41598-018-20602-y
117. de Oliveira JL, Campos EVR, Germano-Costa T, Lima R, Vechia JFD, Soares ST, de Andrade DJ, Gonçalves KC, do Nascimento J, Polanczyk RA, Fraceto LF (2019) Association of zein nanoparticles with botanical compounds for effective pest control systems. Pest Manag Sci 75. https://doi.org/10.1002/ps.5338
118. Kumar S, Kumar D, Dilbaghi N (2017) Preparation, characterization, and bio-efficacy evaluation of controlled release carbendazim-loaded polymeric nanoparticles. Environ Sci Pollut Res 24. https://doi.org/10.1007/s11356-016-7774-y
119. Maruyama CR, Guilger M, Pascoli M, Bileshy-José N, Abhilash PC, Fraceto LF, de Lima R (2016) Nanoparticles based on chitosan as carriers for the combined herbicides imazapic and imazapyr. Sci Rep 6. https://doi.org/10.1038/srep19768
120. Grillo R, Pereira AES, Nishisaka CS, de Lima R, Oehlke K, Greiner R, Fraceto LF (2014) Chitosan/tripolyphosphate nanoparticles loaded with paraquat herbicide: an environmentally safer alternative for weed control. J Hazard Mater 278. https://doi.org/10.1016/j.jhazmat.2014.05.079
121. Kumar S, Chauhan N, Gopal M, Kumar R, Dilbaghi N (2015) Development and evaluation of alginate-chitosan nanocapsules for controlled release of acetamiprid. Int J Biol Macromol 81. https://doi.org/10.1016/j.ijbiomac.2015.08.062
122. Sathiyabama M, Manikandan A (2016) Chitosan nanoparticle induced defense responses in fingermillet plants against blast disease caused by *Pyricularia grisea* (Cke.) Sacc. Carbohydr Polym 154. https://doi.org/10.1016/j.carbpol.2016.06.089

Chapter 10
Chitosan as Tools to Combat COVID-19

10.1 Taxonomy of SARS-CoV-2

Over the last twenty years, there have been three variants of coronavirus outbreaks observed over the globe, with several deadly incidences. Coronaviruses are RNA viruses, specifically positive-sensed encapsulated, unsegmented, single-stranded RNA viruses. The distinguishing electron microscopy appearance delivered as a significant indicator for amplifying and sequencing nucleic acids from the respiratory specimen of Dr. Urban, one of the healthcare providers who succumbed in 2003 due to severe acute respiratory syndrome (SARS). By using a consensus coronavirus primer, this can be accomplished. Since febrile patients could develop pneumonia, lower respiratory symptoms such as cough and dyspnea, as well as the severe acute respiratory syndrome, the virus has been subsequently given the name SARS-CoV. The SARS-CoV outbreak started in Guangdong, China, and then expanded to a number of locations in Southeast Asia, North America, Europe, and South Africa. After infecting over 8000 people and 774 expiries with a case fatality rate estimated at 9.5%, the last case of SARS-CoV occurred in 2003 (September). A novel coronavirus that began causing respiratory diseases in the Middle East after nine years was given the name MERS-CoV. Despite the fact that many patients experience significant acute respiratory distress, MERS-CoV signs are unusual. On December 30, 2019, a group of pneumonia patients with an unknown etiology was discovered in China (Wuhan). Later, it has been reported to the Beijing-based World Health Organization (WHO) China office. A week later, on January 7, 2020, a novel coronavirus (SARS-CoV-2) has been identified in these people. On February 11, 2020, the World Health Organization (WHO) officially changed the virus's name from novel coronavirus 2019 (2019-nCoV) to COVID-19 [1, 2]. It is crucial to trace the origin and inhabitant of the SARS-CoV-2 to limit the spread of the virus in humans and other species and to identify if the virus causing a disease belongs to new or

K. M. Zia, *Chitosan*, Smart Nanomaterials Technology,
https://doi.org/10.1007/978-981-96-2629-8_10

Table 10.1 Complete profile of SARS-CoV-2 [6]

Common features of SARS-CoV-2	
Name of disease	Coronavirus disease 2019 (COVID-19)
Symptoms	Severe acute respiratory syndrome
Year of evolution	2020
Family	Coronaviridae
Genus	Betacoronavirus
Subgenus	Sarbecovirus
Specie	SARS-CoV-2
Techniques to detect SARS-CoV-2 in clinical samples	Reverse transcription polymerase chain reaction (RT-PCR)
Structure	Similar to SARS-CoV in amino acid composition and protein structure
Incubation period	Within 14 days
Median incubation period	3–4 days
Source of infection	Bat origin

existing species. The newly emerged viruses are assigned taxonomic position based on the information of sequence-based family classification [3]. Single-stranded RNA viruses with a positive sense that are widely recognized as coronaviruses can infect multiple kinds of animals. Strong evidence suggests that bats, mice, or domestic animals may have contributed to the spread of human coronaviruses in the past times [4]. In fact, more evidence specifies bats as the ancestral source of all HCoVs, which are well-flexible, non-pathogenic, but exhibit a variating genetic range. By tracing the zoonotic origin of coronaviruses, it is possible to comprehend the natural history, motivation, and obstacles to interspecies transmission [5]. The International Committee on Taxonomy of Viruses gave the severe acute respiratory syndrome coronavirus 2 (SARS-CoV-2) its name after doing extensive study and sequencing analysis. A complete profile of SARS-CoV-2 is listed in Table 10.1 [6]. CoVs are widely found in humans and other species. The first infectious bronchitis virus (IBV) isolation came from an outbreak in chicken herds in 1937. Since then, more viral isolations in rodents, domestic animals, and people have shown related CoVs. These CoVs typically cause respiratory, neurologic, and hepatic problems [5, 7]. The symptoms of the common cold in immunocompetent adults are caused by four out of six human pathogenic CoVs, including HCoV-229E, HCoV-OC43, HCoV-NL63, and HKU1 [4]. The others, severe acute respiratory syndrome coronavirus (SARS-CoV) and Middle East respiratory syndrome coronavirus (MERS-CoV), are zoonotic in origin and are fatal respiratory ailments [8, 9].

Aside from above cited common features of SARS-CoV-2, the following are the remaining profiles of SAR-CoV-2:

Routes of transmission	• Short-distance droplet transmission • Contact transmission • Aerosols transmission
Pathological characteristics	• Pathological interstitial pneumonia and prominent pulmonary edema, with protein exudation and minor inflammatory cell infiltration
Clinical characteristics	• Respiratory symptoms: low fever, and temperature does not rise in some cases, cough • Gastrointestinal symptoms: uncommon; can be the initial symptoms • Cardiovascular symptoms: palpitations and chest pain can be the initial symptoms • Concomitant diseases: higher frequency including cardio-cerebrovascular diseases, hypertension, diabetes • Imaging features
Susceptible population	• Susceptibility to the entire population, mostly young adults

Letovirinae and Orthocoronavirinae are the two subfamilies that make up the family Coronaviridae. *Alphacoronavirus*, *Betacoronavirus*, *Deltacoronavirus*, and *Gammacoronavirus* are the four genera belonging to the subfamily named Orthocoronavirinae. This family of 39 species of viruses consists of enclosed viruses having a single-stranded, positive-sensed RNA genome. The severe acute respiratory syndrome-related coronavirus, which includes SARS and SARS-CoV-2, is a member of the genus Betacoronavirus [10, 11]. Figure 10.1 shows that all the coronaviruses accountable for the epidemic of pneumonia belong to Betacoronavirus genus [12–14]. These coronavirus genera, which are thought to have been separated between 2400 and 3000 BC, were known to infect a variety of animal species. *Alphacoronaviruses* and *Betacoronaviruses* are mainly found in mammals, whereas *Delta coronaviruses* are found mainly in birds. *Gama coronavirus* infects birds and some other animals such as cetaceans [15–17]. The taxonomy and classification of families was developed by Coronaviridae Study Group of International Committee on Taxonomy of Viruses (ICTV) [18]. *Betacoronavirus* and SARS-CoV genus formed a cluster that included SARS-CoV-2 [19, 20].

Depending on the measurements being employed and the genomic area preferred for study, the distance between SARS-CoV-2 and the coronaviruses that are most closely related varies.

In terms of innovation, SARS-CoV-2 varies from the SARS-CoV and MERS coronaviruses that propagated in human populations in the earlier twenty-first century. The *Coronaviridae* Study Group CSG previously determined that these viruses represent a new species within a new subgroup of the *Betacoronavirus* genus named these subgroups as *Sarbecovirus* and *Merbecovirus* [22–24]. The SARS-CoV-2 can be classified as a member of an existing species that also comprises hundreds of viruses that have chiefly been isolated from human kind and bats (Fig. 10.2).

Even though it is obvious that SARS-CoV-2 is not a descendant of SARS-CoV and that each of these viruses likely entered human bodies from unrelated and unknown external sources, these two viruses are genetically similar enough that their shared evolutionary history and characteristics can be used to understand them.

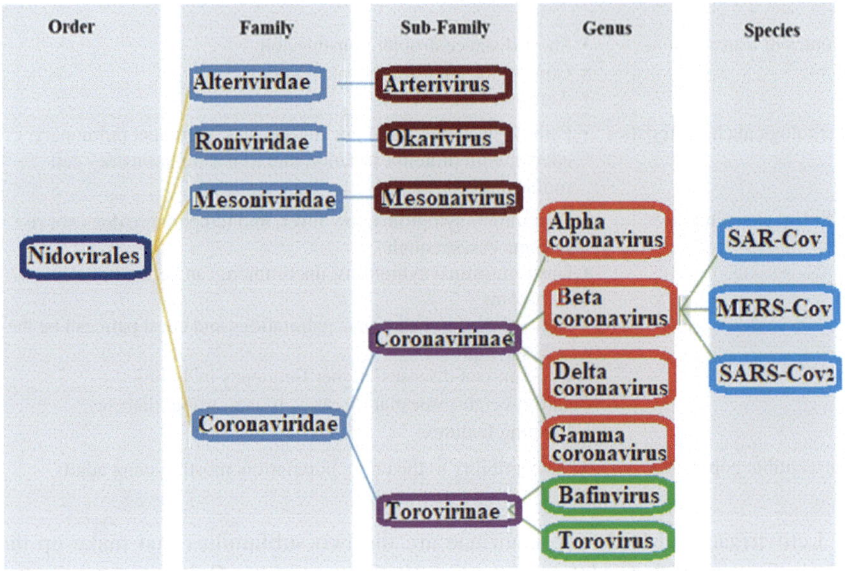

Fig. 10.1 Taxonomy of SARS-Cov-2 [21]

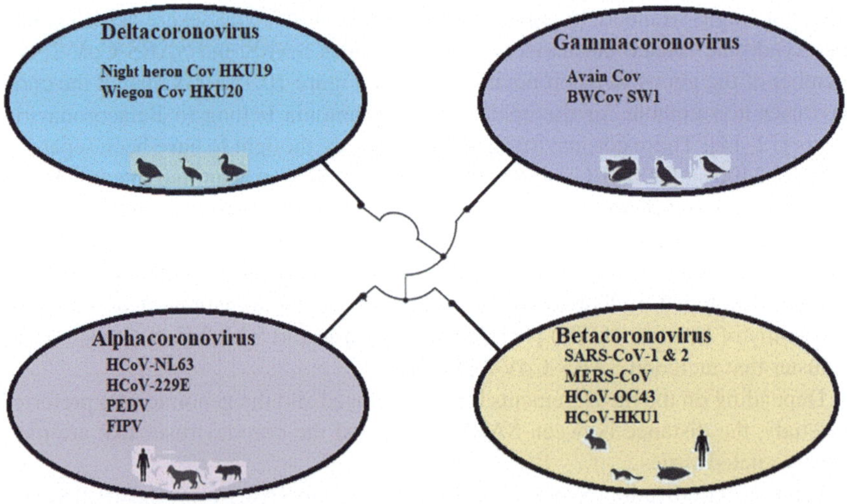

Fig. 10.2 Schematic depiction of the four genera of coronaviruses their evolutionary relationship and their animal hosts [24]

10.2 Immunology and Immunopathology of COVID-19

10.2.1 Innate Immunity to SARS-CoV-2

Coronaviruses have a genome consisting of about 30,000 nucleotide units of positive-sense RNA strands. The innate immune system detects the presence of viruses using a range of pattern recognition receptors (PRRs). There are two distinct methods for detecting RNA viruses, like under discussed coronaviruses. One method involved the usage of Toll-like receptor 7 (TLR7) by plasmacytoid dendritic cells (pDCs) that are specialized immune cells, to identify incoming viral genomic RNA in the endosome. A range of cell types including myeloid cells express endosomal TLR3 and TLR8, which are also capable of identifying endocytosed double-stranded RNA (dsRNA) and single-stranded RNA (ssRNA), respectively. Alveolar macrophages are the first line of defense, scavenging the respiratory tract lumen. Viral recognition within infected cells is the other tactic. The RIG-I-like receptors (RLRs) and other cytosolic RNA sensors can detect dsRNA intermediates during viral propagation. The transcription of type-I and type-III interferons (IFNs), as well as nuclear factor kB (NF-kB)-dependent pro-inflammatory cytokines and chemokines, is activated by signaling downstream following TLR and RLR interaction [25].

The innate immunity must first identify SARS-COV-2 as a pathogen in order to establish an immunological defense against it. The SARS-CoV-2 virus enters cells via interacting with the ACE2 receptor, also known as the angiotensin-converting enzyme 2, through its spike protein. An immunological reaction is triggered when a virus infects a host because the host's immune system recognizes the virus through its surface epitopes. Transmembrane serine protease 2 (TMPRSS2) enzyme is the major protease associated with activation of SARS-CoV-2 [26–28]. Recent data from single cell RNA-sequencing analysis has indicated that other proteases such as cathepsin B and L are also involved in this process. During infection, pattern recognition receptors (PRRs), primarily Toll-such Receptors (TLRs) such as TLR3, TLR7, and TLR8, detect the viral pathogen-associated molecular patterns (PAMPs). This leads to the activation of several immune responses of the host involving nuclear factor-kappa B-κB (NF-κB), phosphoinositide 3-kinase, and interferon regulatory factor-3 (IRF-3) [29–31]. Finally, the activation of these immune responses triggers the innate immune response such as type-I interferon (IFN-1) and pro-inflammatory cytokines production, e.g., IL-6, IL-18, TNF-α, and IL-1β [32]. Endosomal TLR3 and TLR7 PRRs get activated after SARS-CoV-2 infection and detect the virus dsRNA and ssRNA [33, 34]. Type-I IFNs and cytokines are then used to stimulate the natural killer (NK) cells. The early immune response that restricts SARS-CoV-2 pathogenesis in the early stages of infection depends on these immune cells [35]. Appropriate synthesis of the IFNs restricts the replication of virus by inducing immune response in the infected cells. Several SARS-CoV-2 proteins, including open reading frame 6 (ORF6) and ORF3b, have been demonstrated to inhibit antiviral type-I IFN (IFN-I) synthesis and signaling [36, 37]. Due to delayed response of IFN-I, the virus replicates and spreads unrestrictedly in the infected host resulting in an eventual increase in IFN-I which can worsen hyperinflammation as the disease progresses to severe conditions [38]. The respiratory tract's epithelial cells are among the first effector cells to engage immune cells in the lung tissue through the production of immune cytokines. Type-1 macrophages in the innate immune system mount an antiviral immunological response upon recognition of the virus by producing immune cytokines [39]. The production of type-I interferons (IFNs) and interleukin-6 (IL-6) by dendritic cells (DCs), which are found in the respiratory mucosa, plays an essential role in the suppression of viral infections. The DCs such as antigen-presenting cells (APCs) are essential for initiating adaptive immune responses. In critically ill individuals, unregulated type-I IFN production by DCs promotes acute inflammation and acute respiratory distress syndrome [40]. Type-I IFNs such as IFN-α, IFN-β, and IFN-λ are one of the crucial immune signals in innate immunity which have important antiviral activity. Interferon regulatory factor (IRF)-3 controls the production of type-I interferon (IFN-I) [41]. Compared with SARS-CoV and MERS virus, SARS-CoV-2 infection compels a lower antiviral transcriptional response characterized by low-level type-I IFN and type-III land high expression of chemokines. This may explain the pro-inflammatory disease condition associated with COVID-19 [42]. The IRF-3 is induced within the macrophage after the virus has been identified, and this causes the creation of IFN-1. The IFN-1 triggers a virus-specific adaptive immune response by triggering downstream signaling pathways. The IFN-induced transmembrane family proteins are activated by type-I IFN, one of the processes by which it contributes to antiviral immunity and effectively prevents the virus from entering the host cell. This behavior makes it more difficult for the SARS-CoV-2 virus to replicate and infect more cells.

10.2.2 Adaptive Immunity to SARS-CoV-2

Due to the substantial contribution of adaptive immunity to the provision of defense against viral infection and disease, the scientific community has aggressively investigated the research of humoral antibody (Ab) and cell-mediated immunity. This study intends to find elements that indicate protection, develop therapies based on the immune system, and enhance the design and delivery of vaccines [25].

Adaptive immunity has played a significant role in COVID-19 infection by activating cytotoxic T cells and B cells. The T cells and B cells destroy the infected cells and produce the antibodies against the viral antigens. Helper T cells get the antigens from antigen-presenting cells such DCs and monocytes. Due to the same pathophysiology of these two coronaviruses, the SARS-CoV-2's antigen presentation method resembles SARS-CoVs. Patients with severe SARS-CoV-2 infection frequently have lymphopenia, which is characterized by lack in lymphocytes and an abnormal innate response with low IFN-I. Other mechanisms which may contribute to COVID-19 lymphopenia are infection of T cells [43], tissue damage of lymphatic organs and pyroptosis [44], and apoptosis of lymphocytes [45]. The CD^{4+} T cells, namely helper T cell 1 (Th1), play a crucial role in directing cellular immune responses against SARS-CoV-2. The release of cytotoxic T cells (CTLs) activating cytokines, including TNF-α, IFN-γ, and IL-2, is observed in patients with COVID-19, specifically in Th1 cells. Cytotoxic T lymphocytes (CTLs) engage in the elimination of infected cells through the process of perforin and granzyme-mediated cytotoxicity. Th2 cells facilitate the presentation of the viral antigen to B lymphocytes, resulting in the production of neutralizing antibodies specifically targeting the viral spike (S) protein. The function of these antibodies is to inhibit viral replication within the host organism and elicit humoral immunity. In contrast, the production of antibody molecules by plasma cells promotes the humoral immune response. The humoral immune system is a biological mechanism that allows B cells to produce antibodies, which then help to destroy extracellular pathogens and stop the propagation of intracellular infections. Helper T cells are frequently required for the activation of B cells and their subsequent transformation into plasma cells that release antibodies, which are typically triggered by the presence of antigens. The designation "helper T cell" is commonly employed to refer to a cell originating from the TH2 subclass of CD^{4+} T cells. It is important to keep in mind that a particular subset of TH1 cells can also help to activate B cells. The foundational idea of humoral immunity served as the basis for the development of SARS-CoV-2 vaccines [46]. Initiating the adaptive immune response against SARS-CoV-2 is significantly influenced by the humoral immune response, which is mostly mediated by B cells. Humoral response in COVID-19 has been found to be the same as earlier reported coronavirus infections, with the development of the typical Immunoglobulin G (IgG) and Immunoglobulin M (IgM) antibodies. After the SARS-CoV-2 infection, the B cells generate response against the nucleocapsid (N) protein of the virus. The IgM and IgG antibodies, as well as other antibodies that target the viral S-protein, are also produced by B cells [47, 48]. About nine days after the onset of the first symptoms, SARS-CoV-2-IgM antibodies were found in the serum of COVID-19 patients, whereas SARS-CoV-2-IgG antibodies were found about 2 weeks after the start of the infection [49]. In order to stop the virus from infecting the patient again, it is essential that the right number of antibodies is produced. On the other hand, a lack of sufficient antibodies raises the danger of the host becoming re-infected [50]. One prominent characteristic of severe COVID-19 is the heightened synthesis of cytokines, sometimes referred to as cytokine storm. In severe COVID-19 instances, there is an increased production of inflammatory cytokines, particularly IL-6, IL-8, IL-10, and granulocyte macrophage-colony-stimulating factor (GM-CSF) [51, 52]. From mild to severe inflammation, cytokines have a substantial impact on how it develops [53]. A crucial element in the impaired immune response seen in people with COVID-19 is the decrease in T cells (Fig. 10.3). The upregulation of programmed cell death protein, T cell immunoglobulin, and mucin domain 3 (Tim-3) expressions on T cells has been seen in individuals diagnosed with severe COVID-19 [54].

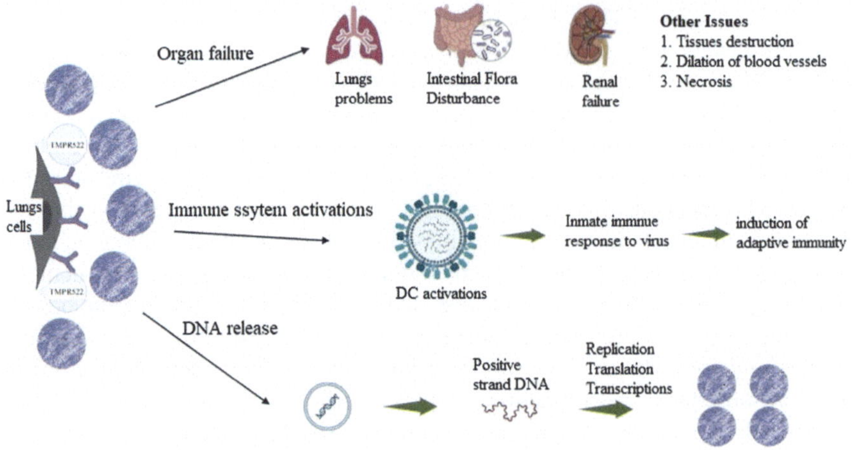

Fig. 10.3 Manifestations of COVID-19 in body [54]

10.2.3 Immunopathology of COVID-19

The first line defense against COVID-19 infection is a quick and harmonized innate immune response; however, unregulated and uncontrolled immune responses can cause immunopathology [55]. The general consensus among experts is that SARS-CoV-2 infection is restricted by adaptive immune responses, particularly those mediated by cells. This is made possible by controlling important cytokines and other anti-inflammatory proteins. Additionally, the rapid evolution of SARS-CoV-2 into various variations marked by several mutations has led researchers to reevaluate the function of cell-mediated immune response [56]. Studies show that SARS-CoV-2 alters the immune response, resulting in weakened immunity and unchecked inflammatory responses. By evading innate immune responses and replicating indiscriminately in predominantly infected tissue, the SARS-CoV-2 can cause more severe illness and a poorer prognosis in some patients. The increased levels of IL-2, IL-6, IL-7, IL-10, monocyte chemoattractant protein-1 (MCP1), IFN, granulocyte colony-stimulating factor (GCSF), macrophage inflammatory protein-1 (MIP1), and TNF in these patients are positively correlated with disease severity and mortality [57, 58].

Also, in the delayed response of IFN-I, the virus replicates and spreads unrestrictedly in the infected host resulting in an eventual increase in IFN-I which can worsen hyperinflammation as the disease progresses to severe conditions [38]. Lymphopenia is a major characteristic of patients with severe SARS-CoV-2 infection. By producing viral-specific antibodies, CD^{4+} T and CD^{8+} T lymphocytes play an essential role in controlling chronic viral infections. In COVID-19-infected individuals, the number of CD^{4+} T, CD^{8+} T, NK, and B cells decreases significantly, indicating an impaired immune response. Patients' antiviral immunity is weakened as a result of their decreased number. Lymphopenia is more prevalent in severe cases, occurring in 63% of patients who are critically ill. The SARS-CoV-2 virus can also suppress the immune system by inducing the demise of programmed T cells. As the number of lymphocytes decreases in COVID-19 patients, inflammatory cytokines increase and the severity of the disease worsens. In addition, multiple reports suggest that in severe COVID-19 cases, regulatory T cells are decreased, resulting in lung immunopathology [39, 59]. Several studies have demonstrated that in addition to lymphopenia, T cell exhaustion and functional diversity decrease as a result of the expression of programmed cell death protein-1 (PD-1), which promotes the progression of COVID-19 [39]. Downregulation and stimulation of ACE2 excretion following viral infection are additional hypothesized pathways by which SARS-CoV-2 can generate inflammatory responses [60].

Because the renin-angiotensin system (RAS) is disrupted when ACE2 is missing from cells, it affects blood pressure, electrolyte balance, inflammation regulation, and vascular permeability in the airways [61]. The alveolar capillary barrier and alveolar edema in the lungs are damaged as a result of an increase in the neutrophil to lymphocyte ratio (NLR) and acute inflammatory processes following a high viral burden and cytokine storm. This can ultimately result in shock, organ failure, particularly in organs with high ACE2 expression, and mortality.

10.3 Common Antiviral Approaches and Limitations

All viruses, including influenza viruses, must invade a host cell in order to replicate, release, and disseminate. This procedure involves multiple steps [62]. Viruses are pathogens that can replicate in host cells using protein or nucleic acid synthesis machinery of cell. Hence, treating viral infections is more difficult than treating bacteria or other pathogenic microorganisms. Treatment of the infected host cell with the antiviral drugs can be toxic to the infected cell. Currently, the most common ways to control viral diseases are through vaccination and antiviral drugs. Current antiviral comprises nucleoside analogs, neuraminidase, transcription blocking compounds, proteinase, and immunoglobulins [63]. All the antiviral drugs have a different mechanism of action. Some antiviral drugs target the viral proteins, metabolism, and transcription, while some drugs target the proteins of the host cell. Several antiviral drugs boost the defense mechanism of the host [64].

In the first phase of life cycle, the virus interacts with the host receptors; this phase is termed as fusion and entry. In order to treat respiratory syncytial virus (RSV) and human immunodeficiency virus (HIV), antiviral drugs that prevent the fusion of viruses have been frequently used [65–67]. The interaction between chemokine receptors and glycoprotein receptors offers an attractive strategy for blocking viral fusion and entry, a crucial step in preventing viral infection (Fig. 10.4). Uncoating antiviral drugs/inhibitors have been developed to combat viral infections, including influenza. Upon entering the endosome, the virus encounters an acidic environment, activating the M2 proton channels. This activation leads to a reduction in the strength of the electrostatic contacts between the virus and the host cell, facilitating the process of viral uncoating. Nevertheless, the emergence of resistance to these drugs has been seen, hence prompting worries over their widespread utilization [68, 69]. The third stage in the viral life cycle involves the process of nucleic acid synthesis, which is facilitated by viral enzymes such as DNA polymerase, RNA polymerase, and reverse transcriptase. The viral enzymes mentioned in this context are frequently utilized as targets for a multitude of antiviral medications, which are employed in the treatment of diverse viral diseases like herpes simplex virus-1/2 and hepatitis B virus [70]. The inhibitors that specifically target the enzymes of the virus can be classified into three distinct categories: non-nucleoside reverse transcriptase inhibitors (NNRTIs), nucleoside reverse transcriptase inhibitors (NRTIs), and transcriptase inhibitors (NtRTIs). A significant drawback associated with these types of inhibitors is the emergence of resistance to the drugs [71, 72]. Targeted host proteins and several steps of the viral life cycle are presented in Fig. 10.4.

Understanding the molecular biology and life cycle of virus aids in the development of vaccines and drug to combat them. Nevertheless, it is important to acknowledge that viruses possess the capacity to acquire resistance against the administered treatment, presenting a significant obstacle that researchers must contend with [74].

Antiviral medications should possess the following characteristics:

- The drugs should not be mutagenic, carcinogenic, and toxic.

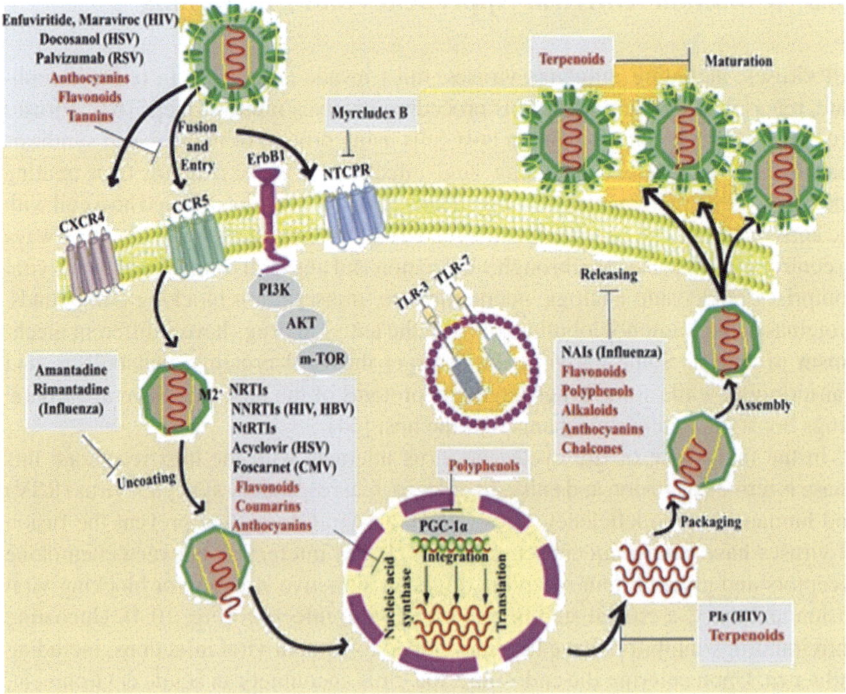

Fig. 10.4 Targeting host proteins and several steps of the viral life cycle by current therapeutic and alternative natural products [73]

- The drug should not harm the host cell and should target the virus only.
- It should be stable, effective, properly absorbed, clinically safe, and act solely in the target organ [64].

Antiviral drugs, on the other hand, also have several limitations. High rate of replication and high mutation rate in the genome of virus lead to drug resistance generations. Antiviral drugs usually kill only those viruses which are active and replicating whereas leaving the viruses that are hidden inside the host cells. Some viruses may escape from the effect of the antiviral drug. The inhibition of host immunological responses, which weakens the immune system by stopping or lowering the proliferation of dividing immune cells, is one of the harmful effects of various antiviral drugs [75]. Various drugs stop the DNA synthesis, thus having deadly and toxic effects on the cell. Another major limitation of the antiviral drugs is that these drugs are usually specific and selective inhibiting just a limited number of viruses.

Antiviral therapy should be initiated as soon as possible before the virus causes damage to tissues, and delayed use of these drugs will result in no effective treatment. Drug resistance highlights the value of combining medications that are more expensive and have more negative side effects for the patient. Due to these limitations, alternative antiviral approaches are being investigated such as ribozymes, peptide aptamers, and antisense technology [64]. A financial strategy and the allocated funds, as well as a combination of knowledge and the right research instruments, are all required to tackle the challenging process of creating and manufacturing drugs. Detailed knowledge of the molecular biology, epidemiology, life cycle, and pathogenesis of virus as well as the development of immune-informatic and understanding of host–pathogen interactions facilitates the development and selection of appropriate antiviral interventions. Another valuable perspective is drug repurposing, which aids in the quick development of drugs for newly discovered and unknown viral infections that harm public health. Effective treatment of the newly developed coronavirus necessitates substantial antiviral medication development [74].

10.4 Antiviral Activities of Chitosan

Chitosan (deacetylated chitin) has a wide range of biological activities such as antiviral, antibacterial, and antitumor, allowing for its widespread use in medicine and pharmacy. The ability of chitosan to boost the immune system and afterward be used to protect organisms from stress and disease is of great interest [76]. The potential of chitosan as vaccine adjuvant is increasingly being investigated [77]. Chitosan is a linear polysaccharide composed of two monosaccharides β-d-glucosamine N-acetyl-D-glucosamine which are linked by β-1,4 glycosidic bonds (Fig. 10.5). Molecular weight, degree of acetylation, and degree of substitution, among other factors, all have an impact on the antiviral properties of chitosan and its derivatives. In relation to the mechanism of action of chitosan-based antiviral therapeutics, the literature has examined and evaluated the effects of chitosan-based therapeutic characteristics on antiviral activity [78]. Chitosan and its derivatives have the ability to modulate the immune system of the host and the virus-infected organism, or to directly influence the virus. Tobacco mosaic virus (TMV), MS2 (bacteriophage), phiX174, Autographa californica multicapsid nucleopolyhedrovirus, *Bombyx mori* Nuclear Polyhedrosis, and moderately large to large molecular weight chitosan (53–1150 kDa, 9.1–32.5% degree of acetylation) all exhibit growth inhibition (Fig. 10.5). The Newcastle Disease Virus (NDV) was discovered to be active against smaller chitosan with a molecular weight of 3.375–10.491 kDa and 10% degree of acetylation [78].

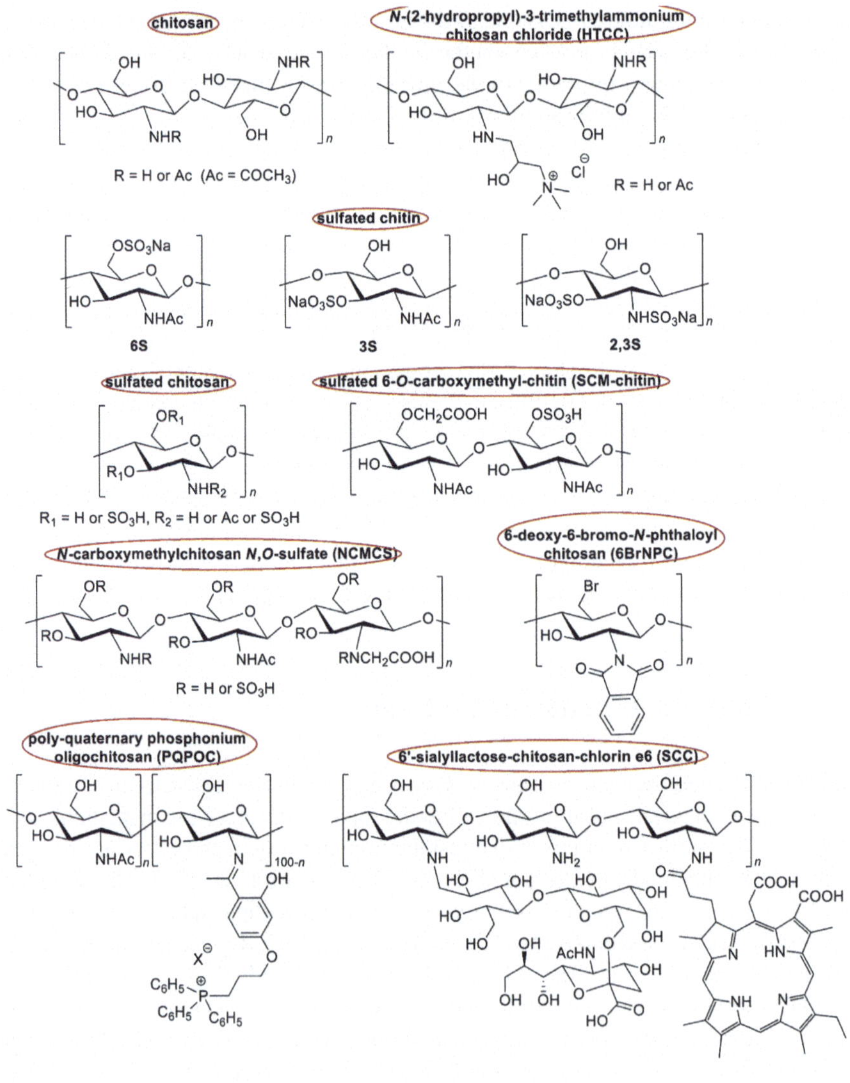

Fig. 10.5 Chemical structures of chitosan and its antiviral derivatives [78]

Chitosan was discovered in 1859 by Rouget. The size of chitosan polymer differs widely and is described by mean degree of polymerization, its mean molecular weight, deacetylation degree, and pKa values [79]. Chitosan is a low-cost product made from deacetylation of chitin, which is widely dispersed in nature and serves as a structural component of the exoskeleton of arthropods and the fungal cell wall. Chitosan also has several favorable biological properties; the most important are low toxicity and biodegradability. It is easily degraded by specific and non-specific enzymes such as chitosanases, cellulases, chitinases, lysozymes, proteases, and lipases [80]. Due to its efficacy as an adjuvant, chitosan is employed in the production of antiviral vaccines [81]. The potential use of chitosan in the management of viral infections is summarized in Fig. 10.6 [82].

Antiviral activity has previously been demonstrated in marine polysaccharides and oligosaccharides produced from chitosan. Chitosan in plants has been carefully studied for its antiviral properties. Chitosan has been proven to affect hypersensitivity to viral infection in plants and prevention of systemic transmission of the virus in plants [82, 83]. Chitosan does not have direct antiviral effect; rather, it enhances the plant's immune response and strengthens its defense mechanisms [84, 85].

Chitosan has been used to treat the viral infection caused by *B. mori* nucleopolyhedrosis virus in larvae of *Bombyx mori* L. Chitosan significantly suppressed the mortality rate of the larvae. Chitosan can also inhibit the viral infections caused by *Autographa californica* nucleopolyhedrovirus [86]. Chitosan also has potential inhibitory effects against the viral infections in human. It can significantly inhibit

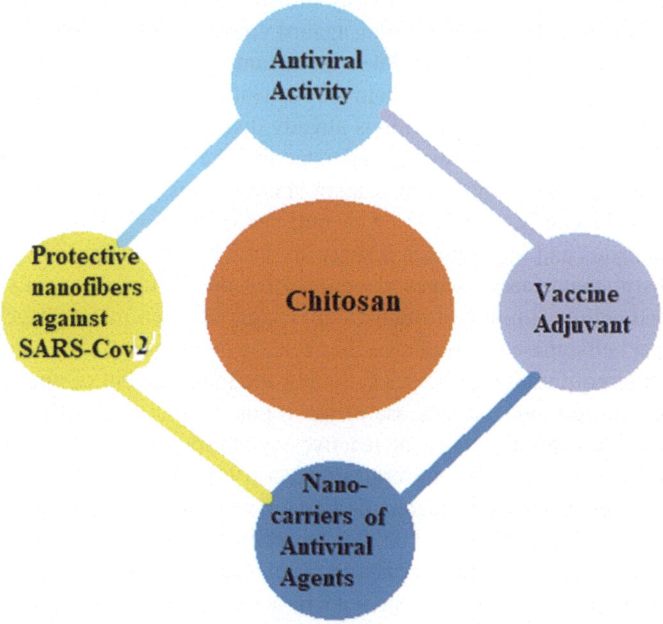

Fig. 10.6 Schematic diagram illustrating the potential role of chitosan in controlling viral infections

H_1N_1 influenza "A" virus infection and human cytomegalovirus strain AD169 [87, 88]. Chitosan has also shown antiviral activity against Herpes Simplex-1, Coxsackie viruses, and Rift Valley fever virus [89]. The molecular weight, chitosan concentration, acetylation level, and chemical modification of the chitosan polymer represent a few of the characteristics that affect chitosan's capability to prevent viral growth [90]. It has been shown that the low concentration of chitosan inhibits the tobacco mosaic virus and potato virus X (PVX) in one-third of treated tomato and potato plants while high concentration of chitosan inhibits 80% of the virus infection in treated plants. These findings show how antiviral action of chitosan is affected by its concentration [91]. Some studies demonstrate that the antiviral activity of chitosan increases when the molecular weight decreases [83, 84]. High molecular weight chitosan, according to other studies, provides stronger antiviral effects against PVX infection [92]. Although chitosan's degree of acetylation has an insignificant effect on antiviral activity, deamination of the substance has a significant impact [87, 90]. Numerous chemical modifications have been made to the backbone of chitosan, resulting in a variety of polymer derivatives with novel antiviral properties. The 6-amine chitosan, sulfonated chitosan, and fluorinated derivative of chitosan increase the antiviral activity [93–95].

10.4.1 Mechanism of Action of Chitosan

Chitosan can inhibit the viral infection by directly entering the infected host cells and by strengthening the immune system against virus. Various studies demonstrated that chitosan does not show direct link with the antiviral activity in some diseases; instead, it modulates the immune system thus activating the innate immune response toward them; this kind of modulation is already discussed [84]. Understanding the stages of viral infection is essential to understanding the antiviral activity of chitosan and its derivatives. The antiviral mechanism of chitosan is shown in Fig. 10.7 [89].

Chitosan and its derivatives have a wide range of applications as preventive antivirals. Granulocytes and macrophage activity are both heavily regulated by chitosan. As previously stated macrophages are crucial for the immune system's reaction. Humoral and cellular immune responses are triggered when macrophages interact with T-helper cells; these macrophages then release interleukin-1. The interleukin-1 stimulates the proliferation of helper cells. For the management of viral infections in animals, chitosan-induced increase of macrophage functional activity may also be important. Chitosan stimulates the reactive oxygen species production in alveolar macrophages and stimulate g-interferon synthesis [96]. The increase in synthesis of interferon by chitosan is a vital feature of antiviral resistance. The ability of genomic RNAs and early viral mRNAs to translate has been demonstrated to be reduced by interferon, which prevents virus replication [90]. Chitosan has also been proved to induce neutrophil chemotaxis. The chemotactic activities of the neutrophil are related to the chitosan molecular weight and shown to be many times greater than that of chitin [97, 98]. Chitosan increases the production the IgA and IgG antibodies in

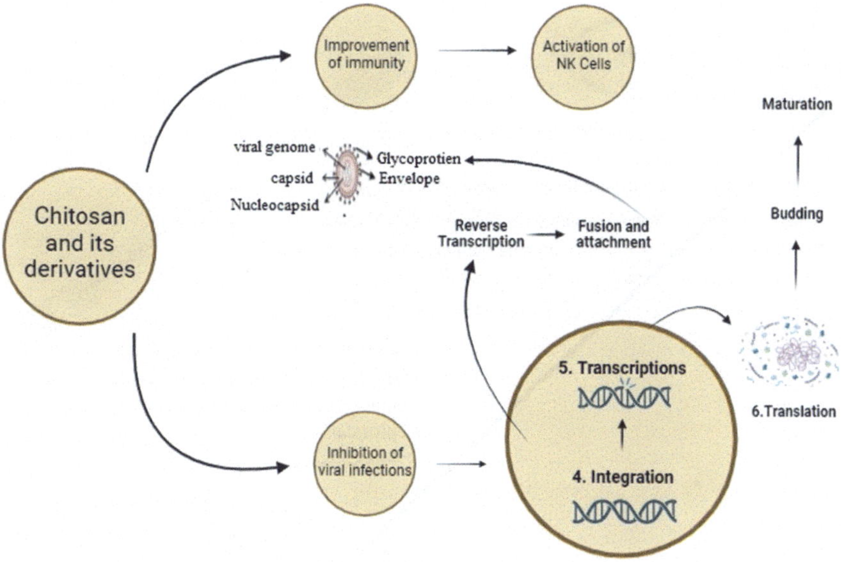

Fig. 10.7 Mechanism of action of chitosan [89]

influenza (H_1N_1) A and B viral infections. Chitosan can alter the induction phase of immunological responses in animals, as well as various immune system effector mechanisms [99].

10.5 Direct Antiviral Effects of Chitosan

As previously indicated, substantial research has been done on the antibacterial properties of chitosan biopolymers and their derivatives. These products have the ability to inhibit the growth of many types of bacteria and fungi. Researchers have examined the effects of direct antimicrobial agents such as chitosan on a range of pathogens [100]. Various polysaccharides may be able to directly enter the host cell and kill viruses there, according to some investigations. By rupturing the virus's protective membrane, the interaction between the positive charge of chitosan and the negative charge of the virus surface kills the virus. A chitosan derivative called N-(2-hydroxypropyl)-3-trimethylchitosan ammonium chloride (HTCC) can effectively stop SARS-CoV-2 and MERS-CoV replication (Fig. 10.8). When the viral S-protein and HTCC polymer connect, a complex is generated that prevents the virus from entering host cells, according to analysis of the interaction mechanism.

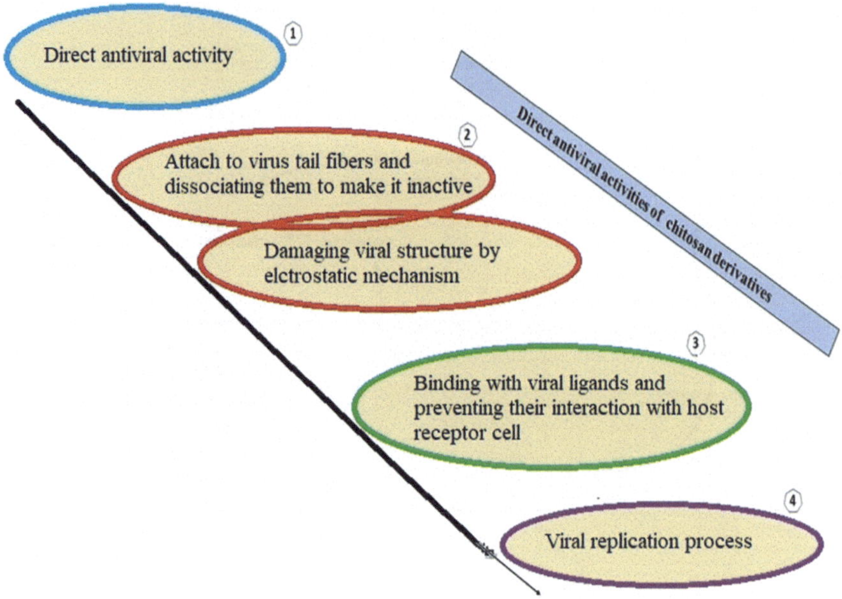

Fig. 10.8 Direct antiviral effects of chitosan [100]

10.6 Immune-Boosting Effects of Chitosan

Chitosan, a linear polysaccharide compound, bears a high biodegradability and cytocompatibility; due to these properties, it is used in a range of biomedical applications and various drug delivery assays [101]; these are also used as tributary for vaccines [102]. It has been observed in various studies that chitosan is not only an immune booster but also an immunomodulatory; it acts by immunostimulatory activity. This immunostimulant works by activating the innate immune cells which leads to the release of various growth factors, chemokines, bioactive lipids, pro-inflammatory, and anti-inflammatory cytokines release of the immune mediators is influenced by chitosan [103].

Chitosan is a chitin derivative, and it is not a single polysaccharide. It belongs to a group of linear polymers with varying amounts of N-acetylglucosamine (GlcNAc) and -O-(1–4)-linked glucosamine (GlcN). Chitosan is derived from deacetylation of chitin by enzyme which results in production of 40–100% GlcN content [104]. After this chemical deacetylation, the GlcNAc residues are dispersed unevenly throughout the structure, but these can be arranged in case of chitosan using deacetylation followed by reacetylation using DDA. Not only the acetylation patterns but the molecular weight of chitosan highly effects its physiochemical properties and its interactions with the biological systems in which it is induced [105].

As the chitosan primary structure is highly dense ranging a high molecular weight, the small oligomers of chitosan can be used. As these fall in range of very low molecular weight so can be used as a target for Beta hydrolysis by biological body enzymes and acids, this low pass hydrolysis does not affect the DDA levels of these oligomers of chitosan [106]. At this point, the major concern is that different DDA polydispersity molecular weight and the purity of chitosan all affect the innate immune system and its immune system. This makes it a highly potent immune booster. The immune-modulating activities of chitosan are stimulation of natural killer cells and the macrophage cellular population. These activities of chitosan make it a highly potent immune booster. In past three decades, there have been a detailed study about immune-modulating properties of chitosan but the cellular response of chitosan inside the biological systems are considered recently [107]. Yet only the cGAS-STING and NLRP3 are the pathways that are studied extensively; this study gives an elaborated understanding of their working [108]. The response of these pathways triggers the macrophage exposure to chitosan molecules with molecular weight of 3–400 kDa and about 90% DDA. The type-I, INF response-mediated downstream action is observed which leads to further expression of the different chemokines, namely CXCL10/IP-10. Not only this response but STAT1/STAT2 are also activated; it has been observed that activation of these responses induces the production and release of, IL-1ra which is an active anti-inflammatory factor. Thus, this mode of action is involved in creating a highly potent immune-boosting response in biological systems [109]. Whereas the NLRP3 pathway does not require any intermediate response rather it directly activates the inflammasomes and release of IL-1B and PGE2 the pro-inflammatory factors.

These two types of signaling pathways are seen by chitosan modulation in response of mycobacterium intracellular infection. The early intracellular pathogenic infections lead to the production of macrophages. This paracrine signaling results in the production of IL-10 and IL-ra, which have an effect on the nearby cells and trigger their own inflammatory response. These macrophages also produce low levels of type-I IFN. Pro-inflammatory cytokines are inhibited by IL-10, whereas IL-1ra blocks IL-1B autocrine signals, promoting the pathogens' demise through pyknosis [110]. High spread of infections leads to the production of high level of PEG-2 and the activation of inflammasomes. Few chitosan dependent studies which included chitosan's of different properties like consisting of different glycine residue mimic the model of macrophage infection by intracellular pathogens. Different dose of chitosan's is administered.

At low dose of chitosan consisting of 3000 Da glycine residue a high potential type-I IFN macrophagic response has been induced without activation of inflammasomes, whereas at high dose, chitosan-mediated inflammasomes have been activating without triggering-1 IFN response [109]. Based on these findings, it can be concluded that chitosan mimics macrophages and has inherent properties that are essential for the release of pro- and anti-inflammatory signals. The dose adjustment is still a debate which can be altered accordingly; this will lead to mold immunomodulating properties as per the application response. Chitosan at low doses causes type-I IFN response, increased levels of IL-1ra and CXCL10/IP-10 production, and lower levels of PEG-2 and IL-1B. In contrast, high dose of chitosan leads to the higher production of IL-1B and PEG-2 and activation of inflammasomes. Chitosan causes the production of pro-inflammatory cytokines when the cell is in the M1 state, but it causes the release of anti-inflammatory cytokines when the cell is polarized toward the M2 state [111]. The type of cellular response induction is dependent on type of chitosan's used (different DDA, MW, polydispersity), different chitosan doses, and type of cellular models used (murine or human origin and cell lines), and the most considerate agent is the type of pro-inflammatory co-stimulators present [112]. This indicates that prior to administer chitosan's, the full characterization of chitosan's must be considered. Chitosan acts as an immune booster by influencing the immune cell cytokines production. It directly regulates the functioning of macrophages, neutrophil, mast cells, and NK cells and their involvement with other immune cells [113]. Chitosan triggers a number of innate immune responses that have an impact on how the host and pathogen interact. Chitosan interacts with extracellular serum proteins which help in pathogen recognizing mechanisms. Chitosan is the only naturally occurring cationic polysaccharide with adaptable properties. The immunological potential of the individuals is affected by an extensive range of cytokines with anti-inflammatory and pro-inflammatory responses as a result of its immune modulation properties [114].

10.7 Chitosan as a Drug Delivery Vehicle for the Treatment of Viral Infections

Chitosan is a naturally occurring abundant organic polysaccharide which is used for both veterinary and human medicinal purposes [109]. Chitosan bearing such pharmacological properties makes it widely studied thus revealing that chitosan effects the immune system of biological system as already discussed. Chitosan and its derivatives directly affect the immune system affecting the immune cells like macrophages and granulocytes. The N-acetylglucosamine residue of chitosan is more effective than N-acetylmannosamine in case of peripheral leukocytosis [90]. The stimulation of macrophages by chitosan is very important for effective immune response by suppressing the viral infection. When chitosan was given subcutaneously to mice in different instances, the alveolar macrophages phagocytosed the CH, which resulted in a significant amount of reactive oxygen species being produced. High levels of interferons are secreted by mouse splenocytes; these interferons prevent viral replication by changing the activity of early viral mRNA [96]. This ability of chitosan makes it under consideration for antiviral therapies, as chitosan works by suppressing the intracellular infection and by suppressing the uptake of pathogens by neighboring cells as well [115]. Chitosan increases the potency of antiviral immunity by production of various antibodies like IgG and Ig A [90]. Each of the derivatives of chitosan is designed with respect to particular viral agent, like sulfated chitosan's derivatives work particularly for inhibiting retro viral replication, whereas N-carboxymethyl chitosan works with inhibition of HIV replication. Likewise, N-carboxymethylated chitosan prevents the interaction of gp120 the viral glycoprotein with host T lymphocytes receptors. This kind of interaction prevents the entry of viruses in CD^{4+} cells. Along with disruption of the interaction, the viral reverse transcriptase activity which is the key role for viral genome replication is also disrupted which influence the life cycle of virus. Investigation so far suggests that the highest potent derivative of chitosan is ones with sulfated at O_2 and O_3 positions. These compounds hinder MT-4 cells from producing HIV. Anionic chitosan is believed to prevent retroviral infections. The two principal functional groups that exhibit and affect its biological activity are the amino and hydroxyl groups. The mechanism of biological activity is also influenced by the molecular weight and degree of deacetylation [116]. Sialyl lactose-chitosan derivative used for inhibiting influenza virus is presented (Fig. 10.9).

In drug delivery system, chitosan plays an important role for targeted drug delivery at specific cellular and tissue level. The primary goal of this naturally occurring polymer-mediated drug transport is the much lower cytotoxicity of chitosan, a derivative that can be used for a wide range of treatments. The biological attributes of chitosan make it anticarcinogenic, anticholesteremic, and bacteriostatic [117]. Using

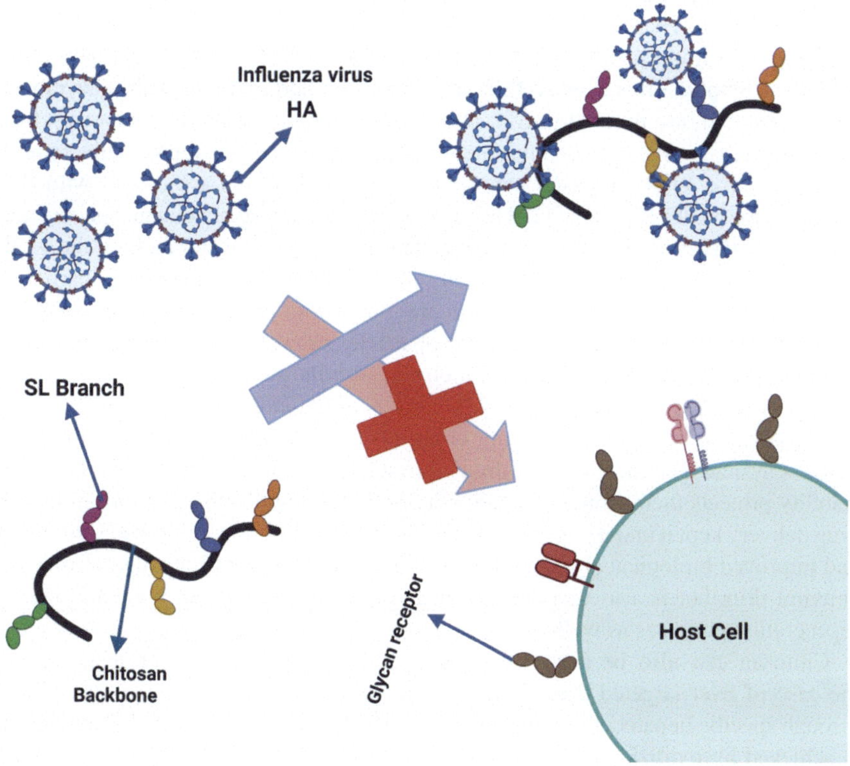

Fig. 10.9 Sialyl lactose-CH (SLCC1) chitosan derivative used for inhibiting influenza virus [68, 69]

chitosan for drug delivery is effective because the less molecular weight leads to its effective degradation by safe enzymatic means and its clearance from the human body by normal renal function like urination.

Chitosan drug delivery system is preferred due to its cationic nature which helps its interaction with other anionic drugs or other considerant compounds. The small structure of chitosan helps its penetration inside the tissues and its mucoadhesive properties make it stable; along with these attributes, it allows the cellular transfection of nucleotides allows it to target the pathogenic genome [118]. In case of diseases like HIV where a combination of drugs is used as a high potential antiretroviral therapy, optimizing drug selection and dosage is crucial to improving patient outcomes and minimizing resistance development. Different types of drugs are combined which are used for multiple life stages of viruses [119]. The FDA has approved eight different kinds of protease inhibitor; one of them is saquinavir but its bioavailability has led a serious concern. Its lower intracellular concentration makes it more convenient for drug efflux; it can be metabolized in liver and intestine. The lower bioavailability of this protease leads to excessive consumption which makes the cell resistant toward it. To resolve this issue, various studies have been done to investigate the usage of chitosan nanoparticles. In recent studies, chitosan-based nanoparticles can be loaded with saquinavir. These loaded molecules control the viral proliferation and thus targeting two different HIV strains in two different kinds of T cells [120]. In the case of Herpes Simplex Virus Type I (HSV-1), different antiherpes drugs, such as acyclovir and famciclovir, are used. However, acyclovir, being a hydrophilic compound, has difficulty crossing the fatty cellular membrane. To enhance its targeted delivery and cellular uptake, chitosan-based nanospheres are used, which are coated with the drug to improve its efficiency [121]. It is observed that these novel nanospheres are excessively targeting the viral pathogens cleaving its genome and making the viral load less and thus reducing the infection ratio. Not only the chitosan nanoparticles but chitosan-based microparticles which are linked with tripolyphosphate make the delivery of drugs more efficient. The microparticles showed more enhanced biocompatibility and bioadhesive properties which increases the penetration of these targeted microparticles to penetrate through skin barriers of the individuals.

The targeted drug delivery using the chitosan molecules not only enhances the drug delivery system but also makes the drug more stable and more efficient; this chemical stability protects them from degradation as well. The application of chitosan-based drug delivery kept it under consideration for its high penetration, improved solubility, and improved biological absorption. Chitosan is not only used for drug delivery of antiviral drug but is also vaccine carrier and can also be used to bone and nerve regeneration purposes as well.

Chitosan can also be used for organ-specific drug delivery. For instance, in the case of liver-targeted drug delivery, a modified system enhances drug uptake through specific hepatocellular receptors Fig. 10.10 [114]. This targeted conjugation is achieved by oxidizing the drug, allowing it to interact effectively with chitosan molecules.

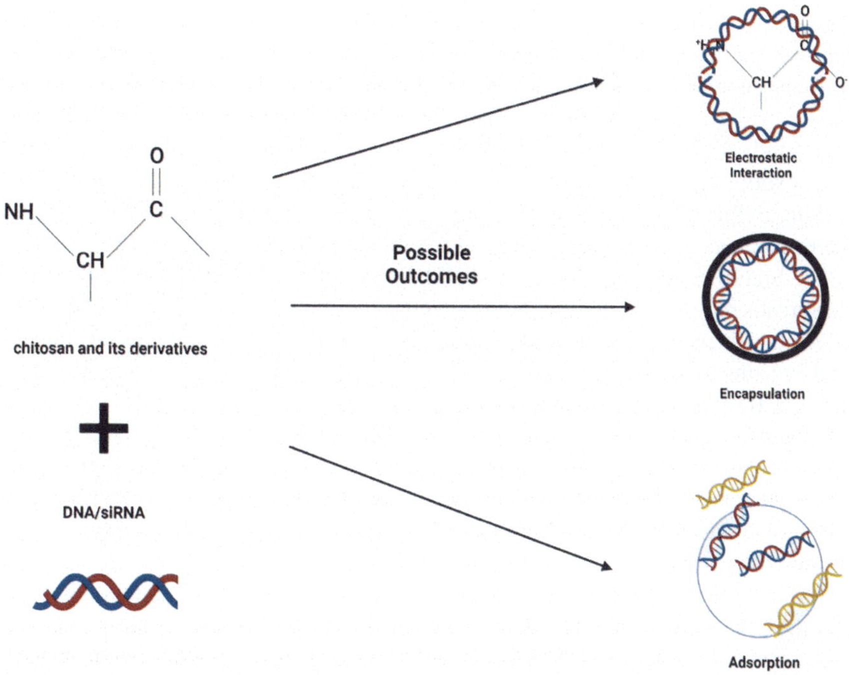

Fig. 10.10 Use of chitosan nanoparticle for drug delivery and gene editing [114]

10.8 Chitosan Potential Antiviral Effects for SARS-CoV-2

Numerous researches have been carried out recently to look at the potential use of chitosan for medicine delivery against viral diseases. The consideration of chitosan as a drug delivery medium and as an antiviral drug is due to its pharmacologically favorable traits [122]. For recent years, chitosan is considered as potent drug delivery molecule due to its biocompatible qualities which allow it to provide multiple reaction receptor sites for various bioactive compounds; the chitosan molecule protects the drug which are highly unstable from the blood flow and strong acids produced by the digestive track and their mucoadhesive properties allow them to adhere to the mucosal tissue for their better absorption [123]. Chitosan is not only considered as safe biomolecule but also a low-cost and abundant natural molecule which makes it readily available for pharmaceutical industry.

The cationic charge on chitosan molecule makes it visibly more effective; thus, this ionic difference allows it to interact with various proteins, phospholipids, and nucleic acids, i.e., DNA and RNA. The presence of hydroxyl and amine group both makes it easily modulable as per the desired targets. While considering organ-specific diseases, the inability of drugs to localize effectively at the target site remains a major concern, necessitating the development of advanced drug delivery systems for improved therapeutic outcomes. Like delivery of drugs into the lungs of infected induvial in case of SARS-CoV-2 is a concern so the delivery of drugs by chitosan nano and microparticles is under consideration [124]. In a study, it has been observed that chitosan polymers, encapsulated with an antitubercular drug, showed more stability and drug release up to 24 h without any cytotoxic effects on the organs or cells. In a recent study, Debnath and his coworkers worked on particularly chitosan nanoparticles for various respiratory diseases [125]. The field of nanotechnology is particularly working in this era of a deadly pandemic for better drug delivery in SARS-CoV-2. The researchers are using nanochitosan for anti-COVID therapeutics [126].

The highly contagious disease COVID-19 exhibits a high transmit ability ratio and spreads rapidly and widely throughout the community. This spread is due to the presence of its pathogenic aerosol particles. Not only the aerosol particles are inhaled but the resting of these particles on surfaces makes the contact more vulnerable [127]. The COVID particles are of nanosize and possessing high zeta potential. This unique zeta potential links it with the capsid it bears. The capsid of COVID-19 virus closely links it with the earlier SARS viruses by arginine rich motifs on the capsid link with unique properties which differentiate it from rest of SARS viruses as well [128]. The impact of chitosan on the cell lines of SARS-CoV-2 infected lungs has been studied by numerous groups.

The chitosan derivatives in current study efficiently disrupted the infection ratio inhibiting the growth of SARS-CoV-2 pathogens [129]. The use of these chitosan derivatives conjugated with DNA molecules suppresses the inflammation in lung tissues and results in enhancement of the antiviral response. In experimental studies conducted in 2020, a DNA-based vaccine was used that expressed the nucleocapsid protein of the SARS-CoV-2 virus in combination with chitosan nanoparticles to enhance its delivery and effectiveness. The vaccine has been administered intranasally; the aim of such administration has direct target for the nasal presented DCs and the cells that have T cells antigen present over them. The delivery of chitosan showed the increased production of IgG and IgA antibodies that are present in intranasal region; this shows the immune-boosting effect of chitosan and targeted therapeutic showing effective antiviral properties against SARS-CoV-2 molecule. The chitosan-based antiviral response is enhanced as compared to the trivial drugs used which loses the biostability and does not pass the mucosal lungs line. The absorption and redirection of these drugs is highly compromised whereas chitosan directs the drug delivery and increases the stability. The potential antiviral treatments for SARS-CoV-2 include chitosan and its derivatives are presented (Fig. 10.11).

The chitosan-based nanoparticles used as vaccine assessor or the mode of drug delivery using the internasal mode of action is considered as a promising approach to combat the SARS-CoV-2 infection [130]. Chitosan can be thought of as an appropriate defense against COVID-19 due to its capability to stimulate the immune system, pass through tight junctions between mucosal epithelial cells, and potential to physiochemically alter antigen molecules in order to improve immunogenicity. After being administered intravenously, biotinylated chitosan nanoparticles can be functionalized by a fusion protein vector to selectively target dendritic cells (DCs) and deliver the plasmid DNA encoding the SARS-CoV-2 N-protein. This increased mucosal IgA concentration and improved systemic IgG against the N-protein. The

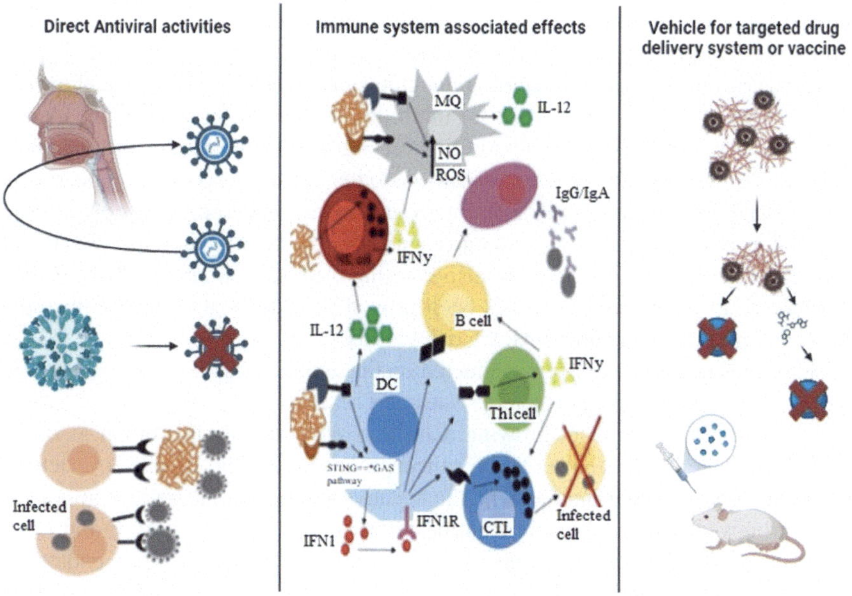

Fig. 10.11 Chitosan and its derivatives-based antiviral therapy against SARS-CoV-2 [130]

potential of Bioavanta-Bosti's chitosan nanoparticle technology for creating aerosol anti-COVID-19 medications was announced in 2020. Any prospective anti-COVID-19 medication can be delivered to the lungs of critically ill patients using aerosol formulations of Novochizol nanoparticles. Chitosan is the raw ingredient used to make Novochizol. The Novochizol method produces chitosan nanoparticles that are spherical and have distinct, adaptable physicochemical properties. Simple aerosol formulations of Novochizol nanoparticles are available for administration at all illness phases, such as in ventilators. The encapsulated active ingredient is gradually released from immobilized Novochizol nanoparticles in the respiratory system due to diffusion and slow chitosan degradation. Novochizol nanoparticles cling to mucosal membranes in the respiratory tract. Due to the lack of systemic distribution, this localized pharmacokinetics assures a high local concentration of the medication [131].

10.9 Chitosan Derivatives; A Novel Inhibitory Discovery Against Wild Type and Variants of SARS-CoV-2

The compounds originating from marine resources are just as effective antivirals in this regard. Chitosan is one such bioactive macromolecule that primarily comes from marine sources. Although marine resources as prospective medication therapy molecules have not received as much attention, new studies of chitosan as a potential molecule to treat COVID-19 disease are extensively documented. The N-(2-hydroxypropyl)-3-trimethylammonium chitosan chloride (HTCC), a quaternized chitosan derivative, has been noted for inhibiting the growth of a variety of human coronaviruses, with the degree of inhibition varying with different substitutions [132, 133]. Table 10.2 lists the function of various other chitosan-derived composites in anti-COVID-19 applications. The antimicrobial and antiviral properties of chitosan have led a way toward its use for a carrier for vaccine bearing antiviral [134]. Not only treating the viral diseases but also creating a protection against them is also considered. In recent studies, it has been proven that chitosan is an effective and highly potent therapy against SARS-CoV-2 [135].

Numerous studies have been conducted on the antibacterial capabilities of chitin and chitosan biopolymers and their derivatives. These have the ability to stop a variety of bacteria and fungus from growing. Chitin or chitosan administration as direct antibacterial agent has been studied for their protective benefits against a variety of infections. Because their interact with the negative surfaces of bacterial membranes, NH^{3+} groups enable them to exert antibacterial actions. By attaching to the DNA molecule in the nucleus, it can also stop transcription, which prevents pathogens from synthesizing proteins. Furthermore, research has demonstrated the antiviral efficacy of chitin and chitosan. Both direct antiviral action and eliciting antiviral immune responses are thought to be effective ways for chitin and chitosan polymers to combat viral infections. Chitosan prevents viral infections in plants, animals, and microbes (phage infection), according to observations. This polymer can make plant hosts resistant to viral infection and can stop phages from spreading in microbial cultures. The antiphage properties of chitosan are affected by a variety of structural variables. Its performance may be influenced by the degree of polymerization, the degree of deacetylation, the positive charge, and the kind of chemical alterations that the molecules undergo. The presence of positive charges on the molecule is effective because chitosan anionic derivatives like sulfate and succinate sulfate do not have antiphage properties. Inhibiting phages more well than polymeric precursors are oligomers. The impact of inhibition increases with polymerization degree [138]. The presence of positive charges on the molecule is effective because chitosan anionic derivatives like sulfate and succinate sulfate do not have antiphage properties. Inhibiting phages more well than polymeric precursors are oligomers. The impact of inhibition increases with polymerization degree [138]. The detailed and elaborated immune-boosting patterns of chitosan are also studied. For treatment of various diseases, before the spread of SARS-CoV-2 chitosan has been used as a potential antiviral drug as in case of HIV, herpes in human; not only this but many plant-based viral diseases have also been targeted using chitosan as an intermediate for efficient drug delivery [90]. Few scientists found that chitosan showed an enhanced hypersensitive response for plant-based viral infections. Various studies demonstrated that chitosan does not show direct link with the antiviral activity in some diseases; instead, it modulates the immune system thus activating the innate immune response toward them; this kind of modulation is already discussed [84]. Chitosan and its derivatives have paved a broad way toward the therapeutics field if considering human-based

Table 10.2 Role of chitosan-derived composites in anti-COVID-19 applications

Chitosan-derived composite	Mechanism of antiviral studies	Anti-COVID application	References
Cationically modified chitosan derivative	Electrostatic interaction exerted by sphere complex with viral nucleic acid	Antiviral activity against HCoV-NL63 and HCoV-OC43	[136]
Hydrophobically modified HTCC and N-(2-hydroxypropyl)-3-trimethylammonium chitosan chloride (HTCC)	Inhibition of HCoV-NL63 replication	Coronavirus NL63 inhibitors	[137]
Aerosol formulations for anti-COVID-19 drugs (Novochizol™) based on chitosan nanoparticles	Sluggish degradation of chitosan responsible for slow release of anti-COVID-19 drug	Chitosan NPs in Novochizol™	[138]
Chitosan hydrogels blended with dicarboxylic acids (malic and glutaric acid)	Risk control of SARS-CoV-2 infection	Viral spike protein to the ACE2 receptors	[139]
N-(2-hydroxypropyl)-3-trimethylammonium chitosan chloride (HTCC), and hydrophobically modified derivative (HM-HTCC) Modifications of chitosan in cationic mode	HTCC and HM-HTCC inhibit interaction of a virus with its receptor	Coronavirus HCoV-NL63 inhibitors	[140]
Nanoparticles of chitosan	Study of mechanism	CNPs encapsulated with silymarin (Sil–CNPs)	[141]
Zero-valent nanosilver/titania-chitosan	MS2 bacteriophages testing responsible for reduction of infection caused by COVID-19	Removal of viral aerosols to minimize the risk of airborne transmission	[142]
Nanocomposite of chitosan-Zeolite-ZnO	Antiviral activity chitosan–zeolite and ZnO against SARS-CoV-2 The SARS-CoV-2 ligand and receptor	Identification of SARS-CoV-2 glycoprotein receptors RCSB and PDB interaction by molecular docking	[143]

(continued)

Table 10.2 (continued)

Chitosan-derived composite	Mechanism of antiviral studies	Anti-COVID application	References
Modification of chitosan derivative in cationic mode (HTCC)	Receptor protein inhibition	Chitosan derivative HTCC as inhibitors for coronavirus MERS-CoV and SARS-CoV-2 infection	[129]
Chitosan nanofibers derived of N,N,N-trimethyl chitosan (TMC), the single N-quaternized (QCS) and the double N-diquaternized (DQCS)	Viral repulsion of SARS-CoV-2 by cationic amino group	Design of Personal Protective Equipment (PPE)	[144]
Nanoparticles of chitosan derivatives	Activation of T cells for the production of IgG and nasal IgA antibodies against SARS-CoV	Carrier molecules for DNA vaccine expressing SARS-CoV nucleocapsid protein	[145]
In silico studies by chitosan and its derivatives	Inhibition of viral spike protein binding to the ACE2 receptors	Inhibition of SARS-CoV-2 viral binding with host cells	[146]
Nanochitosan studies	Vaccine delivery via the intranasal route to fight against SARS-CoV-2	Application as vaccine adjuvant	[130]
Amphiphilic chitosan N-palmitoyl-N-monomethyl-N,N-dimethyl-N,N,N-trimethyl-6-O-glycol chitosan (GCPQ)	Reduction in the infectivity of SARS-CoV-2 in A549ACE2+ and Vero E6 cells and in human airway epithelial cells	As a COVID-19 prophylactic agent	[147]
β-chitosan analysis	Interaction due to binding between SARS-CoV-2 S-RBD and ACE2 receptors	Inhibition of SARS-CoV-2 viral binding with host cells	[148]

diseases multiple diseases like influenza, Rift Valley fever, and herpes simplex. The mode of action of chitosan derivatives is sometimes specific to the type of causative strain [128]. It is proven so far that chitosan derivatives bear antiviral properties but the different kind of polymers and degree of acetylation and molecular weight of these derivatives also plays a key role in efficacy of this antiviral property [91].

Studies indicate that the antiviral properties of chitosan derivatives are enhanced by their low molecular weight. The lower the molecular weight, the higher the antiviral activity, making them more effective against viral infections [83]. The degree of acetylation has a very weak link with the antiviral capacity. But amino acid repeat backbone addressing the content of amino acid residue in different chitosan derivatives directly impacts its antiviral properties [90].

This debate regarding the properties has led to development of various modulated derivatives. The modulated derivatives usually are dependent on type of pathogen used against. The chemical modification on the backbone of chitosan is main target for developing differently potent chitosan derivatives. Use of different functional groups on chitosan has led to the development of derivatives as well. Different studies have shown that chitosan derivatives with bromine ion possess high potential properties against respiratory tract diseases, whereas the sulfate group attached chitosan derivatives are more effective toward immune deficiency virus HIV [82].

Figure 10.12 gives a brief review of approved chitosan derivatives which are specially modulated as per the type of pathogen treated [134]. Chitosan is used in preparation of vaccines against SAR-CoV-2 and its different strain that emerge due to any spontaneous mutation in nucleocapsid proteins. In recent years, the urge for development of different vaccines against different strains of SARS-CoV-2 has increased drastically. The demand increased due to unstoppable transmits ability. As the earlier studies have clearly suggested that chitosan naturally occurring polymer is a successful candidate for development of biodegradable vaccine delivery adjuvant. The use of chitosan as a vaccine addressor in the management of infectious bronchitis brought on by coronavirus has been investigated by Bande et al. [134]. The immunogenic potential of the DNA vaccines and chitosan-saponin nanoparticles has been evaluated for two distinct strains, i.e., avian infectious bronchitis virus (M41) and CR88. A rise in mucosal immunoglobulin A (IgA) and an expression of interferon-gamma-related genes has been observed while discussing the immunogenic response in detail, indicating an improvement in antibody- and cell-mediated responses. Their findings supported the use of chitosan and saponin as adjuvants or immunomodulators to boost the host's immunological responses [134].

Chitosan is a promising means of delivering genes, as evidenced by the rise in mucosal IgA and systemic Immunoglobulin G (IgG) against N-protein [149, 150]. Sun et al. created bovine coronavirus N-protein-coated biotinylated chitosan nanoparticles [150]. The effectiveness of their vaccine against SARS-CoV-2 has been tested. The alteration of dose manner its reduction can still increase the antiviral penetration and enhance the intracellular antiviral activity. The decrease in dose increases the bioavailability of the vaccine, and along with this, it impacts the solubility of the drug. This increased availability provides more times for interaction of the drug with

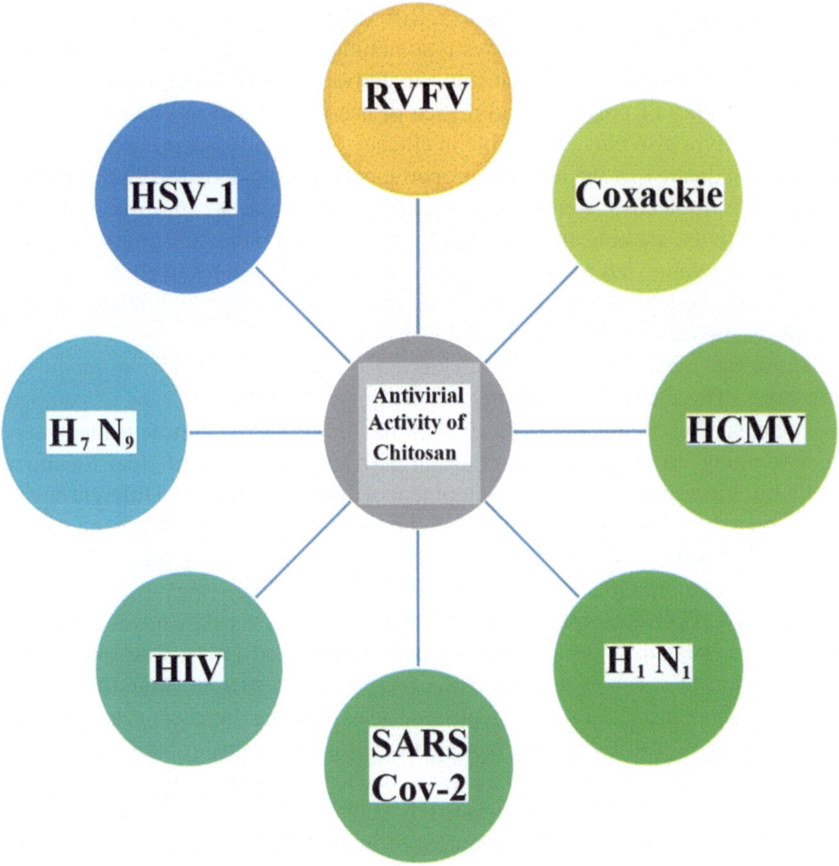

Fig. 10.12 Tested and proven antiviral activity of chitosan against several human viruses [134]

the infected cells thus provoking an enhanced immune response in the case of SARS-CoV-2 and its different variants. The targeted delivery of plasmid DNA expressing nucleocapsids specifically for SARS-CoV-2 is made possible by the chitosan-based nanoparticles that are designed to attack the intranasal cells. The SARS-CoV-2 spike proteins are secreted and transported as a result of this type of vaccine administration, where these compete with live coronavirus for binding to human ACE2 receptors. According to the findings, mice produced significant amounts of IgA and IgG. Chitosan's cationic composition enhances binding to negatively charged DNA and reduces DNA vaccine breakdown. Chitosan's mucoadhesive characteristics also support its application as a delivery adjuvant to mucosal surfaces [151, 152].

10.10 Conclusion

The COVID-19 pandemic, caused by the SARS-CoV-2 virus, has ravaged the world with its unprecedented severity and disruptive impact. This chapter provides an exhaustive overview of the novel applications of chitosan derivatives and nanoparticles in combating COVID-19. COVID-19, caused by SARS-CoV-2, is a member of the coronavirus family, characterized by a positive-sense RNA-strand genome of approximately 30,000 bases. The emergence of SARS-CoV-2 in humans is believed to be facilitated by unknown external factors. The innate immune response serves as the first line of defense against COVID-19. However, unregulated immune responses can lead to immunopathology. Pattern recognition receptors (PRRs) detect viral presence, triggering an immune response. Adaptive immunity to SARS-CoV-2 provides protection against viral infection and disease. Conventional antiviral strategies and their limitations are discussed. Chitosan's role in enhancing immune response and direct antiviral effects is elaborated. Chitosan's immune-boosting and antiviral properties make it an attractive compound for studying its effects on the immune system. Its potential as a drug delivery vehicle for viral infections is highlighted. Chitosan derivatives and nanoparticles demonstrate promise in combating COVID-19. Their immunomodulatory and antiviral effects, combined with their potential as drug delivery vehicles, position chitosan as a valuable tool in enhancing defense mechanisms against SARS-CoV-2.

References

1. Zhu N, Zhang D, Wang W, Li X, Yang B, Song J, Zhao X, Huang B, Shi W, Lu R, Niu P, Zhan F, Ma X, Wang D, Xu W, Wu G, Gao GF, Tan W (2020) A novel coronavirus from patients with pneumonia in China, 2019. N Engl J Med 382. https://doi.org/10.1056/nejmoa2001017
2. Guarner J (2020) Three emerging coronaviruses in two decades: the story of SARS, MERS, and now COVID-19. https://doi.org/10.1093/ajcp/aqaa029
3. Zheng DP, Ando T, Fankhauser RL, Beard RS, Glass RI, Monroe SS (2006) Norovirus classification and proposed strain nomenclature. Virology 346. https://doi.org/10.1016/j.virol.2005.11.015
4. Ye ZW, Yuan S, Yuen KS, Fung SY, Chan CP, Jin DY (2020) Zoonotic origins of human coronaviruses. https://doi.org/10.7150/ijbs.45472
5. Alluwaimi AM, Alshubaith IH, Al-Ali AM, Abohelaika S (2020) The coronaviruses of animals and birds: their zoonosis, vaccines, and models for SARS-CoV and SARS-CoV2. https://doi.org/10.3389/fvets.2020.582287
6. Heinz FX, Stiasny K (2020) Profile of SARS-CoV-2. Wien Klin Wochenschr 132:635–644. https://doi.org/10.1007/s00508-020-01763-1
7. Weiss SR, Leibowitz JL (2011) Coronavirus pathogenesis. Adv Virus Res. https://doi.org/10.1016/B978-0-12-385885-6.00009-2
8. Cui J, Li F, Shi ZL (2019) Origin and evolution of pathogenic coronaviruses. https://doi.org/10.1038/s41579-018-0118-9
9. Hasöksüz M, Kiliç S, Saraç F (2020) Coronaviruses and SARS-CoV-2. https://doi.org/10.3906/sag-2004-127
10. Siddell SG, Walker PJ, Lefkowitz EJ, Mushegian AR, Adams MJ, Dutilh BE, Gorbalenya AE, Harrach B, Harrison RL, Junglen S, Knowles NJ, Kropinski AM, Krupovic M, Kuhn

JH, Nibert M, Rubino L, Sabanadzovic S, Sanfaçon H, Simmonds P, Varsani A, Zerbini FM, Davison AJ (2019) Additional changes to taxonomy ratified in a special vote by the International Committee on Taxonomy of Viruses (October 2018). Arch Virol 164. https://doi.org/10.1007/s00705-018-04136-2
11. Gorbalenya AE, Baker SC, Baric RS, de Groot RJ, Drosten C, Gulyaeva AA, Haagmans BL, Lauber C, Leontovich AM, Neuman BW, Penzar D, Perlman S, Poon LLM, Samborskiy DV, Sidorov IA, Sola I, Ziebuhr J (2020) The species severe acute respiratory syndrome-related coronavirus: classifying 2019-nCoV and naming it SARS-CoV-2. https://doi.org/10.1038/s41564-020-0695-z
12. Forni D, Cagliani R, Clerici M, Sironi M (2017) Molecular evolution of human coronavirus genomes. https://doi.org/10.1016/j.tim.2016.09.001
13. Dudas G, Carvalho LM, Rambaut A, Bedford T (2018) MERS-CoV spillover at the camel-human interface. Elife 7. https://doi.org/10.7554/eLife.31257
14. Woo PCY, Lau SKP, Lam CSF, Lau CCY, Tsang AKL, Lau JHN, Bai R, Teng JLL, Tsang CCC, Wang M, Zheng B-J, Chan K-H, Yuen K-Y (2012) Discovery of seven novel mammalian and avian coronaviruses in the genus deltacoronavirus supports bat coronaviruses as the gene source of alphacoronavirus and betacoronavirus and avian coronaviruses as the gene source of gammacoronavirus and deltacoronavirus. J Virol 86. https://doi.org/10.1128/jvi.06540-11
15. King AMQ, Lefkowitz EJ, Mushegian AR, Adams MJ, Dutilh BE, Gorbalenya AE, Harrach B, Harrison RL, Junglen S, Knowles NJ, Kropinski AM, Krupovic M, Kuhn JH, Nibert ML, Rubino L, Sabanadzovic S, Sanfaçon H, Siddell SG, Simmonds P, Varsani A, Zerbini FM, Davison AJ (2018) Changes to taxonomy and the International Code of Virus Classification and Nomenclature ratified by the International Committee on Taxonomy of Viruses (2018). Arch Virol 163. https://doi.org/10.1007/s00705-018-3847-1
16. Ma Y, Zhang Y, Liang X, Lou F, Oglesbee M, Krakowka S, Li J (2015) Origin, evolution, and virulence of porcine deltacoronaviruses in the United States. mBio 6. https://doi.org/10.1128/mBio.00064-15
17. Woo PCY, Lau SKP, Lam CSF, Tsang AKL, Hui S-W, Fan RYY, Martelli P, Yuen K-Y (2014) Discovery of a novel bottlenose dolphin coronavirus reveals a distinct species of marine mammal coronavirus in gammacoronavirus. J Virol 88. https://doi.org/10.1128/jvi.02351-13
18. King AMQ, Adams MJ, Carstens EB, Lefkowitz EJ (2011) Virus taxonomy. In: Ninth report of the international committee on taxonomy of viruses
19. Lin X, Chen S (2020) Major concerns on the identification of bat coronavirus strain RaTG13 and quality of related nature paper. Preprints (Basel)
20. Ramanathan K, Antognini D, Combes A, Paden M, Zakhary B, Ogino M, Maclaren G, Brodie D (2020) Genomic characterisation and epidemiology of 2019 novel coronavirus: implications for virus origins and receptor binding. Lancet 395
21. Rehman SU, Shafique L, Ihsan A, Liu Q (2020) Evolutionary trajectory for the emergence of novel coronavirus SARS-CoV-2. Pathogens 9. https://doi.org/10.3390/pathogens9030240
22. Gorbalenya AE, Snijder EJ, Spaan WJM (2004) Severe acute respiratory syndrome coronavirus phylogeny: toward consensus. J Virol 78. https://doi.org/10.1128/jvi.78.15.7863-7866.2004
23. de Groot RJ, Baker SC, Baric RS, Brown CS, Drosten C, Enjuanes L, Fouchier RAM, Galiano M, Gorbalenya AE, Memish ZA, Perlman S, Poon LLM, Snijder EJ, Stephens GM, Woo PCY, Zaki AM, Zambon M, Ziebuhr J (2013) Commentary: Middle East respiratory syndrome coronavirus (MERS-CoV): announcement of the coronavirus study group. J Virol 87. https://doi.org/10.1128/jvi.01244-13
24. Singh D, Yi SV (2021) On the origin and evolution of SARS-CoV-2. Exp Mol Med 53(4):537–547. https://doi.org/10.1038/s12276-021-00604-z
25. Merad M, Blish CA, Sallusto F, Iwasaki A (2022) The immunology and immunopathology of COVID-19. https://doi.org/10.1126/science.abm8108
26. Mahmoud IS, Jarrar YB, Alshaer W, Ismail S (2020) SARS-CoV-2 entry in host cells-multiple targets for treatment and prevention. https://doi.org/10.1016/j.biochi.2020.05.012

27. Heurich A, Hofmann-Winkler H, Gierer S, Liepold T, Jahn O, Pöhlmann S (2014) TMPRSS2 and ADAM17 cleave ACE2 differentially and only proteolysis by TMPRSS2 augments entry driven by the severe acute respiratory syndrome coronavirus spike protein. J Virol 88. https://doi.org/10.1128/jvi.02202-13
28. Hoffmann M, Kleine-Weber H, Schroeder S, Krüger N, Herrler T, Erichsen S, Schiergens TS, Herrler G, Wu NH, Nitsche A, Müller MA, Drosten C, Pöhlmann S (2020) SARS-CoV-2 cell entry depends on ACE2 and TMPRSS2 and is blocked by a clinically proven protease inhibitor. Cell 181. https://doi.org/10.1016/j.cell.2020.02.052
29. Thiel V, Weber F (2008) Interferon and cytokine responses to SARS-coronavirus infection. https://doi.org/10.1016/j.cytogfr.2008.01.001
30. Le Bon A, Tough DF (2002) Links between innate and adaptive immunity via type I interferon. https://doi.org/10.1016/S0952-7915(02)00354-0
31. Akira S, Uematsu S, Takeuchi O (2006) Pathogen recognition and innate immunity. https://doi.org/10.1016/j.cell.2006.02.015
32. Liu B, Li M, Zhou Z, Guan X, Xiang Y (2020) Can we use interleukin-6 (IL-6) blockade for coronavirus disease 2019 (COVID-19)-induced cytokine release syndrome (CRS)? J Autoimmun 111
33. Li SW, Wang CY, Jou YJ, Huang SH, Hsiao LH, Wan L, Lin YJ, Kung SH, Lin CW (2016) SARS coronavirus papain-like protease inhibits the TLR7 signaling pathway through removing Lys63-linked polyubiquitination of TRAF3 and TRAF6. Int J Mol Sci 17. https://doi.org/10.3390/ijms17050678
34. Totura AL, Whitmore A, Agnihothram S, Schäfer A, Katze MG, Heise MT, Baric RS (2015) Toll-like receptor 3 signaling via TRIF contributes to a protective innate immune response to severe acute respiratory syndrome coronavirus infection. mBio 6. https://doi.org/10.1128/mBio.00638-15
35. Florindo HF, Kleiner R, Vaskovich-Koubi D, Acúrcio RC, Carreira B, Yeini E, Tiram G, Liubomirski Y, Satchi-Fainaro R (2020) Immune-mediated approaches against COVID-19. https://doi.org/10.1038/s41565-020-0732-3
36. Konno Y, Kimura I, Uriu K, Fukushi M, Irie T, Koyanagi Y, Sauter D, Gifford RJ, Nakagawa S, Sato K (2020) SARS-CoV-2 ORF3b is a potent interferon antagonist whose activity is increased by a naturally occurring elongation variant. Cell Rep 32. https://doi.org/10.1016/j.celrep.2020.108185
37. Xia H, Cao Z, Xie X, Zhang X, Chen JYC, Wang H, Menachery VD, Rajsbaum R, Shi PY (2020) Evasion of type I interferon by SARS-CoV-2. Cell Rep 33. https://doi.org/10.1016/j.celrep.2020.108234
38. Tian W, Zhang N, Jin R, Feng Y, Wang S, Gao S, Gao R, Wu G, Tian D, Tan W, Chen Y, Gao GF, Wong CCL (2020) Immune suppression in the early stage of COVID-19 disease. Nat Commun 11. https://doi.org/10.1038/s41467-020-19706-9
39. Vabret N, Britton GJ, Gruber C, Hegde S, Kim J, Kuksin M, Levantovsky R, Malle L, Moreira A, Park MD, Pia L, Risson E, Saffern M, Salomé B, Esai Selvan M, Spindler MP, Tan J, van der Heide V, Gregory JK, Alexandropoulos K, Bhardwaj N, Brown BD, Greenbaum B, Gümüş ZH, Homann D, Horowitz A, Kamphorst AO, Curotto de Lafaille MA, Mehandru S, Merad M, Samstein RM, Agrawal M, Aleynick M, Belabed M, Brown M, Casanova-Acebes M, Catalan J, Centa M, Charap A, Chan A, Chen ST, Chung J, Bozkus CC, Cody E, Cossarini F, Dalla E, Fernandez N, Grout J, Ruan DF, Hamon P, Humblin E, Jha D, Kodysh J, Leader A, Lin M, Lindblad K, Lozano-Ojalvo D, Lubitz G, Magen A, Mahmood Z, Martinez-Delgado G, Mateus-Tique J, Meritt E, Moon C, Noel J, O'Donnell T, Ota M, Plitt T, Pothula V, Redes J, Reyes Torres I, Roberto M, Sanchez-Paulete AR, Shang J, Schanoski AS, Suprun M, Tran M, Vaninov N, Wilk CM, Aguirre-Ghiso J, Bogunovic D, Cho J, Faith J, Grasset E, Heeger P, Kenigsberg E, Krammer F, Laserson U (2020) Immunology of COVID-19: current state of the science. https://doi.org/10.1016/j.immuni.2020.05.002
40. Chiappelli F, Khakshooy A, Greenberg G (2020) Immunopathology and immunotherapy
41. De Wit E, Van Doremalen N, Falzarano D, Munster VJ (2016) SARS and MERS: recent insights into emerging coronaviruses. https://doi.org/10.1038/nrmicro.2016.81

42. Blanco-Melo D, Nilsson-Payant BE, Liu WC, Uhl S, Hoagland D, Møller R, Jordan TX, Oishi K, Panis M, Sachs D, Wang TT, Schwartz RE, Lim JK, Albrecht RA, tenOever BR (2020) Imbalanced host response to SARS-CoV-2 drives development of COVID-19. Cell 181. https://doi.org/10.1016/j.cell.2020.04.026
43. Helal MA, Shouman S, Abdelwaly A, Elmehrath AO, Essawy M, Sayed SM, Saleh AH, El-Badri N (2022) Molecular basis of the potential interaction of SARS-CoV-2 spike protein to CD147 in COVID-19 associated-lymphopenia. J Biomol Struct Dyn 40. https://doi.org/10.1080/07391102.2020.1822208
44. Liu Q, Shi Y, Cai J, Duan Y, Wang R, Zhang H, Ruan Q, Li J, Zhao L, Ping Y, Chen R, Ren L, Fei X, Zhang H, Tang R, Wang X, Luo T, Liu X, Huang X, Liu Z, Ao Q, Ren Y, Xiong J, He Z, Wu H, Fu W, Zhao P, Chen X, Qu G, Wang Y, Wang X, Liu J, Xiang D, Xu S, Zhou X, Li Q, Ma J, Li H, Zhang J, Huang S, Yao X, Zhou Y, Wang C, Zhang D, Wang G, Liu L, Bian XW (2020) Pathological changes in the lungs and lymphatic organs of 12 COVID-19 autopsy cases. Natl Sci Rev. 7. https://doi.org/10.1093/nsr/nwaa247
45. Moon C (2020) Fighting COVID-19 exhausts T cells. Nat Rev Immunol 20. https://doi.org/10.1038/s41577-020-0304-7
46. Li X, Geng M, Peng Y, Meng L, Lu S (2020) Molecular immune pathogenesis and diagnosis of COVID-19. https://doi.org/10.1016/j.jpha.2020.03.001
47. Tan YJ, Goh PY, Fielding BC, Shen S, Chou CF, Fu JL, Leong HN, Leo YS, Ooi EE, Ling AE, Lim SG, Hong W (2004) Profiles of antibody responses against severe acute respiratory syndrome coronavirus recombinant proteins and their potential use as diagnostic markers. Clin Diagn Lab Immunol 11. https://doi.org/10.1128/CDLI.11.2.362-371.2004
48. Wu HS, Hsieh YC, Su IJ, Lin TH, Chiu SC, Hsu YF, Lin JH, Wang MC, Chen JY, Hsiao PW, Chang GD, Wang AHJ, Ting HW, Chou CM, Huang CJ (2004) Early detection of antibodies against various structural proteins of the SARS-associated coronavirus in SARS patients. J Biomed Sci 11. https://doi.org/10.1159/000075294
49. Totura AL, Baric RS (2012) SARS coronavirus pathogenesis: Host innate immune responses and viral antagonism of interferon. https://doi.org/10.1016/j.coviro.2012.04.004
50. Zhou L, Liu K, Liu HG (2020) Cause analysis and treatment strategies of "recurrence" with novel coronavirus pneumonia (COVID-19) patients after discharge from hospital. Zhonghua Jie He He Hu Xi Za Zhi 43. https://doi.org/10.3760/cma.j.cn112147-20200229-00219
51. Huang C, Wang Y, Li X, Ren L, Zhao J, Hu Y, Zhang L, Fan G, Xu J, Gu X, Cheng Z, Yu T, Xia J, Wei Y, Wu W, Xie X, Yin W, Li H, Liu M, Xiao Y, Gao H, Guo L, Xie J, Wang G, Jiang R, Gao Z, Jin Q, Wang J, Cao B (2020) Clinical features of patients infected with 2019 novel coronavirus in Wuhan, China. Lancet 395. https://doi.org/10.1016/S0140-6736(20)30183-5
52. Chen N, Zhou M, Dong X, Qu J, Gong F, Han Y, Qiu Y, Wang J, Liu Y, Wei Y, Xia J, Yu T, Zhang X, Zhang L (2020) Epidemiological and clinical characteristics of 99 cases of 2019 novel coronavirus pneumonia in Wuhan, China: a descriptive study. Lancet 395. https://doi.org/10.1016/S0140-6736(20)30211-7
53. Zhou Y, Fu B, Zheng X, Wang D, Zhao C, Qi Y, Sun R, Tian Z, Xu X, Wei H (2020) Pathogenic T-cells and inflammatory monocytes incite inflammatory storms in severe COVID-19 patients. https://doi.org/10.1093/nsr/nwaa041
54. Diao B, Wang C, Tan Y, Chen X, Liu Y, Ning L, Chen L, Li M, Liu Y, Wang G, Yuan Z, Feng Z, Zhang Y, Wu Y, Chen Y (2020) Reduction and functional exhaustion of T cells in patients with coronavirus disease 2019 (COVID-19). Front Immunol 11. https://doi.org/10.3389/fimmu.2020.00827
55. Channappanavar R, Fehr AR, Vijay R, Mack M, Zhao J, Meyerholz DK, Perlman S (2016) Dysregulated type I interferon and inflammatory monocyte-macrophage responses cause lethal pneumonia in SARS-CoV-infected mice. Cell Host Microbe 19. https://doi.org/10.1016/j.chom.2016.01.007
56. Dhawan M, Rabaan AA, Alwarthan S, Alhajri M, Halwani MA, Alshengeti A, Najim MA, Alwashmi ASS, Alshehri AA, Alshamrani SA, AlShehail BM, Garout M, Al-Abdulhadi S, Al-Ahmed SH, Thakur N, Verma G (2023) Regulatory T cells (Tregs) and COVID-19: unveiling the mechanisms, and therapeutic potentialities with a special focus on long COVID. Vaccines (Basel) 11:699. https://doi.org/10.3390/vaccines11030699

57. Tay MZ, Poh CM, Rénia L, MacAry PA, Ng LFP (2020) The trinity of COVID-19: immunity, inflammation and intervention. https://doi.org/10.1038/s41577-020-0311-8
58. Jamilloux Y, Henry T, Belot A, Viel S, Fauter M, El Jammal T, Walzer T, François B, Sève P (2020) Should we stimulate or suppress immune responses in COVID-19? Cytokine and anti-cytokine interventions. https://doi.org/10.1016/j.autrev.2020.102567
59. Qin C, Zhou L, Hu Z, Zhang S, Yang S, Tao Y, Xie C, Ma K, Shang K, Wang W, Tian D-S (2020) Dysregulation of immune response in patients with coronavirus 2019 (COVID-19) in Wuhan, China. Clin Infect Dis 71(15):762–768
60. Nikolich-Zugich J, Knox KS, Rios CT, Natt B, Bhattacharya D, Fain MJ (2020) SARS-CoV-2 and COVID-19 in older adults: what we may expect regarding pathogenesis, immune responses, and outcomes. https://doi.org/10.1007/s11357-020-00186-0
61. Fu Y, Cheng Y, Wu Y (2020) Understanding SARS-CoV-2-mediated inflammatory responses: from mechanisms to potential therapeutic tools. https://doi.org/10.1007/s12250-020-00207-4
62. Batool S, Chokkakula S, Song MS (2023) Influenza treatment: limitations of antiviral therapy and advantages of drug combination therapy. https://doi.org/10.3390/microorganisms11010183
63. Lok ASF, McMahon BJ, Brown RS, Wong JB, Ahmed AT, Farah W, Almasri J, Alahdab F, Benkhadra K, Mouchli MA, Singh S, Mohamed EA, Abu Dabrh AM, Prokop LJ, Wang Z, Murad MH, Mohammed K (2016) Antiviral therapy for chronic hepatitis B viral infection in adults: a systematic review and meta-analysis. Hepatology 63:284–306. https://doi.org/10.1002/hep.28280
64. Saxena SK, Saxena S, Saxena R, Arvinda Swamy M, Gupta A, Nair MP (2010) Emerging trends, challenges and prospects in antiviral therapeutics and drug development for infectious diseases. Electron J Biol 6
65. Soriano V, Perno CF, Kaiser R, Calvez V, Gatell JM, Di Perri G, Pillay D, Rockstroh J, Geretti AM (2009) When and how to use maraviroc in HIV-infected patients. https://doi.org/10.1097/QAD.0b013e328332d32d
66. Tilton JC, Doms RW (2010) Entry inhibitors in the treatment of HIV-1 infection. https://doi.org/10.1016/j.antiviral.2009.07.022
67. Hu J, Robinson JL (2010) Treatment of respiratory syncytial virus with palivizumab: a systematic review. World J Pediatr 6. https://doi.org/10.1007/s12519-010-0230-z
68. Pinto LH, Holsinger LJ, Lamb RA (1992) Influenza virus M2 protein has ion channel activity. Cell 69. https://doi.org/10.1016/0092-8674(92)90452-I
69. Martin K, Heleniust A (1991) Nuclear transport of influenza virus ribonucleoproteins: the viral matrix protein (M1) promotes export and inhibits import. Cell 67. https://doi.org/10.1016/0092-8674(91)90576-K
70. De Clercq E (2015) Current treatment of hepatitis B virus infections. https://doi.org/10.1002/rmv.1849
71. Miller WH, Miller RL (1982) Phosphorylation of acyclovir diphosphate by cellular enzymes. Biochem Pharmacol 31. https://doi.org/10.1016/0006-2952(82)90305-7
72. Elion GB (1993) Acyclovir: discovery, mechanism of action, and selectivity. J Med Virol 41. https://doi.org/10.1002/jmv.1890410503
73. Pour PM, Fakhri S, Asgary S, Farzaei MH, Echeverría J (2019) The signaling pathways, and therapeutic targets of antiviral agents: focusing on the antiviral approaches and clinical perspectives of anthocyanins in the management of viral diseases. https://doi.org/10.3389/fphar.2019.01207
74. Asboe D, Pozniak A (2010) Antiviral therapy. Infect Dis
75. Bean B (1992) Antiviral therapy: current concepts and practices. https://doi.org/10.1128/CMR.5.2.146
76. Crini G (2022) Chitin and chitosan: production, properties, and applications. In: Chitin and chitosan. https://doi.org/10.1016/b978-0-323-96119-6.00004-9
77. Idris F, Ting DHR, Alonso S (2021) An update on dengue vaccine development, challenges, and future perspectives. https://doi.org/10.1080/17460441.2020.1811675

78. Tan RSL, Hassandarvish P, Chee CF, Chan LW, Wong TW (2022) Chitosan and its derivatives as polymeric anti-viral therapeutics and potential anti-SARS-CoV-2 nanomedicine. Carbohydr Polym 290:119500. https://doi.org/10.1016/j.carbpol.2022.119500
79. Shigemasa Y, Minami S (1996) Applications of chitin and chitosan for biomaterials. Biotechnol Genet Eng Rev 13. https://doi.org/10.1080/02648725.1996.10647935
80. Amor G, Sabbah M, Caputo L, Idbella M, De Feo V, Porta R, Fechtali T, Mauriello G (2021) Basil essential oil: composition, antimicrobial properties, and microencapsulation to produce active chitosan films for food packaging. Foods 10. https://doi.org/10.3390/foods10010121
81. Van der Weken H, Cox E, Devriendt B (2021) Advances in oral subunit vaccine design. https://doi.org/10.3390/vaccines9010001
82. He X, Xing R, Liu S, Qin Y, Li K, Yu H, Li P (2021) The improved antiviral activities of amino-modified chitosan derivatives on Newcastle virus. Drug Chem Toxicol 44. https://doi.org/10.1080/01480545.2019.1620264
83. Davydova VN, Nagorskaya VP, Gorbach VI, Kalitnik AA, Reunov AV, Solov'eva TF, Ermak IM (2011) Chitosan antiviral activity: dependence on structure and depolymerization method. Appl Biochem Microbiol 47. https://doi.org/10.1134/S0003683811010042
84. Iriti M, Varoni EM (2015) Chitosan-induced antiviral activity and innate immunity in plants. Environ Sci Pollut Res 22. https://doi.org/10.1007/s11356-014-3571-7
85. Iriti M, Varoni EM (2016) Chitosan-elicited plant innate immunity: focus on antiviral activity. In: Research progress in oligosaccharins. https://doi.org/10.1007/978-1-4939-3518-5_6
86. Ai H, Wang F, Xia Y, Chen X, Lei C (2012) Antioxidant, antifungal and antiviral activities of chitosan from the larvae of housefly, *Musca domestica* L. Food Chem 132. https://doi.org/10.1016/j.foodchem.2011.11.033
87. Divya K, Vijayan S, George TK, Jisha MS (2017) Antimicrobial properties of chitosan nanoparticles: mode of action and factors affecting activity. Fibers Polym 18. https://doi.org/10.1007/s12221-017-6690-1
88. Jarach N, Dodiuk H, Kenig S (2020) Polymers in the medical antiviral front-line. https://doi.org/10.3390/POLYM12081727
89. Hassan M, Mohamed A, Taher F, Kamel M (2016) Antimicrobial activities of chitosan nanoparticles prepared from *Lucilia cuprina* maggots (Diptera: Calliphoridae). J Egypt Soc Parasitol 46. https://doi.org/10.21608/jesp.2016.88258
90. Chirkov SN (2002) The antiviral activity of chitosan (review). https://doi.org/10.1023/A:1013206517442
91. Kulikov SN, Chirkov SN, Il'ina AV, Lopatin SA, Varlamov VP (2006) Effect of the molecular weight of chitosan on its antiviral activity in plants. Appl Biochem Microbiol 42. https://doi.org/10.1134/S0003683806020165
92. Chirkov SN, Il'ina AV, Surgucheva NA, Letunova EV, Varitsev YA, Tatarinova NY, Varlamov VP (2001) Effect of chitosan on systemic viral infection and some defense responses in potato plants. Russ J Plant Physiol 48. https://doi.org/10.1023/A:1012508625017
93. Mori Y, Ono T, Miyahira Y, Nguyen VQ, Matsui T, Ishihara M (2013) Antiviral activity of silver nanoparticle/chitosan composites against H1N1 influenza A virus. Nanoscale Res Lett 8. https://doi.org/10.1186/1556-276x-8-93
94. Wang W, Wang SX, Guan HS (2012) The antiviral activities and mechanisms of marine polysaccharides: an overview. https://doi.org/10.3390/md10122795
95. Zhou J, Hu Z, Zabihi F, Chen Z, Zhu M (2020) Progress and perspective of antiviral protective material. https://doi.org/10.1007/s42765-020-00047-7
96. Shibata Y, Foster LA, Metzger WJ, Myrvik QN (1997) Alveolar macrophage priming by intravenous administration of chitin particles, polymers of N-acetyl-D-glucosamine, in mice. Infect Immun 65. https://doi.org/10.1128/iai.65.5.1734-1741.1997
97. Minami S, Masuda M, Suzuki H, Okamoto Y, Matsuhashi A, Kato K, Shigemasa Y (1997) Subcutaneous injected chitosan induces systemic activation in dogs. Carbohydr Polym 33. https://doi.org/10.1016/S0144-8617(97)00024-6
98. Usami Y, Okamoto Y, Takayama T, Shigemasa Y, Minami S (1998) Effect of N-acetyl-D-glucosamine and D-glucosamine oligomers on canine polymorphonuclear cells in vitro. Carbohydr Polym 36. https://doi.org/10.1016/S0144-8617(98)00019-8

99. Bacon A, Makin J, Sizer PJ, Jabbal-Gill I, Hinchcliffe M, Illum L, Chatfield S, Roberts M (2000) Carbohydrate biopolymers enhance antibody responses to mucosally delivered vaccine antigens. Infect Immun 68. https://doi.org/10.1128/IAI.68.10.5764-5770.2000
100. Okawa Y, Kobayashi M, Suzuki S, Suzuki M (2003) Comparative study of protective effects of chitin, chitosan, and N-acetyl chitohexaose against *Pseudomonas aeruginosa* and *Listeria monocytogenes* infections in mice. Biol Pharm Bull 26. https://doi.org/10.1248/bpb.26.902
101. Stanish WD, McCormack R, Forriol F, Mohtadi N, Pelet S, Desnoyers J, Restrepo A, Shive MS (2013) Novel scaffold-based BST-CarGel treatment results in superior cartilage repair compared with microfracture in a randomized controlled trial. J Bone Joint Surg 95. https://doi.org/10.2106/JBJS.L.01345
102. Carroll EC, Jin L, Mori A, Muñoz-Wolf N, Oleszycka E, Moran HBT, Mansouri S, McEntee CP, Lambe E, Agger EM, Andersen P, Cunningham C, Hertzog P, Fitzgerald KA, Bowie AG, Lavelle EC (2016) The vaccine adjuvant chitosan promotes cellular immunity via DNA sensor cGAS-STING-dependent induction of type I interferons. Immunity 44. https://doi.org/10.1016/j.immuni.2016.02.004
103. Hoemann CD, Fong D (2017) Immunological responses to chitosan for biomedical applications. In: Chitosan based biomaterials. https://doi.org/10.1016/B978-0-08-100230-8.00003-0
104. Buschmann MD, Merzouki A, Lavertu M, Thibault M, Jean M, Darras V (2013) Chitosans for delivery of nucleic acids. https://doi.org/10.1016/j.addr.2013.07.005
105. Vårum KM, Antohonsen MW, Grasdalen H, Smidsrød O (1991) Determination of the degree of N-acetylation and the distribution of N-acetyl groups in partially N-deacetylated chitins (chitosans) by high-field n.m.r. spectroscopy. Carbohydr Res 211. https://doi.org/10.1016/0008-6215(91)84142-2
106. Lavertu M, Méthot S, Tran-Khanh N, Buschmann MD (2006) High efficiency gene transfer using chitosan/DNA nanoparticles with specific combinations of molecular weight and degree of deacetylation. Biomaterials 27. https://doi.org/10.1016/j.biomaterials.2006.04.029
107. Nishimura K, Nishimura S, Nishi N, Saiki I, Tokura S, Azuma I (1984) Immunological activity of chitin and its derivatives. Vaccine 2. https://doi.org/10.1016/S0264-410X(98)90039-1
108. Bueter CL, Lee CK, Wang JP, Ostroff GR, Specht CA, Levitz SM (2014) Spectrum and mechanisms of inflammasome activation by chitosan. J Immunol 192. https://doi.org/10.4049/jimmunol.1301695
109. Fong D, Grégoire-Gélinas P, Cheng AP, Mezheritsky T, Lavertu M, Sato S, Hoemann CD (2017) Lysosomal rupture induced by structurally distinct chitosans either promotes a type 1 IFN response or activates the inflammasome in macrophages. Biomaterials 129. https://doi.org/10.1016/j.biomaterials.2017.03.022
110. McNab F, Mayer-Barber K, Sher A, Wack A, O'Garra A (2015) Type I interferons in infectious disease. https://doi.org/10.1038/nri3787
111. Bueter CL, Lee CK, Rathinam VAK, Healy GJ, Taron CH, Specht CA, Levitz SM (2011) Chitosan but not chitin activates the inflammasome by a mechanism dependent upon phagocytosis. J Biol Chem 286. https://doi.org/10.1074/jbc.M111.274936
112. Mori A, Oleszycka E, Sharp FA, Coleman M, Ozasa Y, Singh M, O'Hagan DT, Tajber L, Corrigan OI, McNeela EA, Lavelle EC (2012) The vaccine adjuvant alum inhibits IL-12 by promoting PI3 kinase signaling while chitosan does not inhibit IL-12 and enhances Th1 and Th17 responses. Eur J Immunol 42. https://doi.org/10.1002/eji.201242372
113. Guzmán-Morales J, Ariganello MB, Hammami I, Thibault M, Jolicoeur M, Hoemann CD (2011) Biodegradable chitosan particles induce chemokine release and negligible arginase-1 activity compared to IL-4 in murine bone marrow-derived macrophages. Biochem Biophys Res Commun 405. https://doi.org/10.1016/j.bbrc.2011.01.063
114. Hoemann CD, Guzmán-Morales J, Tran-Khanh N, Lavallée G, Jolicoeur M, Lavertu M (2013) Chitosan rate of uptake in HEK293 cells is influenced by soluble versus microparticle state and enhanced by serum-induced cell metabolism and lactate-based media acidification. Molecules 18. https://doi.org/10.3390/molecules18011015

115. Pawar D, Goyal AK, Mangal S, Mishra N, Vaidya B, Tiwari S, Jain AK, Vyas SP (2010) Evaluation of mucoadhesive PLGA microparticles for nasal immunization. AAPS J 12. https://doi.org/10.1208/s12248-009-9169-1
116. Chen JL, Zhao Y (2012) Effect of molecular weight, acid, and plasticizer on the physicochemical and antibacterial properties of β-chitosan based films. J Food Sci 77. https://doi.org/10.1111/j.1750-3841.2012.02686.x
117. Saikia C, Gogoi P (2015) Chitosan: a promising biopolymer in drug delivery applications. J Mol Genet Med s4. https://doi.org/10.4172/1747-0862.s4-006
118. Ngo DH, Vo TS, Ngo DN, Kang KH, Je JY, Pham HND, Byun HG, Kim SK (2015) Biological effects of chitosan and its derivatives. https://doi.org/10.1016/j.foodhyd.2015.05.023
119. Gupta U, Jain NK (2010) Non-polymeric nano-carriers in HIV/AIDS drug delivery and targeting. https://doi.org/10.1016/j.addr.2009.11.018
120. Ramana LN, Sharma S, Sethuraman S, Ranga U, Krishnan UM (2014) Evaluation of chitosan nanoformulations as potent anti-HIV therapeutic systems. Biochim Biophys Acta Gen Subj 1840. https://doi.org/10.1016/j.bbagen.2013.10.002
121. Donalisio M, Leone F, Civra A, Spagnolo R, Ozer O, Lembo D, Cavalli R (2018) Acyclovir-loaded chitosan nanospheres from nano-emulsion templating for the topical treatment of herpesviruses infections. Pharmaceutics 10. https://doi.org/10.3390/pharmaceutics10020046
122. Iqbal M, Lin W, Jabbal-Gill I, Davis SS, Steward MW, Illum L (2003) Nasal delivery of chitosan-DNA plasmid expressing epitopes of respiratory syncytial virus (RSV) induces protective CTL responses in BALB/c mice. Vaccine 21. https://doi.org/10.1016/S0264-410X(02)00662-X
123. Mao S, Sun W, Kissel T (2010) Chitosan-based formulations for delivery of DNA and siRNA. https://doi.org/10.1016/j.addr.2009.08.004
124. Ruge CC, Kirch J, Lehr CM (2013) Pulmonary drug delivery: from generating aerosols to overcoming biological barriers-therapeutic possibilities and technological challenges. https://doi.org/10.1016/S2213-2600(13)70072-9
125. Debnath SK, Saisivam S, Debanth M, Omri A (2018) Development and evaluation of chitosan nanoparticles based dry powder inhalation formulations of prothionamide. PLoS ONE 13. https://doi.org/10.1371/journal.pone.0190976
126. van Doremalen N, Bushmaker T, Morris DH, Holbrook MG, Gamble A, Williamson BN, Tamin A, Harcourt JL, Thornburg NJ, Gerber SI, Lloyd-Smith JO, de Wit E, Munster VJ (2020) Aerosol and surface stability of SARS-CoV-2 as compared with SARS-CoV-1. N Engl J Med 382. https://doi.org/10.1056/nejmc2004973
127. Meselson M (2020) Droplets and aerosols in the transmission of SARS-CoV-2. N Engl J Med 382. https://doi.org/10.1056/nejmc2009324
128. Jaimes J, Millet J, Whittaker G (2020) Proteolytic cleavage of the SARS-CoV-2 spike protein and the role of the novel S1/S2 site. SSRN Electron J. https://doi.org/10.2139/ssrn.3581359
129. Milewska A, Chi Y, Szczepanski A, Barreto-Duran E, Liu K, Liu D, Guo X, Ge Y, Li J, Cui L, Ochman M, Urlik M, Rodziewicz-Motowidlo S, Zhu F, Szczubialka K, Nowakowska M, Pyrc K (2020) HTCC as a highly effective polymeric inhibitor of SARS-CoV-2 and MERS-CoV. bioRxiv
130. Itani R, Tobaiqy M, Al Faraj A (2020) Optimizing use of theranostic nanoparticles as a life-saving strategy for treating COVID-19 patients. Theranostics 10:5932–5942. https://doi.org/10.7150/thno.46691
131. Zhou H, Ikeuchi-Takahashi Y, Hattori Y, Onishi H (2020) Nanogels of a succinylated glycol chitosan-succinyl prednisolone conjugate: release behavior, gastrointestinal distribution, and systemic absorption. Int J Mol Sci 21. https://doi.org/10.3390/ijms21072376
132. Kantak MN, Bharate SS (2022) Analysis of clinical trials on biomaterial and therapeutic applications of chitosan: a review. https://doi.org/10.1016/j.carbpol.2021.118999
133. Gopal J, Muthu M, Pushparaj SSC, Sivanesan I (2023) Anti-COVID-19 credentials of chitosan composites and derivatives: future scope? Antibiotics 12:665. https://doi.org/10.3390/antibiotics12040665

134. Bande F, Arshad SS, Bejo MH, Omar AR, Moeini H, Khadkodaei S, Wei TS, Keong YS, Abba Y, Anka IA (2020) Development and immunogenic potentials of chitosan-saponin encapsulated DNA vaccine against avian infectious bronchitis coronavirus. Microb Pathog 149. https://doi.org/10.1016/j.micpath.2020.104560
135. Gurunathan S, Qasim M, Choi Y, Do JT, Park C, Hong K, Kim JH, Song H (2020) Antiviral potential of nanoparticles—can nanoparticles fight against coronaviruses? https://doi.org/10.3390/nano10091645
136. Ciejka J, Wolski K, Nowakowska M, Pyrc K, Szczubiałka K (2017) Biopolymeric nano/microspheres for selective and reversible adsorption of coronaviruses. Mater Sci Eng C 76. https://doi.org/10.1016/j.msec.2017.03.047
137. Milewska A, Ciejka J, Kaminski K, Karewicz A, Bielska D, Zeglen S, Karolak W, Nowakowska M, Potempa J, Bosch BJ, Pyrc K, Szczubialka K (2013) Novel polymeric inhibitors of HCoV-NL63. Antiviral Res 97. https://doi.org/10.1016/j.antiviral.2012.11.006
138. Safarzadeh M, Sadeghi S, Azizi M, Rastegari-Pouyani M, Pouriran R, Haji Molla Hoseini M (2021) Chitin and chitosan as tools to combat COVID-19: a triple approach. https://doi.org/10.1016/j.ijbiomac.2021.04.157
139. Vörös-Horváth B, Živković P, Bánfai K, Bővári-Biri J, Pongrácz J, Bálint G, Pál S, Széchenyi A (2022) Preparation and characterization of ACE2 receptor inhibitor-loaded chitosan hydrogels for nasal formulation to reduce the risk of COVID-19 viral infection. ACS Omega 7. https://doi.org/10.1021/acsomega.1c05149
140. Milewska A, Kaminski K, Ciejka J, Kosowicz K, Zeglen S, Wojarski J, Nowakowska M, Szczubiałka K, Pyrc K (2016) HTCC: broad range inhibitor of coronavirus entry. PLoS ONE 11. https://doi.org/10.1371/journal.pone.0156552
141. Rezaei FS, Khorshidian A, Beram FM, Derakhshani A, Esmaeili J, Barati A (2021) 3D printed chitosan/polycaprolactone scaffold for lung tissue engineering: hope to be useful for COVID-19 studies. RSC Adv 11. https://doi.org/10.1039/d1ra03410c
142. Wang IJ, Chen YC, Su C, Tsai MH, Shen WT, Bai CH, Yu KP (2021) Effectiveness of the nanosilver/TiO_2-chitosan antiviral filter on the removal of viral aerosols. J Aerosol Med Pulm Drug Deliv 34. https://doi.org/10.1089/jamp.2020.1607
143. Maharani DK, Sanjaya IGM, Amaria A, Anggraeni MA, Jannah LR (2021) Molecular docking analysis chitosan-zeolite-ZnO nanocomposite and its potency against SARS-CoV-2. IOP Conf Ser Mater Sci Eng 1125. https://doi.org/10.1088/1757-899x/1125/1/012006
144. Hathout RM, Kassem DH (2020) Positively charged electroceutical spun chitosan nanofibers can protect health care providers from COVID-19 infection: an opinion. Front Bioeng Biotechnol 8. https://doi.org/10.3389/fbioe.2020.00885
145. Raghuwanshi D, Mishra V, Das D, Kaur K, Suresh MR (2012) Dendritic cell targeted chitosan nanoparticles for nasal DNA immunization against SARS CoV nucleocapsid protein. Mol Pharm 9. https://doi.org/10.1021/mp200553x
146. Kalathiya U, Padariya M, Mayordomo M, Lisowska M, Nicholson J, Singh A, Baginski M, Fahraeus R, Carragher N, Ball K, Haas J, Daniels A, Hupp TR, Alfaro JA (2020) Highly conserved homotrimer cavity formed by the SARS-CoV-2 spike glycoprotein: a novel binding site. J Clin Med 9. https://doi.org/10.3390/jcm9051473
147. Pyrć K, Milewska A, Duran EB, Botwina P, Lopes R, Arenas-Pinto A, Badr M, Mellor R, Kalber TL, Fernandes-Reyes D, Schätzlein AG, Uchegbu IF (2020) SARS-CoV-2 inhibition in human airway epithelial cells using a mucoadhesive, amphiphilic chitosan that may serve as an anti-viral nasal spray. https://doi.org/10.1101/2020.12.10.413609
148. Alitongbieke G, Li X, Wu Q-C, Lin Z-C, Huang J-F, Xue Y, Liu J-N, Lin J-M, Pan T, Chen Y-X, Su Y, Zhang G-G, Leng B, Liu S-W, Pan Y-T (2020) Effect of β-chitosan on the binding interaction between SARS-CoV-2 S-RBD and ACE2. bioRxiv
149. Mohammadi Z, Eini M, Rastegari A, Tehrani MR (2021) Chitosan as a machine for biomolecule delivery: a review. https://doi.org/10.1016/j.carbpol.2020.117414
150. Sun Q, Li X, Che X, Chang J, Yang Y, Yu L (2009) Evaluation on the immune efficiency of bovine coronavirus N protein-loaded chitosan microspheres. Zhongguo Yufang Shouyi Xuebao/Chin J Prev Vet Med 31:882–886

151. Tatlow D, Tatlow C, Tatlow S, Tatlow S (2020) A novel concept for treatment and vaccination against COVID-19 with an inhaled chitosan-coated DNA vaccine encoding a secreted spike protein portion. Clin Exp Pharmacol Physiol 47. https://doi.org/10.1111/1440-1681.13393
152. Ejeromedoghene O, Oderinde O, Egejuru G, Adewuyi S (2020) Chitosan-drug encapsulation as a potential candidate for COVID-19 drug delivery systems: a review. https://doi.org/10.18596/jotcsa.773780

Chapter 11
Recent Challenges and Future Prospectus of Green Derivatized Chitosan Nanoparticles

11.1 Isolation, Preparation of Chitin, Chitosan, Their Functionalized Polymers and Bio-nanocomposites

Utilization of shellfish industrial waste for isolation, purification, and characterizations of chitin from crustacean's sources has been reported in the literature [1]. Molecular engineering of novel, biodegradable, thermally stable chitin-based polyurethane elastomers (PUEs) [2, 3], and synthesis and characterization of chitin/curcumin blended PUEs have also been well documented [4]. XRD studies of chitin-based polyurethane elastomers and morphological and thermal studies of chitin-curcumin blends derived polyurethanes have been reported [5, 6]. Surface characteristics of chitin-based polyurethanes using hydroxyl-terminated polybutadiene and UV-irradiated chitin-based polyurethane elastomers have been documented [7, 8]. Molecular engineering and properties of chitin-based shape memory polyurethanes and XRD studies of polyurethane elastomers based on chitin/1,4-butane diol blends have been reported [9, 10]. XRD pattern and cytotoxicity and mechanical behavior of chitin–bentonite clay-based polyurethane bio-nanocomposites have been studied [11, 12]. Molecular engineering and morphological studies of chitin-based polyurethanes using hydroxyl-terminated polybutadiene have been well documented [13, 14]. Surface characteristics of chitin-based shape memory polyurethane elastomers (SMPU) and UV-irradiated chitin-based SMPU have been compared with pristine [15, 16]. Effect of diisocyanate structure on the evaluation of biocompatibility and mechanical behavior of chitin-based polyurethane elastomers and effect of chitin/1,4-butane diol blends on the evaluation of biocompatibility and mechanical behavior of polyurethane elastomers have been well studied [17, 18]. Synthesis and characterization of stable and biological active chitin-based polyurethane elastomers and chitin–bentonite clay-based polyurethane bio-nanocomposites [19, 20].

K. M. Zia, *Chitosan*, Smart Nanomaterials Technology,
https://doi.org/10.1007/978-981-96-2629-8_11

11.2 Recent Challenges in Green Derivatization of Chitosan Nanoparticles (CSNPs)

Green derivatization is a beneficial however a difficult technology to adopt and implement. Despite of a lot of research in this area, and numerous advantages of CSNPs, there are significant challenges associated with this newly emerged material. A few recent challenges have been enlisted here such as:

1. The CSNPs exhibit poor solubility at alkaline and neutral pH [21–23].
2. Reportedly, the CSNPs have shown very high cytotoxicity against malignant tumor cells [24].
3. Although CS is approved for use in diet, wound dressings applications, and formulations of cartilage, it is not currently approved for mass-scale production [25].
4. One of the major challenges is associated with chemical production of CSNPs; the productions are not eco-friendly resulting in the significant quantity of the organic waste [26].
5. Over all the production is not economically favorable due to the additional cost of effluent treatment. The effluents are mainly generated after acidic and basic treatment [27].
6. The continuous hydrolysis of the CSNPs throughout the alkali treatment leads to the decline of molecular weight [28–30].
7. Mechanical resistance of the end product is poor [31].
8. The ultimate control of pore size is complex process.
9. The process of electrospinning is complicated for CSNPs [32, 33].
10. Lack of comprehensive data of in vivo study is a serious question mark on the efficiency and protection of CSNP platforms [34].
11. Preparation of CSNPs by cross-linking can have a serious influence on the intrinsic properties of CS [35].

11.3 Recent Strategies and Practical Aspects

Due to numerous challenges faced in the productions and safe applications of CSNPs, much of the attention has been diverted to develop CSNPs in cost-effective and eco-friendly mode. Nowadays, CSNPs have gained immense significance in developing new curative drug delivery systems with improved nanomedicines. These CSNPs play significant role in biomedical engineering applications. One of the major features is the bioavailability directly linked with enhanced sensitivity and reduced pharmacological toxicity. Some activities may rely on the shape and size of the CSNPs as well [36].

The modified form of novel CSNPs, which consists of clusters of NPs with a size of 9.9–80 nm, has indicated significant possibility in the fields of nanomedicine, biomedical engineering, industry, and pharmaceutical companies [37]. In many cases, CS-based nanosystems have displayed excellent physicochemical and biological properties because they can change the protein load and adjust the value of every parameter throughout the production, making them an advanced drug delivery system. In addition to these features, it is highly stable having high protein encapsulation efficiency that can be synthesized as a lyophilized powder; and its storage and transportation is much convenient [38].

In recent years, the class of cationic polysaccharides (CSNPs) has been extensively explored for targeted deliverance of antitumor agents owing to their distinctive biological and physicochemical properties, such as biocompatibility, biological degradability, and mucosal adhesion properties. Moreover, these CSNPs have also active groups for chemical transformations [39].

11.4 Chitosan Nanoparticles Issues and Opportunities

The CSNP has attracted much consideration in nanomedicine-based techniques for malignant cells treatment owing to its exceptional physicochemical qualities. The CSNPs have been extensively studied for malignant tumor cells targeting and have shown promising future; however, the knowledge of pharmacokinetic (PK) and pharmacodynamic (PD) properties and malignant tumor targeting efficiency of CSNPs are still limited. Therefore, much more progress is needed in understanding such sort of issues in NPs technology, which may create further opportunities to translate nanomedicinal techniques into much better clinical outcomes for malignant tumor-targeted delivery of NPs [40].

Another issue for the production of CSNPs is their commercially accessible therapeutics. This matter, on the one side, is linked to their dissimilar molecular masses, and on the other side, it is linked with their degree of deacetylation, which influences the in vivo qualities of CSNPs. Such issues can be practically solved by a variety of analysis in order to manufacture the optimized NPs, which are entirely appropriate for in vivo drug deliverance systems [41].

The development of mucosal vaccines faces a number of challenges and opportunities. Imperative issues for effective mucosal injection include antigen maintenance time that allows contact with the lymphatic system, selection of adjuvants that are non-toxic and immunogenic; and possibly the ability to initiate mimic mucosal pathogens. The CS seems to be one of the most promising polysaccharides as an essential release agent for mucosal vaccines and also improved immune response. Even though CS (in its unmodified form) and its salts are being commercially accessible in elevated purity grades, the toxicity data related to human exposure and mucosal-administration is not convincing enough to make it a pharmaceutically permitted product. In addition, CS is unlikely to accomplish universal adjuvant status as it may be incompatible and ineffective against certain antigens. The chemical modification of CS determines the physicochemical properties; such as charge and solubility, thus allowing for most favorable match to the antigen. The extensive usage and physical nature of formulations (e.g., NPs) are expected to offer great potential in the future. The CS-based derivatives can incorporate particular ligands (antigens; modifiers; such as mannose) and are thus designed to exclusively interact with preferred kinds of cells; it is also expected to open up possibilities and new capabilities in vaccine-targeting. However, there are some challenges in obtaining regulatory endorsement and bringing these new materials to market and, therefore, their inclusion in vaccine products can be considered a long-term objective [42].

11.5 Challenges to Be Addressed

Despite exceptional research endeavors to conquer the problems in oral incorporation of proteins, a clinically powerful and secure deliverance machine stays an elusive challenge. The pH-responsive NPs shelled with CS are evidently very much promising for oral deliverance of protein tablets along with insulin. Though, there are some barriers that need to be addressed for those CSNPs in order to turn oral delivery into a reality. For instance, these CSNPs need to reveal their superiority over injectable insulin in order to gain higher control of the blood glucose and slighter prevalence of diabetic complications. Moreover, their long-time period effectiveness needs to be verified in massive animals and in human beings. Another task in the use of mucoadhesive CSNPs, with extended retention inside the intestinal region, for oral protein transport is keeping the physicochemical and organic balance of the added drugs. Immobilizing of enzyme inhibitors within CSNPs can provide this advantage and increase the bioavailability of orally administered proteins. The attachment of each reactive group (i.e., amino and hydroxyl) on CS offers possibilities for chemical amendment to include diverse sorts of enzyme inhibitors. Moreover, CS is insoluble in the neutral pH medium because of the deprotonation of its amino groups, which leads to the lack of its mucoadhesive feature and its opening activity in jejunum and ileum. It is pertinent to mention here that the primary function of the jejunum is to absorb sugars, amino acids, and fatty acids. Both the jejunum and ileum are peritoneal. The ileum absorbs any remaining nutrients that did not get absorbed by the duodenum or jejunum, in particular vitamin B12, as well as bile acids that will go on to be recycled.

Chemically transformed CSNPs can be manufactured to enhance its physiochemical qualities along with aqueous solubility and mucoadhesive and intestinal penetration abilities. If the above demanding situations are effectively addressed, then the CSNPs system for oral transport of protein drugs might soon become a reality [43].

11.6 Brief Outlook and Concluding Remarks

Chitosan is an excellent biocompatible, biodegradable, and polycationic polymer which has great structural modification ability to create new properties and purposes in diverse fields, particularly in industry and biomedical fields. Most important fundamental technological advances have been utilized to augment the capabilities and improve the qualities and consistency of CS and modified CSNPs. In the ancient times, research into functional biomaterials such as CSNPs has led to the development of innovative and better regenerative drug delivery systems. At present, one of the fastest-growing fields is the health science. These properties evidently reveal that CSNPs have a better future development prospective in various scientific fields, i.e., drug and gene delivery, cell imaging, sensing, as well as in the cure and diagnosis of certain diseases such as malignancy. The CS is also a promising food preservative. The CS-based nanomaterials have better physical and chemical properties, i.e., elevated surface area, porosity, tensile strength, electrical conductance, and luminescence as well as enhanced mechanical properties in contrast with pure CS. Despite of a lot of qualities of CSNPs, some limitations have been reported which should be addressed in the near future.

11.7 Open Questions

- How the solubility of CSNPs can be increased specially in neutral or alkaline pH? What further steps are needed to be taken?
- How to make much more cost-effective, eco-friendly, and chemically modified form of CSNPs? What should we do in chemical transformation of CSNPs; in order to make modified form of CSNPs?
- How to make electrospinning of CSNPs uncomplicated? What sort of experiments should we needed?
- How to make the much more effective use of CSNPs in agricultural field?
- How to use CSNPs in malignant tumor treatment in much better way in near future?

11.8 Future Prospects and Probabilities

In near future, we look forward to observe supplementary work in in vivo studies to estimate the antibacterial phenomenon and biocompatibility. Furthermore, their action against a wide diversity of microbes, fungi, and viruses, presently cannot seem to be clarified, although these things are likely to come into view in the coming years owing to their promising studies [34].

The CS-based morphine nasal formulation (RylomineTM) is presently in phase-2 (UK and EU) and phase-3 clinical testing in the USA. When it hits the market, it will pave the way for alike products in the near era and will help to differentiate unintended effects in humans [25].

The CSNPs have the absence of high colloidal constancy in biological serum and restricted response to blood cells (at very extremely high NPs dose) are the serious problems that attract attention and may be addressed in the near future by efficient and better surface modification and preparation techniques of NPs [41]. Supplementary, preclinical, and clinical evaluations are required in near future to move ionically cross-linked CSNPs toward clinical approval [44].

In the near future, based on investigation on CSNPs, with the arrival of new techniques to overcome CSNPs drawbacks through better formulation techniques for wide assortment of drugs and even macromolecules, it is inferred to see more applications particularly in nasal and pneumonic drug administration with anticipation that future studies on CSNPs will concentrates on the toxicity analysis in humans [34].

The CSNPs have better prospective for future advances in diverse areas of science that is drug delivery, agriculture, cell imaging, and also in the cure as well as in the diagnosis of various ailments like malignancy [45–47]. Such capabilities of CSNPs count the considerable versatility and attention to CSNPs, and it is anticipated that this biomaterial has a brilliant future. Consequently, this wide panel of CS qualities and processed nanomaterials given to this bio-based polymer can serve as an excellent imminent material. With innovative inspirations recommended by novel characterization of CS and modified CSNPs and vast innovative opportunities brought by nano-based technology, it is realistic to anticipate that the considerable advances that will revolutionize nanotechnology field is not so distant future [31]. Additional research to design stable and effective CS-based NPs formulation is also needed in near future.

References

1. Ilyas HN, Zia KM, Rehman S, Ilyas R, Sultana S (2021) Utilization of shellfish industrial waste for isolation, purification, and characterizations of chitin from crustacean's sources in Pakistan. J Polym Environ 29:2337–2348. https://doi.org/10.1007/s10924-020-02037-7
2. Mahmood K, Zia KM, Aftab W, Zuber M, Tabasum S, Noreen A, Zia F (2018) Synthesis and characterization of chitin/curcumin blended polyurethane elastomers. Int J Biol Macromol 113:150–158
3. Zia KM, Barikani M, Bhatti IA, Zuber M, Bhatti HN (2008) Synthesis and characterization of novel, biodegradable, thermally stable chitin-based polyurethane elastomers. J Appl Polym Sci 110(2):769–776. https://doi.org/10.1002/app.28533
4. Zia KM, Barikani M, Zuber M, Bhatti IA, Sheikh MA (2008) Molecular engineering of chitin based polyurethane elastomers. Carbohydr Polym 74(2):149–158. https://doi.org/10.1016/j.carbpol.2008.03.013
5. Mahmood K, Zia KM, Zuber M, Tabasum S, Rehman S, Zia F, Noreen A (2017) Morphological and thermal studies of chitin-curcumin blends derived polyurethanes. Int J Biol Macromol 105:1180–1191. https://doi.org/10.1016/j.ijbiomac.2017.07.149

6. Zia KM, Bhatti IA, Barikani M, Zuber M, Sheikh MA (2008) XRD studies of chitin-based polyurethane elastomers. Int J Biol Macromol 43(2):136–141. https://doi.org/10.1016/j.ijbiomac.2008.04.009
7. Zia KM, Zuber M, Saif MJ, Jawaid M, Mahmood K, Shahid M et al (2013) Chitin based polyurethanes using hydroxyl terminated polybutadiene, part III: surface characteristics. Int J Biol Macromol 62:670–676. https://doi.org/10.1016/j.ijbiomac.2013.10.001
8. Zia KM, Barikani M, Khalid AM, Honarkar H (2009) Surface characteristics of UV-irradiated chitin-based polyurethane elastomers. Carbohydr Polym 77(3):621–627. https://doi.org/10.1016/j.carbpol.2009.02.005
9. Barikani M, Zia KM, Bhatti IA, Zuber M, Bhatti HN (2008) Molecular engineering and properties of chitin based shape memory polyurethanes. Carbohydr Polym 74(3):621–626. https://doi.org/10.1016/j.carbpol.2008.04.023
10. Zia KM, Bhatti IA, Barikani M, Zuber M, Bhatti HN (2009) XRD studies of polyurethane elastomers based on chitin/1,4-butane diol blends. Carbohydr Polym 76(2):183–187. https://doi.org/10.1016/j.carbpol.2008.10.005
11. Zia KM, Zuber M, Barikani M, Jabbar A, Khosa MK (2010) XRD pattern of chitin based polyurethane bio-nanocomposites. Carbohydr Polym 80(2):539–543. https://doi.org/10.1016/j.carbpol.2009.12.017
12. Zia KM, Zuber M, Barikani M, Hussain R, Jamil T, Anjum S (2011) Cytotoxicity and mechanical behavior of chitin–bentonite clay based polyurethane bio-nanocomposites. Int J Biol Macromol 49(5):1131–1136. https://doi.org/10.1016/j.ijbiomac.2011.09.010
13. Zia KM, Mahmood K, Zuber M, Jamil T, Shafiq M (2013) Chitin based polyurethanes using hydroxyl terminated polybutadiene. Part I: molecular engineering. Int J Biol Macromol 59:320–327. https://doi.org/10.1016/j.ijbiomac.2013.04.069
14. Zia KM, Qureshi NA, Mujahid M, Mahmood K, Zuber M (2013) Chitin based polyurethanes using hydroxyl terminated polybutadiene. Part II: morphological studies. Int J Biol Macromol 59:313–319. https://doi.org/10.1016/j.ijbiomac.2013.04.042
15. Zia KM, Zuber M, Barikani M, Bhatti IA, Khan MB (2009) Surface characteristics of chitin-based shape memory polyurethane elastomers. Colloids Surf B Biointerfaces 72(2):248–252. https://doi.org/10.1016/j.colsurfb.2009.04.011
16. Zia KM, Zuber M, Mahboob S, Sultana T, Sultana S (2010) Surface characteristics of UV-irradiated chitin-based shape memory polyurethanes. Carbohydr Polym 80(1):229–234. https://doi.org/10.1016/j.carbpol.2009.11.015
17. Zia KM, Zuber M, Bhatti IA, Barikani M, Sheikh MA (2009) Evaluation of biocompatibility and mechanical behavior of chitin-based polyurethane elastomers. Part-II: effect of diisocyanate structure. Int J Biol Macromol 44(1):23–28. https://doi.org/10.1016/j.ijbiomac.2008.11.002
18. Zia KM, Zuber M, Bhatti IA, Barikani M, Sheikh MA (2009) Evaluation of biocompatibility and mechanical behavior of polyurethane elastomers based on chitin/1,4-butane diol blends. Int J Biol Macromol 44(1):18–22. https://doi.org/10.1016/j.ijbiomac.2008.09.011
19. Zuber M, Zia KM, Mahboob S, Hassan M, Bhatti IA (2010) Synthesis of chitin–bentonite clay based polyurethane bio-nanocomposites. Int J Biol Macromol 47(2):196–200. https://doi.org/10.1016/j.ijbiomac.2010.04.022
20. Amjed N, Bhatti IA, Zia KM, Iqbal J, Jamil Y (2020) Synthesis and characterization of stable and biological active chitin-based polyurethane elastomers. Int J Biol Macromol 154:1149–1157. https://doi.org/10.1016/j.ijbiomac.2019.11.097
21. Nasirimoghaddam S, Zeinali S, Sabbaghi S (2015) Chitosan coated magnetic nanoparticles as nano-adsorbent for efficient removal of mercury contents from industrial aqueous and oily samples. J Ind Eng Chem 27. https://doi.org/10.1016/j.jiec.2014.12.020
22. Bodnar M, Hartmann JF, Borbely J (2005) Preparation and characterization of chitosan-based nanoparticles. Biomacromolecules 6. https://doi.org/10.1021/bm0502258
23. Cheung RCF, Ng TB, Wong JH, Chan WY (2015) Chitosan: an update on potential biomedical and pharmaceutical applications. https://doi.org/10.3390/md13085156
24. Jimtaisong A, Saewan N (2014) Utilization of carboxymethyl chitosan in cosmetics. https://doi.org/10.1111/ics.12102

25. Bellich B, D'Agostino I, Semeraro S, Gamini A, Cesàro A (2016) "The good, the bad and the ugly" of chitosans. https://doi.org/10.3390/md14050099
26. Yadav M, Goswami P, Paritosh K, Kumar M, Pareek N, Vivekanand V (2019) Seafood waste: a source for preparation of commercially employable chitin/chitosan materials. https://doi.org/10.1186/s40643-019-0243-y
27. Hamed I, Özogul F, Regenstein JM (2016) Industrial applications of crustacean by-products (chitin, chitosan, and chitooligosaccharides): a review. https://doi.org/10.1016/j.tifs.2015.11.007
28. Tsaih ML, Chen RH (2003) Effect of degree of deacetylation of chitosan on the kinetics of ultrasonic degradation of chitosan. J Appl Polym Sci 90. https://doi.org/10.1002/app.13027
29. Li L, Hsieh YL (2006) Chitosan bicomponent nanofibers and nanoporous fibers. Carbohydr Res 341. https://doi.org/10.1016/j.carres.2005.11.028
30. Baxter S, Zivanovic S, Weiss J (2005) Molecular weight and degree of acetylation of high-intensity ultrasonicated chitosan. Food Hydrocoll 19. https://doi.org/10.1016/j.foodhyd.2004.11.002
31. Shukla SK, Mishra AK, Arotiba OA, Mamba BB (2013) Chitosan-based nanomaterials: a state-of-the-art review. https://doi.org/10.1016/j.ijbiomac.2013.04.043
32. Lee KY, Jeong L, Kang YO, Lee SJ, Park WH (2009) Electrospinning of polysaccharides for regenerative medicine. https://doi.org/10.1016/j.addr.2009.07.006
33. Zomer Volpato F, Almodóvar J, Erickson K, Popat KC, Migliaresi C, Kipper MJ (2012) Preservation of FGF-2 bioactivity using heparin-based nanoparticles, and their delivery from electrospun chitosan fibers. Acta Biomater 8. https://doi.org/10.1016/j.actbio.2011.12.023
34. Landriscina A, Rosen J, Friedman AJ (2015) Biodegradable chitosan nanoparticles in drug delivery for infectious disease. https://doi.org/10.2217/nnm.15.7
35. Garg U, Chauhan S, Nagaich U, Jain N (2019) Current advances in chitosan nanoparticles based drug delivery and targeting. https://doi.org/10.15171/apb.2019.023
36. Sharifi-Rad J, Quispe C, Butnariu M, Rotariu LS, Sytar O, Sestito S, Rapposelli S, Akram M, Iqbal M, Krishna A, Kumar NVA, Braga SS, Cardoso SM, Jafernik K, Ekiert H, Cruz-Martins N, Szopa A, Villagran M, Mardones L, Martorell M, Docea AO, Calina D (2021) Chitosan nanoparticles as a promising tool in nanomedicine with particular emphasis on oncological treatment. https://doi.org/10.1186/s12935-021-02025-4
37. Ghadi A, Mahjoub S, Tabandeh F, Talebnia F (2014) Synthesis and optimization of chitosan nanoparticles: potential applications in nanomedicine and biomedical engineering. Casp J Intern Med 5
38. Wang X, Li J, Wang Y, Cho KJ, Kim G, Gjyrezi A, Koenig L, Giannakakou P, Shin HJC, Tighiouart M, Nie S, Chen Z, Shin DM (2009) HFT-T, a targeting nanoparticle, enhances specific delivery of paclitaxel to folate receptor-positive tumors. ACS Nano 3. https://doi.org/10.1021/nn900649v
39. Zhang X, Yang X, Ji J, Liu A, Zhai G (2016) Tumor targeting strategies for chitosan-based nanoparticles. https://doi.org/10.1016/j.colsurfb.2016.09.020
40. Jee JP, Na JH, Lee S, Kim SH, Choi K, Yeo Y, Kwon IC (2012) Cancer targeting strategies in nanomedicine: design and application of chitosan nanoparticles. https://doi.org/10.1016/j.cossms.2013.01.002
41. Narmani A, Jafari SM (2021) Chitosan-based nanodelivery systems for cancer therapy: recent advances. https://doi.org/10.1016/j.carbpol.2021.118464
42. Jabbal-Gill I, Watts P, Smith A (2012) Chitosan-based delivery systems for mucosal vaccines. https://doi.org/10.1517/17425247.2012.697455
43. Sung HW, Sonaje K, Liao ZX, Hsu LW, Chuang EY (2012) pH-responsive nanoparticles shelled with chitosan for oral delivery of insulin: from mechanism to therapeutic applications. Acc Chem Res 45. https://doi.org/10.1021/ar200234q
44. Desai KGH (2016) Chitosan nanoparticles prepared by ionotropic gelation: an overview of recent advances. https://doi.org/10.1615/CritRevTherDrugCarrierSyst.2016014850
45. Inamuddin, Asiri AM, Mohammad A (2018) Applications of nanocomposite materials in drug delivery. https://doi.org/10.1016/C2016-0-05075-1

46. Upadhyaya L, Singh J, Agarwal V, Tewari RP (2014) The implications of recent advances in carboxymethyl chitosan based targeted drug delivery and tissue engineering applications. https://doi.org/10.1016/j.jconrel.2014.04.043
47. Patra JK, Das G, Fraceto LF, Campos EVR, Rodriguez-Torres MDP, Acosta-Torres LS, Diaz-Torres LA, Grillo R, Swamy MK, Sharma S, Habtemariam S, Shin HS (2018) Nano based drug delivery systems: recent developments and future prospects. https://doi.org/10.1186/s12951-018-0392-8

The manufacturer's authorised representative in the EU is Springer Nature Customer Service Centre GmbH, Europaplatz 3, 69115 Heidelberg, Germany. If you have any concerns regarding our products, please contact ProductSafety@springernature.com

Printed and bound by CPI Group (UK) Ltd, Croydon, CR0 4YY
01/10/2026
02225928-0008